space
sciences

space
sciences

VOLUME **2**
Planetary Science and Astronomy

Pat Dasch, Editor in Chief

MACMILLAN
REFERENCE
USA™

THOMSON
GALE

New York • Detroit • San Diego • San Francisco • Cleveland • New Haven, Conn. • Waterville, Maine • London • Munich

Macmillan Reference USA
300 Park Avenue South
New York, NY 10010

Gale Group
27500 Drake Rd.
Farmington Hills, MI 48331-3535

Library of Congress Cataloging-in-Publication Data
Space sciences / Pat Dasch, editor in chief.
    p. cm.
Includes bibliographical references and indexes.
ISBN 0-02-865546-X (set : alk. paper)
  1. Space sciences. I. Dasch, Pat.
     QB500 .S63 2002
     500.5—dc21

                                    2002001707

Volume 1: ISBN 0-02-865547-8
Volume 2: ISBN 0-02-865548-6
Volume 3: ISBN 0-02-865549-4
Volume 4: ISBN 0-02-865550-8

Printed in the United States of America
1 2 3 4 5 6 7 8 9 10

Preface

Astronomers have studied the heavens for more than two millennia, but in the twentieth century, humankind ventured off planet Earth into the dark vacuum void of space, forever changing our perspective of our home planet and on our relationship to the universe in which we reside.

Our explorations of space—the final frontier in our niche in this solar system—first with satellites, then robotic probes, and finally with humans, have given rise to an extensive space industry that has a major influence on the economy and on our lives. In 1998, U.S. space exports (launch services, satellites, space-based communications services, and the like) totaled $64 billion. As we entered the new millennium, space exports were the second largest dollar earner after agriculture. The aerospace industry directly employs some 860,000 Americans, with many more involved in subcontracting companies and academic research.

Beginnings

The Chinese are credited with developing the rudiments of rocketry—they launched rockets as missiles against invading Mongols in 1232. In the nineteenth century William Congrieve developed a rocket in Britain based on designs conceived in India in the eighteenth century. Congrieve extended the range of the Indian rockets, adapting them specifically for use by armies. Congrieve's rockets were used in 1806 in the Napoleonic Wars.

The Birth of Modern Space Exploration

The basis of modern spaceflight and exploration came with the writings of Konstantin Tsiolkovsky (1857–1935), a Russian mathematics teacher. He described multi-stage rockets, winged craft like the space shuttle developed in the 1970s, space stations like Mir and the International Space Station, and interplanetary missions of discovery.

During the same period, space travel captured the imagination of fiction writers. Jules Verne wrote several novels with spaceflight themes. His book, *From the Earth to the Moon* (1865), describes manned flight to the Moon, including a launch site in Florida and a spaceship named Columbia—the name chosen for the Apollo 11 spaceship that made the first lunar landing in July 1969 and the first space shuttle, which flew in April 1981. In the twentieth century, Arthur C. Clarke predicted the role of communications satellites and extended our vision of human space exploration while television series such as *Star Trek* and *Dr. Who* challenged the imagination and embedded the idea of space travel in our culture.

The first successful test of the V-2 rocket developed by Wernher von Braun and his team at Peenemünde, Germany, in October 1942 has been described as the "birth of the Space Age." After World War II some of the Peenemünde team under von Braun came to the United States, where they worked at the White Sands Missile Range in New Mexico, while others went to Russia. This sowed the seeds of the space race of the 1960s. Each team worked to develop advanced rockets, with Russia developing the R-7, while a series of rockets with names like Thor, Redstone, and Titan were produced in the United States.

When the Russians lofted Sputnik, the first artificial satellite, on October 4, 1957, the race was on. The flights of Yuri Gagarin, Alan Shepard, and John Glenn followed, culminating in the race for the Moon and the Apollo Program of the 1960s and early 1970s.

The Emergence of a Space Industry

The enormous national commitment to the Apollo Program marked a new phase in our space endeavors. The need for innovation and technological advance stimulated the academic and engineering communities and led to the growth of a vast network of contract supporters of the aerospace initiative and the birth of a vibrant space industry. At the same time, planetary science emerged as a new geological specialization.

Following the Apollo Program, the U.S. space agency's mission remained poorly defined through the end of the twentieth century, grasping at major programs such as development of the space shuttle and the International Space Station, in part, some argue, to provide jobs for the very large workforce spawned by the Apollo Program. The 1980s saw the beginnings of what would become a robust commercial space industry, largely independent of government programs, providing communications and information technology via space-based satellites. During the 1990s many thought that commercialization was the way of the future for space ventures. Commercially coordinated robotic planetary exploration missions were conceived with suggestions that NASA purchase the data, and Dennis Tito, the first paying space tourist in 2001, raised hopes of access to space for all.

The terrorist attacks on the United States on September 11, 2001 and the U.S. recession led to a re-evaluation of the entrepreneurial optimism of the 1990s. Many private commercial space ventures were placed on hold or went out of business. Commentators suggested that the true dawning of the commercial space age would be delayed by up to a decade. But, at the same time, the U.S. space agency emerged with a more clearly defined mandate than it had had since the Apollo Program, with a role of driving technological innovation—with an early emphasis on reducing the cost of getting to orbit—and leading world class space-related scientific projects. And military orders, to fill the needs of the new world order, compensated to a point for the downturn in the commercial space communications sector.

It is against this background of an industry in a state of flux, a discipline on the cusp of a new age of innovation, that this encyclopedia has been prepared.

Organization of the Material

The 341 entries in *Space Sciences* have been organized in four volumes, focusing on the business of space exploration, planetary science and astronomy, human space exploration, and the outlook for the future exploration of space. Each entry has been newly commissioned for this work. Our contributors are drawn from academia, industry, government, professional space institutes and associations, and nonprofit organizations. Many of the contributors are world authorities on their subject, providing up-to-the-minute information in a straightforward style accessible to high school students and university undergraduates.

One of the outstanding advantages of books on space is the wonderful imagery of exploration and achievement. These volumes are richly illustrated, and sidebars provide capsules of additional information on topics of particular interest. Entries are followed by a list of related entries, as well as a reading list for students seeking more information.

Acknowledgements

I wish to thank the team at Macmillan Reference USA and the Gale Group for their vision and leadership in bringing this work to fruition. In particular, thanks to Hélène Potter, Cindy Clendenon, and Gloria Lam. My thanks to Associate Editors Nadine Barlow, Leonard David, and Frank Sietzen, whose expertise, commitment, and patience have made *Space Sciences* possible. My thanks also go to my husband, Julius, for his encouragement and support. My love affair with space began in the 1970s when I worked alongside geologists using space imagery to plan volcanological field work in remote areas of South America, and took root when, in the 1980s, I became involved in systematic analysis of the more than 3,000 photographs of Earth that astronauts bring back at the end of every shuttle mission. The beauty of planet Earth, as seen from space, and the wealth of information contained in those images, convinced me that space is a very real part of life on Earth, and that I wanted to be a part of the exploration of space and to share the wonder of it with the public. I hope that *Space Sciences* conveys the excitement, achievements, and potential of space exploration to a new generation of students.

Pat Dasch
Editor in Chief

For Your Reference

The following section provides information that is applicable to a number of articles in this reference work. Included in the following pages is a chart providing comparative solar system planet data, as well as measurement, abbreviation, and conversion tables.

SOLAR SYSTEM PLANET DATA

	Mercury	Venus[2]	Earth	Mars	Jupiter	Saturn	Uranus	Neptune	Pluto
Mean distance from the Sun (AU): [1]	0.387	0.723	1	1.524	5.202	9.555	19.218	30.109	39.439
Siderial period of orbit (years):	0.24	0.62	1	1.88	11.86	29.46	84.01	164.79	247.68
Mean orbital velocity (km/sec):	47.89	35.04	29.79	24.14	13.06	9.64	6.81	5.43	4.74
Orbital essentricity:	0.206	0.007	0.017	0.093	0.048	0.056	0.047	0.009	0.246
Inclination to ecliptic (degrees):	7.00	3.40	0	1.85	1.30	2.49	0.77	1.77	17.17
Equatorial radius (km):	2439	6052	6378	3397	71492	60268	25559	24764	1140
Polar radius (km):	same	same	6357	3380	66854	54360	24973	24340	same
Mass of planet (Earth = 1):[3]	0.06	0.82	1	0.11	317.89	95.18	14.54	17.15	0.002
Mean density (gm/cm ^{3}):	5.44	5.25	5.52	3.94	1.33	0.69	1.27	1.64	2.0
Body rotation period (hours):	1408	5832.R	23.93	24.62	9.92	10.66	17.24	16.11	153.3
Tilt of equator to orbit (degrees):	0	2.12	23.45	23.98	3.08	26.73	97.92	28.8	96

[1]AU indicates one astronomical unit, defined as the mean distance between Earth and the Sun (~1.495 x 10[8] km).
[2]R indicates planet rotation is retrograde (i.e., opposite to the planet's orbit).
[3]Earth's mass is approximately 5.976 x 10[26] grams.

SI BASE AND SUPPLEMENTARY UNIT NAMES AND SYMBOLS

Physical Quality	Name	Symbol
Length	meter	m
Mass	kilogram	kg
Time	second	s
Electric current	ampere	A
Thermodynamic temperature	kelvin	K
Amount of substance	mole	mol
Luminous intensity	candela	cd
Plane angle	radian	rad
Solid angle	steradian	sr

Temperature

Scientists commonly use the Celsius system. Although not recommended for scientific and technical use, earth scientists also use the familiar Fahrenheit temperature scale (°F). $1°F = 1.8°C$ or K. The triple point of H_2O, where gas, liquid, and solid water coexist, is 32°F.

- To change from Fahrenheit (F) to Celsius (C):
  $°C = (°F - 32)/(1.8)$
- To change from Celsius (C) to Fahrenheit (F):
  $°F = (°C \times 1.8) + 32$
- To change from Celsius (C) to Kelvin (K):
  $K = °C + 273.15$
- To change from Fahrenheit (F) to Kelvin (K):
  $K = (°F - 32)/(1.8) + 273.15$

UNITS DERIVED FROM SI, WITH SPECIAL NAMES AND SYMBOLS

Derived Quantity	Name of SI Unit	Symbol for SI Unit	Expression in Terms of SI Base Units
Frequency	hertz	Hz	s^{-1}
Force	newton	N	$m\ kg\ s^{-2}$
Pressure, stress	Pascal	Pa	$N\ m^{-2}$ $= m^{-1}\ kg\ s^{-2}$
Energy, work, heat	Joule	J	$N\ m$ $= m^2\ kg\ s^{-2}$
Power, radiant flux	watt	W	$J\ s^{-1}$ $= m^2\ kg\ s^{-3}$
Electric charge	coulomb	C	$A\ s$
Electric potential, electromotive force	volt	V	$J\ C^{-1}$ $= m^{-2}\ kg\ s^{-3}\ A^{-1}$
Electric resistance	ohm	—	$V\ A^{-1}$ $= m^2\ kg\ s^{-3}\ A^{-2}$
Celsius temperature	degree Celsius	C	K
Luminous flux	lumen	lm	cd sr
Illuminance	lux	lx	$cd\ sr\ m^{-2}$

UNITS USED WITH SI, WITH NAME, SYMBOL, AND VALUES IN SI UNITS

The following units, not part of the SI, will continue to be used in appropriate contexts (e.g., angtsrom):

Physical Quantity	Name of Unit	Symbol for Unit	Value in SI Units
Time	minute	min	60 s
	hour	h	3,600 s
	day	d	86,400 s
Plane angle	degree	°	$(\pi/180)$ rad
	minute	'	$(\pi/10,800)$ rad
	second	"	$(\pi/648,000)$ rad
Length	angstrom	Å	10^{-10} m
Volume	liter	l, L	1 $dm^3 = 10^{-3}\ m^3$
Mass	ton	t	1 $mg = 10^3$ kg
	unified atomic mass unit	u $(= m_a(^{12}C)/12)$	$\approx 1.66054 \times 10^{-27}$ kg
Pressure	bar	bar	10^5 Pa $= 10^5\ N\ m^{-2}$
Energy	electronvolt	eV $(= e \times V)$	$\approx 1.60218 \times 10^{-19}$ J

CONVERSIONS FOR STANDARD, DERIVED, AND CUSTOMARY MEASUREMENTS

Length

1 angstrom (Å)	0.1 nanometer (exactly) 0.000000004 inch
1 centimeter (cm)	0.3937 inches
1 foot (ft)	0.3048 meter (exactly)
1 inch (in)	2.54 centimeters (exactly)
1 kilometer (km)	0.621 mile
1 meter (m)	39.37 inches 1.094 yards
1 mile (mi)	5,280 feet (exactly) 1.609 kilometers
1 astronomical unit (AU)	1.495979×10^{13} cm
1 parsec (pc)	206,264.806 AU 3.085678×10^{18} cm 3.261633 light-years
1 light-year	9.460530×10^{17} cm

Area

1 acre	43,560 square feet (exactly) 0.405 hectare
1 hectare	2.471 acres
1 square centimeter (cm²)	0.155 square inch
1 square foot (ft²)	929.030 square centimeters
1 square inch (in²)	6.4516 square centimeters (exactly)
1 square kilometer (km²)	247.104 acres 0.386 square mile
1 square meter (m²)	1.196 square yards 10.764 square feet
1 square mile (mi²)	258.999 hectares

MEASUREMENTS AND ABBREVIATIONS

Volume

1 barrel (bbl)*, liquid	31 to 42 gallons
1 cubic centimeter (cm³)	0.061 cubic inch
1 cubic foot (ft³)	7.481 gallons 28.316 cubic decimeters
1 cubic inch (in³)	0.554 fluid ounce
1 dram, fluid (or liquid)	⅛ fluid ounce (exactly) 0.226 cubic inch 3.697 milliliters
1 gallon (gal) (U.S.)	231 cubic inches (exactly) 3.785 liters 128 U.S. fluid ounces (exactly)
1 gallon (gal) (British Imperial)	277.42 cubic inches 1.201 U.S. gallons 4.546 liters
1 liter	1 cubic decimeter (exactly) 1.057 liquid quarts 0.908 dry quart 61.025 cubic inches
1 ounce, fluid (or liquid)	1.805 cubic inches 29.573 milliliters
1 ounce, fluid (fl oz) (British)	0.961 U.S. fluid ounce 1.734 cubic inches 28.412 milliliters
1 quart (qt), dry (U.S.)	67.201 cubic inches 1.101 liters
1 quart (qt), liquid (U.S.)	57.75 cubic inches (exactly) 0.946 liter

Units of mass

1 carat (ct)	200 milligrams (exactly) 3.086 grains
1 grain	64.79891 milligrams (exactly)
1 gram (g)	15.432 grains 0.035 ounce
1 kilogram (kg)	2.205 pounds
1 microgram (μg)	0.000001 gram (exactly)
1 milligram (mg)	0.015 grain
1 ounce (oz)	437.5 grains (exactly) 28.350 grams
1 pound (lb)	7,000 grains (exactly) 453.59237 grams (exactly)
1 ton, gross or long	2,240 pounds (exactly) 1.12 net tons (exactly) 1.016 metric tons
1 ton, metric (t)	2,204.623 pounds 0.984 gross ton 1.102 net tons
1 ton, net or short	2,000 pounds (exactly) 0.893 gross ton 0.907 metric ton

Pressure

1 kilogram/square centimeter (kg/cm²)	0.96784 atmosphere (atm) 14.2233 pounds/square inch (lb/in²) 0.98067 bar
1 bar	0.98692 atmosphere (atm) 1.02 kilograms/square centimeter (kg/cm²)

* There are a variety of "barrels" established by law or usage. For example, U.S. federal taxes on fermented liquors are based on a barrel of 31 gallons (141 liters); many state laws fix the "barrel for liquids" as 31½ gallons (119.2 liters); one state fixes a 36-gallon (160.5 liters) barrel for cistern measurment; federal law recognizes a 40-gallon (178 liters) barrel for "proof spirts"; by custom, 42 gallons (159 liters) comprise a barrel of crude oil or petroleum products for statistical purposes, and this equivalent is recognized "for liquids" by four states.

Milestones in Space History

c. 850	The Chinese invent a form of gunpowder for rocket propulsion.
1242	Englishman Roger Bacon develops gunpowder.
1379	Rockets are used as weapons in the Siege of Chioggia, Italy.
1804	William Congrieve develops ship-fired rockets.
1903	Konstantin Tsiolkovsky publishes *Research into Interplanetary Science by Means of Rocket Power*, a treatise on space travel.
1909	Robert H. Goddard develops designs for liquid-fueled rockets.
1917	Smithsonian Institute issues grant to Goddard for rocket research.
1918	Goddard publishes the monograph *Method of Attaining Extreme Altitudes*.
1921	Soviet Union establishes a state laboratory for solid rocket research.
1922	Hermann Oberth publishes *Die Rakete zu den Planetenräumen*, a work on rocket travel through space.
1923	Tsiolkovsky publishes work postulating multi-staged rockets.
1924	Walter Hohmann publishes work on rocket flight and orbital motion.
1927	The German Society for Space Travel holds its first meeting.
	Max Valier proposes rocket-powered aircraft adapted from Junkers G23.
1928	Oberth designs liquid rocket for the film *Woman in the Moon*.
1929	Goddard launches rocket carrying barometer.
1930	Soviet rocket designer Valentin Glusko designs U.S.S.R. liquid rocket engine.

1931	Eugene Sänger test fires liquid rocket engines in Vienna.
1932	German Rocket Society fires first rocket in test flight.
1933	Goddard receives grant from Guggenheim Foundation for rocket studies.
1934	Wernher von Braun, member of the German Rocket Society, test fires water-cooled rocket.
1935	Goddard fires advanced liquid rocket that reaches 700 miles per hour.
1936	Glushko publishes work on liquid rocket engines.
1937	The Rocket Research Project of the California Institute of Technology begins research program on rocket designs.
1938	von Braun's rocket researchers open center at Pennemünde.
1939	Sänger and Irene Brendt refine rocket designs and propose advanced winged suborbital bomber.
1940	Goddard develops centrifugal pumps for rocket engines.
1941	Germans test rocket-powered interceptor aircraft Me 163.
1942	V-2 rocket fired from Pennemünde enters space during ballistic flight.
1943	First operational V-2 launch.
1944	V-2 rocket launched to strike London.
1945	Arthur C. Clarke proposes geostationary satellites.
1946	Soviet Union tests version of German V-2 rocket.
1947	United States test fires Corporal missile from White Sands, New Mexico.
	X-1 research rocket aircraft flies past the speed of sound.
1948	United States reveals development plan for Earth satellite adapted from RAND.
1949	Chinese rocket scientist Hsueh-Sen proposes hypersonic aircraft.
1950	United States fires Viking 4 rocket to record 106 miles from USS Norton Sound.
1951	Bell Aircraft Corporation proposes winged suborbital rocket-plane.
1952	Wernher von Braun proposes wheeled Earth-orbiting space station.
1953	U.S. Navy D-558II sets world altitude record of 15 miles above Earth.
1954	Soviet Union begins design of RD-107, RD-108 ballistic missile engines.
1955	Soviet Union launches dogs aboard research rocket on suborbital flight.

1956	United States announces plan to launch Earth satellite as part of Geophysical Year program.
1957	U.S. Army Ballistic Missile Agency is formed.
	Soviet Union test fires R-7 ballistic missile.
	Soviet Union launches the world's first Earth satellite, Sputnik-1, aboard R-7.
	United States launches 3-stage Jupiter C on test flight.
	United States attempts Vanguard 1 satellite launch; rocket explodes.
1958	United States orbits Explorer-1 Earth satellite aboard Jupiter-C rocket.
	United States establishes the National Aeronautics and Space Administration (NASA) as civilian space research organization.
	NASA establishes Project Mercury manned space project.
	United States orbits Atlas rocket with Project Score.
1959	Soviet Union sends Luna 1 towards Moon; misses by 3100 miles.
	NASA announces the selection of seven astronauts for Earth space missions.
	Soviet Union launches Luna 2, which strikes the Moon.
1960	United States launches Echo satellite balloon.
	United States launches Discoverer 14 into orbit, capsule caught in midair.
	Soviet Union launches two dogs into Earth orbit.
	Mercury-Redstone rocket test fired in suborbital flight test.
1961	Soviet Union tests Vostok capsule in Earth orbit with dummy passenger.
	Soviet Union launches Yuri Gagarin aboard Vostok-1; he becomes the first human in space.
	United States launches Alan B. Shepard on suborbital flight.
	United States proposes goal of landing humans on the Moon before 1970.
	Soviet Union launches Gherman Titov into Earth orbital flight for one day.
	United States launches Virgil I. "Gus" Grissom on suborbital flight.
	United States launches first Saturn 1 rocket in suborbital test.

1962	United States launches John H. Glenn into 3-orbit flight.
	United States launches Ranger to impact Moon; craft fails.
	First United States/United Kingdom international satellite launch; Ariel 1 enters orbit.
	X-15 research aircraft sets new altitude record of 246,700 feet.
	United States launches Scott Carpenter into 3-orbit flight.
	United States orbits Telstar 1 communications satellite.
	Soviet Union launches Vostok 3 and 4 into Earth orbital flight.
	United States launches Mariner II toward Venus flyby.
	United States launches Walter Schirra into 6-orbit flight.
	Soviet Union launches Mars 1 flight; craft fails.
1963	United States launches Gordon Cooper into 22-orbit flight.
	Soviet Union launches Vostok 5 into 119-hour orbital flight.
	United States test fires advanced solid rockets for Titan 3C.
	First Apollo Project test in Little Joe II launch.
	Soviet Union orbits Vostok 6, which carries Valentina Tereshkova, the first woman into space.
	Soviet Union tests advanced version of R-7 called Soyuz launcher.
1964	United States conducts first Saturn 1 launch with live second stage; enters orbit.
	U.S. Ranger 6 mission launched towards Moon; craft fails.
	Soviet Union launches Zond 1 to Venus; craft fails.
	United States launches Ranger 7 on successful Moon impact.
	United States launches Syncom 3 communications satellite.
	Soviet Union launches Voshkod 1 carrying three cosmonauts.
	United States launches Mariner 4 on Martian flyby mission.
1965	Soviet Union launches Voshkod 2; first space walk.
	United States launches Gemini 3 on 3-orbit piloted test flight.
	United States launches Early Bird 1 communications satellite.
	United States launches Gemini 4 on 4-day flight; first U.S. space walk.

United States launches Gemini 5 on 8-day flight.

United States launches Titan 3C on maiden flight.

Europe launches Asterix 1 satellite into orbit.

United States Gemini 6/7 conduct first space rendezvous.

1966   Soviet Union launches Luna 9, which soft lands on Moon.

United States Gemini 8 conducts first space docking; flight aborted.

United States launches Surveyor 1 to Moon soft landing.

United States tests Atlas Centaur advanced launch vehicle.

Gemini 9 flight encounters space walk troubles.

Gemini 10 flight conducts double rendezvous.

United States launches Lunar Orbiter 1 to orbit Moon.

Gemini 11 tests advanced space walks.

United States launches Saturn IB on unpiloted test flight.

Soviet Union tests advanced Proton launch vehicle.

United States launches Gemini 12 to conclude two-man missions.

1967   Apollo 1 astronauts killed in launch pad fire.

Soviet Soyuz 1 flight fails; cosmonaut killed.

Britain launches Ariel 3 communications satellite.

United States conducts test flight of M2F2 lifting body research craft.

United States sends Surveyor 3 to dig lunar soils.

Soviet Union orbits anti-satellite system.

United States conducts first flight of Saturn V rocket (Apollo 4).

1968   Yuri Gagarin killed in plane crash.

Soviet Union docks Cosmos 212 and 213 automatically in orbit.

United States conducts Apollo 6 Saturn V test flight; partial success.

Nuclear rocket engine tested in Nevada.

United States launches Apollo 7 in three-person orbital test flight.

Soviet Union launches Soyuz 3 on three-day piloted flight.

United States sends Apollo 8 into lunar orbit; first human flight to Moon.

1969   Soviet Union launches Soyuz 4 and 5 into orbit; craft dock.

Largest tactical communications satellite launched.

United States flies Apollo 9 on test of lunar landing craft in Earth orbit.

United States flies Apollo 10 to Moon in dress rehearsal of landing attempt.

United States cancels military space station program.

United States flies Apollo 11 to first landing on the Moon.

United States cancels production of Saturn V in budget cut.

Soviet lunar rocket N-1 fails in launch explosion.

United States sends Mariner 6 on Mars flyby.

United States flies Apollo 12 on second lunar landing mission.

Soviet Union flies Soyuz 6 and 7 missions.

United States launches Skynet military satellites for Britain.

1970   China orbits first satellite.

Japan orbits domestic satellite.

United States Apollo 13 mission suffers explosion; crew returns safely.

Soviet Union launches Venera 7 for landing on Venus.

United States launches military early warning satellite.

Soviet Union launches Luna 17 to Moon.

United States announces modifications to Apollo spacecraft.

1971   United States flies Apollo 14 to Moon landing.

Soviet Union launches Salyut 1 space station into orbit.

First crew to Salyut station, Soyuz 11, perishes.

Soviet Union launches Mars 3 to make landing on the red planet.

United States flies Apollo 15 to Moon with roving vehicle aboard.

1972   United States and the Soviet Union sign space cooperation agreement.

United States launches Pioneer 10 to Jupiter flyby.

Soviet Union launches Venera 8 to soft land on Venus.

United States launches Apollo 16 to moon.

India and Soviet Union sign agreement for launch of Indian satellite.

United States initiates space shuttle project.

United States flies Apollo 17, last lunar landing mission.

1973	United States launches Skylab space station.
	United States launches first crew to Skylab station.
	Soviet Union launches Soyuz 12 mission.
	United States launches second crew to Skylab space station.
1974	United States launches ATS research satellite.
	Soviet Union launches Salyut 3 on unpiloted test flight.
	Soviet Union launches Soyuz 12, 13, and 14 flights.
	Soviet Union launches Salyut 4 space station.
1975	Soviet Union launches Soyuz 17 to dock with Salyut 4 station.
	Soviet Union launches Venera 9 to soft land on Venus.
	United States and Soviet Union conduct Apollo-Soyuz Test Project joint flight.
	China orbits large military satellite.
	United States sends Viking 1 and 2 towards landing on Martian surface.
	Soviet Union launches unpiloted Soyuz 20.
1976	Soviet Union launches Salyut 5 space station.
	First space shuttle rolls out; Enterprise prototype.
	Soviet Union docks Soyuz 21 to station.
	China begins tests of advanced ballistic missile.
1977	Soyuz 24 docks with station.
	United States conducts atmospheric test flights of shuttle Enterprise.
	United States launches Voyager 1 and 2 on deep space missions.
	Soviet Union launches Salyut 6 space station.
	Soviet Soyuz 25 fails to dock with station.
	Soyuz 26 is launched and docks with station.
1978	Soyuz 27 is launched and docks with Salyut 6 station.
	Soyuz 28 docks with Soyuz 27/Salyut complex.
	United States launches Pioneer/Venus 1 mission.
	Soyuz 29 docks with station.
	Soviet Union launches Progress unpiloted tankers to station.
	Soyuz 30 docks with station.
	United States launches Pioneer/Venus 2.
	Soyuz 31 docks with station.

1979	Soyuz 32 docks with Salyut station.
	Voyager 1 flies past Jupiter.
	Soyuz 33 fails to dock with station.
	Voyager 2 flies past Jupiter.
1980	First Ariane rocket launches from French Guiana; fails.
	Soviet Union begins new Soyuz T piloted missions.
	STS-1 first shuttle mission moves to launching pad.
1981	Soviet Union orbits advanced Salyut stations.
	STS-1 launched on first space shuttle mission.
	United States launches STS-2 on second shuttle flight; mission curtailed.
1982	United States launches STS-5 first operational shuttle flight.
1983	United States launches Challenger, second orbital shuttle, on STS-6.
	United States launches Sally Ride, the first American woman in space, on STS-7.
	United States launches Guion Bluford, the first African-American astronaut, on STS-8.
	United States launches first Spacelab mission aboard STS-9.
1984	Soviet Union tests advanced orbital station designs.
	Shuttle Discovery makes first flights.
	United States proposes permanent space station as goal.
1985	Space shuttle Atlantis enters service.
	United States announces policy for commercial rocket sales.
	United States flies U.S. Senator aboard space shuttle Challenger.
1986	Soviet Union launches and occupies advanced Mir space station.
	Challenger—on its tenth mission, STS-51-L—is destroyed in a launching accident.
	United States restricts payloads on future shuttle missions.
	United States orders replacement shuttle for Challenger.
1987	Soviet Union flies advanced Soyuz T-2 designs.
	United States' Delta, Atlas, and Titan rockets grounded in launch failures.
	Soviet Union launches Energyia advanced heavy lift rocket.

1988	Soviet Union orbits unpiloted shuttle Buran.
	United States launches space shuttle Discovery on STS-26 flight.
	United States launches STS-27 military shuttle flight.
1989	United States launches STS-29 flight.
	United States launches Magellan probe from shuttle.
1990	Shuttle fleet grounded for hydrogen leaks.
	United States launches Hubble Space Telescope.
1992	Replacement shuttle Endeavour enters service.
	United States probe Mars Observer fails.
1993	United States and Russia announce space station partnership.
1994	United States shuttles begin visits to Russian space station Mir.
1995	Europe launches first Ariane 5 advanced booster; flight fails.
1996	United States announces X-33 project to replace shuttles.
1997	Mars Pathfinder lands on Mars.
1998	First elements of International Space Station launched.
1999	First Ocean space launch of Zenit rocket in Sea Launch program.
2000	Twin United States Mars missions fail.
2001	United States cancels shuttle replacements X-33 and X-34 because of space cutbacks.
	United States orbits Mars Odyssey probe around Mars.
2002	First launches of United States advanced Delta IV and Atlas V commercial rockets.

Frank Sietzen, Jr.

Human Achievements in Space

The road to space has been neither steady nor easy, but the journey has cast humans into a new role in history. Here are some of the milestones and achievements.

Oct. 4, 1957 The Soviet Union launches the first artificial satellite, a 184-pound spacecraft named Sputnik.

Nov. 3, 1957 The Soviets continue pushing the space frontier with the launch of a dog named Laika into orbit aboard Sputnik 2. The dog lives for seven days, an indication that perhaps people may also be able to survive in space.

Jan. 31, 1958 The United States launches Explorer 1, the first U.S. satellite, and discovers that Earth is surrounded by radiation belts. James Van Allen, who instrumented the satellite, is credited with the discovery.

Apr. 12, 1961 Yuri Gagarin becomes the first person in space. He is launched by the Soviet Union aboard a Vostok rocket for a two-hour orbital flight around the planet.

May 5, 1961 Astronaut Alan Shepard becomes the first American in space. Shepard demonstrates that individuals can control a vehicle during weightlessness and high gravitational forces. During his 15-minute suborbital flight, Shepard reaches speeds of 5,100 mph.

May 24, 1961 Stung by the series of Soviet firsts in space, President John F. Kennedy announces a bold plan to land men on the Moon and bring them safely back to Earth before the end of the decade.

Feb. 20, 1962 John Glenn becomes the first American in orbit. He flies around the planet for nearly five hours in his Mercury capsule, Friendship 7.

June 16, 1963 The Soviets launch the first woman, Valentina Tereshkova, into space. She circles Earth in her Vostok spacecraft for three days.

Nov. 28, 1964 NASA launches Mariner 4 spacecraft for a flyby of Mars.

Mar. 18, 1965 Cosmonaut Alexei Leonov performs the world's first space walk outside his Voskhod 2 spacecraft. The outing lasts 10 minutes.

Mar. 23, 1965 Astronauts Virgil I. "Gus" Grissom and John Young blast off on the first Gemini mission and demonstrate for the first time how to maneuver from one orbit to another.

June 3, 1965 Astronaut Edward White becomes the first American to walk in space during a 21-minute outing outside his Gemini spacecraft.

Mar. 16, 1966 Gemini astronauts Neil Armstrong and David Scott dock their spacecraft with an unmanned target vehicle to complete the first joining of two spacecraft in orbit. A stuck thruster forces an early end to the experiment, and the crew makes America's first emergency landing from space.

Jan. 27, 1967 The Apollo 1 crew is killed when a fire breaks out in their command module during a prelaunch test. The fatalities devastate the American space community, but a subsequent spacecraft redesign helps the United States achieve its goal of sending men to the Moon.

Apr. 24, 1967 Tragedy also strikes the Soviet space program, with the death of cosmonaut Vladimir Komarov. His new Soyuz spacecraft gets tangled with parachute lines during re-entry and crashes to Earth.

Dec. 21, 1968 Apollo 8, the first manned mission to the Moon, blasts off from Cape Canaveral, Florida. Frank Borman, Jim Lovell and Bill Anders orbit the Moon ten times, coming to within 70 miles of the lunar surface.

July 20, 1969 Humans walk on another world for the first time when astronauts Neil Armstrong and Edwin "Buzz" Aldrin climb out of their spaceship and set foot on the Moon.

Apr. 13, 1970 The Apollo 13 mission to the Moon is aborted when an oxygen tank explosion cripples the spacecraft. NASA's most serious inflight emergency ends four days later when the astronauts, ill and freezing, splash down in the Pacific Ocean.

June 6, 1971 Cosmonauts blast off for the first mission in the world's first space station, the Soviet Union's Salyut 1. The crew spends twenty-two days aboard the outpost. During re-entry, however, a faulty valve leaks air from the Soyuz capsule, and the crew is killed.

Jan. 5, 1972 President Nixon announces plans to build "an entirely new type of space transportation system," pumping life into NASA's dream to build a reusable, multi-purpose space shuttle.

Dec. 7, 1972 The seventh and final mission to the Moon is launched, as public interest and political support for the Apollo program dims.

May 14, 1973 NASA launches the first U.S. space station, Skylab 1, into orbit. Three crews live on the station between May 1973 and February 1974. NASA hopes to have the shuttle flying in time to reboost and resupply Skylab, but the outpost falls from orbit on July 11, 1979.

July 17, 1975 In a momentary break from Cold War tensions, the United States and Soviet Union conduct the first linking of American and Russian spaceships in orbit. The Apollo-Soyuz mission is a harbinger of the cooperative space programs that develop between the world's two space powers twenty years later.

Apr. 12, 1981 Space shuttle Columbia blasts off with a two-man crew for the first test-flight of NASA's new reusable spaceship. After two days in orbit, the shuttle lands at Edwards Air Force Base in California.

June 18, 1983 For the first time, a space shuttle crew includes a woman. Astronaut Sally Ride becomes America's first woman in orbit.

Oct. 30, 1983 NASA's increasingly diverse astronaut corps includes an African-American for the first time. Guion Bluford, an aerospace engineer, is one of the five crewmen assigned to the STS-8 mission.

Nov. 28, 1983 NASA flies its first Spacelab mission and its first European astronaut, Ulf Merbold.

Feb. 7, 1984 Shuttle astronauts Bruce McCandless and Robert Stewart take the first untethered space walks, using a jet backpack to fly up to 320 feet from the orbiter.

Apr. 9–11, 1984 First retrieval and repair of an orbital satellite.

Jan. 28, 1986 Space shuttle Challenger explodes 73 seconds after launch, killing its seven-member crew. Aboard the shuttle was Teacher-in-Space finalist Christa McAuliffe, who was to conduct lessons from orbit. NASA grounds the shuttle fleet for two and a half years.

Feb. 20. 1986 The Soviets launch the core module of their new space station, Mir, into orbit. Mir is the first outpost designed as a module system to be expanded in orbit. Expected lifetime of the station is five years.

May 15, 1987 Soviets launch a new heavy-lift booster from the Baikonur Cosmodrome in Kazakhstan.

Oct. 1, 1987 Mir cosmonaut Yuri Romanenko breaks the record for the longest space mission, surpassing the 236-day flight by Salyut cosmonauts set in 1984.

Sept. 29, 1988 NASA launches the space shuttle Discovery on the first crewed U.S. mission since the 1986 Challenger explosion. The shuttle carries a replacement communications satellite for the one lost onboard Challenger.

May 4, 1989 Astronauts dispatch a planetary probe from the shuttle for the first time. The Magellan radar mapper is bound for Venus.

Nov. 15, 1989 The Soviets launch their space shuttle Buran, which means snowstorm, on its debut flight. There is no crew onboard, and unlike the U.S. shuttle, no engines to help place it into orbit. Lofted into orbit by twin Energia heavy-lift boosters, Buran circles Earth twice and lands. Buran never flies again.

Apr. 24, 1990 NASA launches the long-awaited Hubble Space Telescope, the cornerstone of the agency's "Great Observatory" program, aboard space shuttle Discovery. Shortly after placing the telescope in orbit, astronomers discover that the telescope's prime mirror is misshapen.

Dec. 2, 1993 Space shuttle Endeavour takes off for one of NASA's most critical shuttle missions: repairing the Hubble Space Telescope. During an unprecedented five space walks, astronauts install corrective optics. The mission is a complete success.

Feb. 3, 1994 A Russian cosmonaut, Sergei Krikalev, flies aboard a U.S. spaceship for the first time.

Mar. 16, 1995 NASA astronaut Norman Thagard begins a three and a half month mission on Mir—the first American to train and fly on a Russian spaceship. He is the first of seven Americans to live on Mir.

Mar. 22, 1995 Cosmonaut Valeri Polyakov sets a new space endurance record of 437 days, 18 hours.

June 29, 1995 Space shuttle Atlantis docks for the first time at the Russian space station Mir.

Mar. 24, 1996 Shannon Lucid begins her stay aboard space aboard Mir, which lasts 188 days—a U.S. record for spaceflight endurance at that time.

Feb. 24, 1997 An oxygen canister on Mir bursts into flames, cutting off the route to the station's emergency escape vehicles. Six crewmembers are onboard, including U.S. astronaut Jerry Linenger.

June 27, 1997 During a practice of a new docking technique, Mir commander Vasily Tsibliyev loses control of an unpiloted cargo ship and it plows into the station. The Spektr module is punctured, The crew hurriedly seals off the compartment to save the ship.

Oct. 29, 1998 Senator John Glenn, one of the original Mercury astronauts, returns to space aboard the shuttle.

Nov. 20, 1998 A Russian Proton rocket hurls the first piece of the International Space Station into orbit.

Aug. 27, 1999 Cosmonauts Viktor Afanasyev, Sergei Avdeyev, and Jean-Pierre Haignere leave Mir. The station is unoccupied for the first time in almost a decade.

Oct. 31, 2000 The first joint American-Russian crew is launched to the International Space Station. Commander Bill Shepherd requests the radio call sign "Alpha" for the station and the name sticks.

Mar. 23, 2001 The Mir space station drops out of orbit and burns up in Earth's atmosphere.

Apr. 28, 2001 Russia launches the world's first space tourist for a week-long stay at the International Space Station. NASA objects to the flight, but is powerless to stop it.

Irene Brown

Contributors

Richard G. Adair
Allochthon Enterprises
Reno, Nevada

Constance M. Adams
Lockheed Martin Space Operations Company
Houston, Texas

Joseph K. Alexander
Space Studies Board of the National Research Council
Washington, D.C.

Judith H. Allton
Apollo Historian
Houston, Texas

Ulises R. Alvarado
Instrumentation Technology Associates, Inc.
Exton, Pennsylvania

Susan Ames
University of Bonn
Bonn, Germany

Jayne Aubele
New Mexico Museum of Natural History and Science
Albuquerque, New Mexico

Michael Babich
University of Illinois
Rockford, Illinois

Nadine G. Barlow
Northern Arizona University
Flagstaff, Arizona

William E. Barrett
GIO Space
Sydney, Australia

Jill Bechtold
University of Arizona
Tucson, Arizona

James Bell
Cornell University
Ithaca, New York

Gregory R. Bennett
Bigelow Aerospace Co.
Las Vegas, Nevada

James W. Benson
SpaceDev
Poway, California

Joseph T. Bielitzki
Arlington, Virginia

William Bottke
Southwest Research Institute
Boulder, Colorado

Chad Boutin
Chicago, Illinois

Irene Brown
Melbourne, Florida

Lance B. Bush
NASA Headquarters
Washington, D.C.

Vickie Elaine Caffey
Milano, Texas

Sherri Chasin Calvo
Clarksville, Maryland

Len Campaigne
IBM Global Services
Colorado Springs, Colorado

Humberto Campins
University of Arizona
Tucson, Arizona

John M. Cassanto
Instrumentation Technology Associates, Inc.
Exton, Pennsylvania

Michael R. Cerney
Space Age Publishing Company
Kailua–Kona, Hawaii

Charles M. Chafer
Celestic Inc.
Houston, Texas

Clark R. Chapman
Southwest Research Institute
Boulder, Colorado

David Charbonneau
California Institute of Technology
Pasadena, California

Carissa Bryce Christensen
The Tauri Group, LLC
Alexandria, Virginia

Anita L. Cochran
University of Texas
Austin, Texas

Larry S. Crumpler
New Mexico Museum of Natural History and Science
Albuquerque, New Mexico

Neil Dahlstrom
Space Business Archives
Alexandria, Virginia

Thomas Damon
Pikes Peak Community College
Colorado Springs, Colorado

E. Julius Dasch
NASA Headquarters
Washington, D.C.

Pat Dasch
RCS International
Washington, D.C.

Leonard David
SPACE.com
Boulder, Colorado

Dwayne A. Day
Space Policy Institute
Washington, D.C.

Peter J. Delfyett
University of Central Florida
Orlando, Florida

Lawrence J. DeLucas
University of Alabama at Birmingham
Birmingham, Alabama

David Desrocher
The Aerospace Corporation
Colorado Springs, Colorado

Neelkanth G. Dhere
Florida Solar Energy Center
Cocoa, Florida

Peter H. Diamandis
Zero-Gravity Corporation
Santa Monica, California

John R. Dickel
University of Illinois
Urbana, Illinois

Taylor Dinerman
SpaceEquity.com
New York, New York

Dave Dooling
Infinity Technology Inc.
Huntsville, Alabama

Michael B. Duke
Colorado School of Mines
Golden, Colorado

Douglas Duncan
University of Chicago
Chicago, Illinois

Frederick C. Durant III
Raleigh, North Carolina

Steve Durst
Space Age Publishing Company
Kailua–Kona, Hawaii

Peter Eckart
Technische Universität München
Munich, Germany

Stephen J. Edberg
Jet Propulsion Laboratory
Pasadena, California

Meridel Ellis
Orlando, Florida

Bruce G. Elmegreen
IBM T.J. Watson Research Center
Yorktown Heights, New York

Debra Meloy Elmegreen
Vassar College
Poughkeepsie, New York

Kimberly Ann Ennico
NASA Ames Research Center
Moffett Field, California

R. Bryan Erb
Sunsat Energy Council
Friendswood, Texas

Jack D. Farmer
Arizona State University
Tempe, Arizona

Adi R. Ferrara
Bellevue, Washington

Margaret G. Finarelli
International Space University
Washington, D.C.

Rick Fleeter
AeroAstro, Inc.
Herndon, Virginia

Theodore T. Foley II
NASA Johnson Space Center
Houston, Texas

Jeff Foust
Rockville, Maryland

Wendy L. Freedman
Carnegie Observatories
Pasadena, California

Michael Fulda
Fairmont State College
Fairmont, West Virginia

Lori Garver
DFI International
Washington, D.C.

Sarah Gibson
National Center for Atmospheric Research
Boulder, Colorado

John F. Graham
American Military University
Manassas, Virginia

Bernard Haisch
Palo Alto, California

Robert L. Haltermann
Space Transportation Association
Alexandria, Virginia

Heidi B. Hammel
Space Sciences Institute
Boulder, Colorado

Alan Harris
Jet Propulsion Laboratory
Pasadena, California

Albert A. Harrison
University of California, Davis
Davis, California

Mary Kay Hemenway
University of Texas
Austin, Texas

Henry R. Hertzfeld
Space Policy Institute
Washington, D.C.

Adrian J. Hooke
Jet Propulsion Laboratory
Pasadena, California

Brian Hoyle
BDH Science Communications
Bedford, Nova Scotia

Robert P. Hoyt
Tethers Unlimited, Inc.
Seattle, Washington

Edward Hujsak
Mina-Helwig Company
La Jolla, California

Nadine M. Jacobson
Takoma Park, Maryland

Kevin Jardine
Spectrum Astro, Inc.
Gilbert, Arizona

Terry C. Johnson
Kansas State University
Manhattan, Kansa

Brad Jolliff
Washington University
St. Louis, Missouri

Thomas D. Jones
The Space Agency
Houston, Texas

Mark E. Kahn
National Air and Space Museum
Washington, D.C.

Marshall H. Kaplan
Launchspace, Inc.
Rockville, Maryland

Michael S. Kelley
NASA Johnson Space Center
Houston, Texas

Michael S. Kelly
Kelly Space & Technology
San Bernardino, California

Lisa Klink
Cement City, Michigan

Peter Kokh
Lunar Reclamation Society, Inc.
Milwaukee, Wisconsin

Randy L. Korotev
Washington University
St. Louis, Missouri

Roger E. Koss
Ormond Beach, Florida

Lillian D. Kozloski
James Monroe Museum and Memorial Library
Fredericksburg, Virginia

Saunders B. Kramer
Montgomery Village, Maryland

John F. Kross
Ad Astra Magazine
Lincoln University, PA

Timothy B. Kyger
Universal Space Lines
Alexandria, Virginia

Geoffrey A. Landis
NASA John Glenn Research Center
Cleveland, Ohio

Roger D. Launius
NASA History Office
Washington, D.C.

Jennifer Lemanski
Orlando, Florida

Andrew J. LePage
Visidyne, Inc.
Burlington, Masschusetts

Debra Facktor Lepore
Kistler Aerospace Corporation
Kirkland, Washington

David H. Levy
Jarnac Observatory, Inc.
Vail, Arizona

John S. Lewis
University of Arizona
Tucson, Arizona

David L. Lihani
Pierson and Burnett, L.L.P.
Washington, D.C.

Arthur H. Litka
Seminole Community College
Sanford, Florida

Bill Livingstone
GlobalOptions, Inc.
Washington, D.C.

John M. Logsdon
Space Policy Institute
Washington, D.C.

Rosaly M. C. Lopes
Jet Propulsion Laboratory
Pasadena, California

Mark L. Lupisella
NASA Goddard Space Flight Center
Greenbelt, Maryland

Jeff Manber
Space Business Archives
Alexandria, Virginia

John C. Mankins
NASA Headquarters
Washington, D.C.

Robert L. Marcialis
University of Arizona
Tucson, Arizona

Margarita M. Marinova
Massachusetts Institute of Technology
Cambridge, Massachusetts

John T. Mariska
Naval Research Laboratory
Washington, D.C.

Shinji Matsumoto
Shimizu Corporation
Tokyo, Japan

Julie L. McDowell
Baltimore, Maryland

Christopher P. McKay
NASA Ames Research Center
Moffett Field, California

David A. Medek
Palo Verde Nuclear Generating Station
Wintersburg, Arizona

Karen J. Meech
University of Hawaii
Honolulu, Hawaii

Wendell Mendell
NASA Johnson Space Center
Houston, Texas

Douglas M. Messier
SpaceJobs.com
Arlington, Virginia

Mark Miesch
National Center for Atmospheric Research
Boulder, Colorado

Frank R. Mignone
University of Central Florida
Orlando, Florida

Ron Miller
King George, Virginia

Julie A. Moberly
Microgravity News
Hampton, Virginia

Jefferson Morris
Aerospace Daily
Washington, D.C.

Clayton Mowry
Arianespace, Inc.
Washington, D.C.

Peter Norvig
NASA Ames Research Center
Moffett Field, California

Tim Palucka
Pittsburgh, Pennsylvania

Jay Pasachoff
Williams College
Williamstown, Massachusetts

Chris A. Peterson
University of Hawaii
Honolulu, Hawaii

A. G. Davis Philip
ISO & Union College
Schenectady, New York

Deborah Pober
Montgomery County, Maryland

Barbara Poppe
NOAA Space Environment Center
Boulder, Colorado

David S. F. Portree
Flagstaff, Arizona

Jane Poynter
Paragon Space Development Corporation
Tucson, Arizona

Cynthia S. Price
Celestic Inc.
Houston, Texas

Nick Proach
Nick Proach Models
Sechelt, British Columbia, Canada

Margaret S. Race
SETI Institute
Mountain View, California

Sudhakar Rajulu
NASA/NSBRI–Baylor College of Medicine
Houston, Texas

Clinton L. Rappole
University of Houston
Houston, Texas

Martin Ratcliffe
International Planetarium Society
Wichita, Kansas

Pat Rawlings
Science Applications International Corporation
Houston, Texas

Elliot Richmond
Education Consultants
Austin, Texas

Carlos J. Rosas-Anderson
Kissimmee, Florida

John D. Rummel
NASA Headquarters
Washington, D.C.

Graham Ryder
Lunar and Planetary Institute
Houston, Texas

Salvatore Salamone
New York, New York

Craig Samuels
University of Central Florida
Orlando, Florida

Eagle Sarmont
AffordableSpaceFlight.com
Stockton, California

Joel L. Schiff
Meteorite! Magazine and Auckland University
Takapuna-Auckland, New Zealand

Mark A. Schneegurt
Wichita State University
Wichita, Kansas

David G. Schrunk
Quality of Laws Institute
Poway, California

Alison Cridland Schutt
National Space Society
Washington, D.C.

Derek L. Schutt
Carnegie Institution of Washington
Washington, D.C.

Seth Shostak
SETI Institute
Mountain View, California

Frank Sietzen, Jr.
Space Transportation Association
Alexandria, Virginia

Samuel Silverstein
Space News for SPACE.com
Springfield, Virginia

Michael A. Sims
NASA Ames Research Center
Moffett Field, California

Phil Smith
Futron Corporation
Bethesda, Maryland

Amy Paige Snyder
Federal Aviation Administration
Washington, D.C.

Mark J. Sonter
Asteroid Enterprises Pty Ltd
Hawthorne Dene, South Australia

Barbara Sprungman
Space Data Resources & Information
Boulder, Colorado

S. Alan Stern
Southwest Research Institute
Boulder, Colorado

Robert G. Strom
University of Arizona
Tucson, Arizona

Angela Swafford
Miami Beach, Florida

Amy Swint
Houston, Texas

Leslie K. Tamppari
Pasadena, California

Jeffrey R. Theall
NASA Johnson Space Center
Houston, Texas

Frederick E. Thomas
Orlando, Florida

Robert Trevino
NASA Johnson Space Center
Houston, Texas

Roeland P. van der Marel
Space Telescope Science Institute
Baltimore, Maryland

Anthony L. Velocci, Jr.
Aviation Week & Space Technology
New York, New York

Joan Vernikos
Thirdage LLC
Alexandria, Virginia

Ray Villard
*Space Telescope Science
Institute*
Baltimore, Maryland

Matt Visser
Washington University
St. Louis, Missouri

Linda D. Voss
Arlington, Virginia

Charles D. Walker
Annandale, Virginia

Timothy R. Webster
*Science Applications International
Corporation*
Torrance, California

Jean-Marie (J.-M.) Wersinger
Auburn University
Auburn, Alabama

Joshua N. Winn
*Harvard–Smithsonian Center for
Astrophysics*
Cambridge, Massachusetts

Grace Wolf-Chase
*Adler Planetarium & Astronomy
Museum and the University of
Chicago*
Chicago, Illinois

Sidney C. Wolff
*National Optical Astronomy
Observatory*
Tucson, Arizona

Cynthia Y. Young
University of Central Florida
Orlando, Florida

Table of Contents

VOLUME 1: SPACE BUSINESS

PREFACE v
FOR YOUR REFERENCE ix
TIMELINE: MAJOR BUSINESS
   MILESTONES IN U.S. HISTORY xiii
TIMELINE: MILESTONES IN
   SPACE HISTORY xix
TIMELINE: HUMAN ACHIEVEMENTS
   IN SPACE xxix
LIST OF CONTRIBUTORS xxxv

A

Accessing Space 1
Advertising 4
Aerospace Corporations 6
Aging Studies 12
AIDS Research 14
Aldrin, Buzz 16
Artwork 17
Augustine, Norman 19

B

Barriers to Space Commerce 20
Burial 24
Business Failures 25
Business Parks 27

C

Cancer Research 31
Career Astronauts 33
Careers in Business and Program
   Management 37
Careers in Rocketry 40
Careers in Space Law 44
Careers in Space Medicine 45

Careers in Writing, Photography,
   and Filmmaking 48
Clarke, Arthur C. 52
Commercialization 53
Communications Satellite Industry 57
Crippen, Robert 63

D

Data Purchase 64

E

Education 66
Energy from Space 70
Entertainment 72

F

Financial Markets, Impacts on 76

G

Getting to Space Cheaply 77
Global Industry 81
Global Positioning System 85
Goddard, Robert Hutchings 89
Ground Infrastructure 91

H

Human Spaceflight Program 94

I

Insurance 97
International Space Station 100
International Space University 105

L

Launch Industry 107
Launch Services 113

Launch Vehicles, Expendable 116
Launch Vehicles, Reusable 120
Law of Space 125
Legislative Environment 128
Licensing 131
Literature 132
Lucas, George 134
Lunar Development 135

M

Made in Space 138
Made with Space Technology 142
Market Share 145
Marketplace 147
McCall, Robert 149
Military Customers 150
Mueller, George 153

N

Navigation from Space 154

O

Oberth, Hermann 156

P

Payloads and Payload Processing 156
Planetary Exploration 161

R

Reconnaissance 164
Regulation 166
Remote Sensing Systems 168
Rocket Engines 177
Roddenberry, Gene 180

S

Satellite Industry 181
Satellites, Types of 188
Search and Rescue 191
Servicing and Repair 193
Small Satellite Technology 194
Space Shuttle, Private Operations 197
Spaceports 202
Sputnik 205

T

Thompson, David 207
Tourism 207
Toys 210

V

Verne, Jules 212

X

X PRIZE 214

PHOTO AND ILLUSTRATION CREDITS .. 217
GLOSSARY 219
INDEX 243

VOLUME 2: PLANETARY SCIENCE AND ASTRONOMY

PREFACE v
FOR YOUR REFERENCE ix
TIMELINE: MILESTONES IN SPACE
   HISTORY xiii
TIMELINE: HUMAN ACHIEVEMENTS
   IN SPACE xxiii
LIST OF CONTRIBUTORS xxix

A

Age of the Universe 1
Asteroids 3
Astronomer 7
Astronomy, History of 9
Astronomy, Kinds of 10

B

Black Holes 15

C

Careers in Astronomy 16
Careers in Space Science 20
Cassini, Giovanni Domenico 23
Close Encounters 23
Comets 27
Copernicus, Nicholas 30
Cosmic Rays 30
Cosmology 32

E

Earth 35
Einstein, Albert 39
Exploration Programs 40
Extrasolar Planets 47

G

Galaxies 50
Galilei, Galileo 56

Government Space Programs 57
Gravity 63

H

Herschel Family 66
Hubble Constant 66
Hubble, Edwin P. 67
Hubble Space Telescope 68
Huygens, Christiaan 75

J

Jupiter 76

K

Kepler, Johannes 81
Kuiper Belt 81
Kuiper, Gerard Peter 83

L

Life in the Universe, Search for 84
Long Duration Exposure Facility
   (LDEF) 90

M

Mars 92
Mercury 98
Meteorites 103
Microgravity 107
Military Exploration 108
Moon 109

N

Neptune 115
Newton, Isaac 119

O

Observatories, Ground 119
Observatories, Space-Based 126
Oort Cloud 132
Orbits 134

P

Planet X 137
Planetariums and Science Centers 138
Planetary Exploration, Future of 142
Planetesimals 146
Pluto 147
Pulsars 151

R

Robotic Exploration of Space 154
Robotics Technology 160

S

Sagan, Carl 163
Saturn 164
Sensors 168
SETI 170
Shapley, Harlow 173
Shoemaker, Eugene 173
Small Bodies 174
Solar Particle Radiation 177
Solar Wind 180
Space Debris 182
Space Environment, Nature of the 183
Spacecraft Buses 188
Spaceflight, History of 190
Spacelab 194
Stars 198
Sun 201
Supernova 205

T

Tombaugh, Clyde 207
Trajectories 209

U

Uranus 210

V

Venus 214

W

Weather, Space 218
What is Space? 220

PHOTO AND ILLUSTRATION CREDITS . . 223
GLOSSARY 225
INDEX 249

VOLUME 3: HUMANS IN SPACE

PREFACE v
FOR YOUR REFERENCE ix
TIMELINE: MILESTONES IN SPACE
   HISTORY xiii
TIMELINE: HUMAN ACHIEVEMENTS IN
   SPACE xxiii
LIST OF CONTRIBUTORS xxix

A

Animals 1
Apollo 3
Apollo I Crew 9
Apollo Lunar Landing Sites 11
Apollo-Soyuz 15
Armstrong, Neil 18
Astronauts, Types of 19

B

Bell, Larry 21
Biosphere 21

C

Capcom 24
Capsules 25
Careers in Spaceflight 27
Challenger 32
Challenger 7 34
Civilians in Space 37
Closed Ecosystems 38
Collins, Eileen 41
Communications for Human Spaceflight . . . 42
Computers, Use of 44
Cosmonauts 47
Crystal Growth 48

E

Emergencies 50
Environmental Controls 52
Escape Plans 54
External Tank 56

F

Faget, Max 58
Flight Control 58
Food 60

G

G Forces 63
Gagarin, Yuri 64
Gemini 65
Getaway Specials 68
Glenn, John 70
Guidance and Control Systems 71
Gyroscopes 72

H

Habitats 74
Heat Shields 77
History of Humans in Space 79
Human Factors 84
Human Missions to Mars 87
Humans versus Robots 89
Hypersonic Programs 93

I

Inertial Measurement Units 96
International Cooperation 96
International Space Station 100

K

KC-135 Training Aircraft 105
Kennedy, John F. 108
Korolev, Sergei 109

L

Launch Management 110
Launch Sites 112
Leonov, Alexei 115
Life Support 116
Living in Space 119
Long-Duration Spaceflight 122
Lunar Rovers 125

M

Manned Maneuvering Unit 127
Medicine 129
Mercury Program 132
Mir 135
Mission Control 138
Mission Specialists 139
Modules 141

N

NASA 143
Navigation 147

O

Oxygen Atmosphere in Spacecraft 149

P

Payload Specialists 151
Payloads 152
Primates, Non-Human 155

R

Reaction Control Systems 156
Re-entry Vehicles 157
Rendezvous 159
Ride, Sally 161
Rockets 162

S

Sänger, Eugene 166
Sanitary Facilities 166
Schmitt, Harrison 169
Shepard, Alan 170
Simulation 171
Skylab 174
Solid Rocket Boosters 177
Space Centers 180
Space Shuttle 183
Space Stations, History of 187
Space Suits 191
Space Walks 195
Stapp, John 199
Sullivan, Kathryn 200

T

T-38 Trainers 201
Teacher in Space Program 202
Tereshkova, Valentina 205
Tools, Apollo Lunar Exploration 206
Tools, Types of 208
Tracking of Spacecraft 210
Tracking Stations 213
Tsiolkovsky, Konstantin 216

V

Vehicle Assembly Building 216
von Braun, Wernher 218
Voshkod 219
Vostok 221

W

White Room 222
Why Human Exploration? 224
Women in Space 227

Y

Young, John 233

Z

Zero Gravity 234
Zond 235

Photo and Illustration Credits . . 239
Glossary 241
Index 265

VOLUME 4: OUR FUTURE IN SPACE

Preface v
For Your Reference ix
Timeline: Milestones in Space
    History xiii
Timeline: Human Achievements in
    Space xxiii
List of Contributors xxix

A

Antimatter Propulsion 1
Asteroid Mining 3
Astrobiology 6

B

Biotechnology 13
Bonestell, Chesley 16

C

Careers in Space 18
Chang-Díaz, Franklin 22
Comet Capture 22
Communications, Future Needs in 24
Communities in Space 27
Cycling Spacecraft 29

D

Domed Cities 30
Dyson, Freeman John 32
Dyson Spheres 33

E

Earth—Why Leave? 34
Ehricke, Kraft 38
Environmental Changes 39

F

Faster-Than-Light Travel 42
First Contact 43
Food Production 45

G

Glaser, Peter 48
Governance 49

H

Hotels 51

I

Impacts 53
Interplanetary Internet 57
Interstellar Travel 58
Ion Propulsion 61

L

L-5 Colonies 62
Land Grants 64
Laser Propulsion 66
Lasers in Space 68
Launch Facilities 71
Law 76
Ley, Willy 78
Lightsails 79
Living on Other Worlds 81
Lunar Bases 85
Lunar Outposts 89

M

Mars Bases 92
Mars Direct 96
Mars Missions 98
Military Uses of Space 101
Miniaturization 106
Movies 108

N

Nanotechnology 111
Natural Resources 115
Nuclear Propulsion 118
O'Neill Colonies 121
O'Neill, Gerard K. 124

P

Planetary Protection 125
Political Systems 128
Pollution 129
Power, Methods of Generating 132
Property Rights 134

R

Rawlings, Pat 137
Religion 138
Resource Utilization 140
Reusable Launch Vehicles 143

S

Satellites, Future Designs 147
Science Fiction 150
Scientific Research 152
Settlements 156
Social Ethics 160
Solar Power Systems 163
Space Elevators 167
Space Industries 171
Space Resources 173
Space Stations of the Future 177
Space Tourism, Evolution of 180
Star Trek 184
Star Wars 185
Stine, G. Harry 187

T

Teleportation 187
Telepresence 189
Terraforming 190
Tethers 194
Time Travel 197
Traffic Control 199
TransHab 201

U

Utopia 204

V

Vehicles 205

W

Wormholes 207

Z

Zero Point Energy 209
Zubrin, Robert 211

PHOTO AND ILLUSTRATION CREDITS . . 213
GLOSSARY 215
CUMULATIVE INDEX 239

Age of the Universe

The idea that the universe had a beginning is common to various religions and mythologies. However, astronomical evidence that the universe truly has a finite age did not appear until early in the twentieth century. The first clue that the universe has a finite age came at the end of World War I, when astronomer Vesto Slipher noted that a mysterious class of objects, collectively called spiral **nebula**, were all receding from Earth. He discovered that their light was stretched or reddened by their apparent motion away from Earth—the same way an ambulance siren's pitch drops when it speeds away from a stationary observer.

nebula clouds of interstellar gas and/or dust

Hubble's Contribution

In the early 1920s American astronomer Edwin P. Hubble was able to measure the distances to these receding objects by using a special class of milepost marker stars called Cepheid variables. Hubble realized that these spiral nebulae were so far away they were actually galaxies—separate cities of stars—far beyond our own Milky Way.

By 1929, Hubble had made the momentous discovery that the farther away a **galaxy** is, the faster it is receding from Earth. This led him to conclude that galaxies are apparently moving away because space itself is expanding uniformly in all directions. Hubble reasoned that the galaxies must inevitably have been closer to each other in the distant past. Indeed, at some point they all must have occupied the same space. This idea led theoreticians to conceive of the notion of the Big Bang, the theory that the universe ballooned from an initially hot and dense state.

galaxy a system of as many as hundreds of billions of stars that have a common gravitational attraction

Hubble realized that if he could measure the universe's speed of expansion, he could easily calculate the universe's true age. Assuming the universe's expansion rate has not changed much over time, he calculated an age of about 2 billion years. One problem with this estimate, however, was that it was younger than geologists' best estimate for the age of Earth at the time.

Astronomers since then have sought to refine the expansion rate—and the estimate for the universe's age—by more precisely measuring distances to galaxies. Based on uncertainties over the true distances of galaxies, estimates for the universe's age have varied from 10 billion to 20 billion years old.

EINSTEIN'S VIEW

Despite his genius for envisioning the farthest reaches of space and time, even the great theoretician Albert Einstein could not imagine that the universe had a beginning. When, using his general theory of relativity, he predicted that the universe should be collapsing under the pull of gravity among galaxies, Einstein arbitrarily altered his equations to maintain an eternal, static universe.

Astronomers were able to date all of the stars in globular star cluster M80 (as seen through the Hubble Space Telescope) at 15 billion years.

More Recent Estimates

A primary task of the Hubble Space Telescope (HST), launched in 1990, was to break this impasse by observing Cepheid variable stars in galaxies much farther away than can be seen from ground-based telescopes. The HST allowed astronomers to measure precisely the universe's expansion rate and calculate an age of approximately 11 to 12 billion years.

Estimating the age is now complicated, however, by recent observations that show the universe expanded at a slower rate in the past. This is due to some mysterious repulsive force, first envisioned by physicist Albert Einstein as part of his so-called fudge factor in keeping the universe balanced. The presence of such a repulsive force pushing galaxies apart means that the universe is more likely to be 13 to 15 billion years old.

Using Stars to Estimate Age. The universe's age can also be estimated independently by observing the oldest stars. Astronomers know that stars must have started forming quickly after the universe expanded and cooled enough for gas to **coagulate** into stars. So the oldest star must be close to the true age of the universe itself. The oldest stars, which lie inside **globular clusters** that orbit our galaxy, are estimated to be at least 12 billion years old. These estimates are difficult because they rely on complex models and calculations about how a star burns its nuclear fuel and ages.

coagulate to cause to come together into a coherent mass

globular clusters roughly spherical collections of hundreds of thousands of old stars found in galactic haloes

A simpler cosmic clock is a class of star called white dwarfs, which are the burned-out remnants of Sun-like stars. Like dying cinders, it takes a long time for dwarfs to cool to absolute zero—longer than the present age of the universe itself. So the coolest, dimmest dwarfs represent the remnants of the oldest stars. Because they are so dim, these dwarfs are hard to find. Astronomers are using the HST to pinpoint the very oldest white dwarfs in globular clusters.

The HST has uncovered the very faintest and coolest dwarfs in the Milky Way galaxy, with ages of 12.6 billion years, thus giving an age estimate for the universe of 13 to 14 billion years. This is a very successful and entirely independent confirmation of previous age estimates of the universe.

Astronomers now know the age of the universe to within a good degree of accuracy. This is quite an achievement considering that less than a century ago, astronomers did not even realize the universe had a beginning. SEE ALSO Cosmology (VOLUME 2); Hubble, Edwin P. (VOLUME 2); Stars (VOLUME 2).

Ray Villard

Bibliography

Guth, Alan H., and Alan P. Lightman. *The Inflationary Universe: The Quest for a New Theory of Cosmic Origins.* Reading, MA: Addison-Wesley Publishing, 1998.

Hogan, Craig J., and Martin Rees. *The Little Book of the Big Bang: A Cosmic Primer.* New York: Copernicus Books, 1998.

Livio, Mario, and Allan Sandage. *The Accelerating Universe: Infinite Expansion, the Cosmological Constant, and the Beauty of the Cosmos.* New York: John Wiley & Sons, 2000.

Asteroids

Asteroids are small bodies in space—the numerous leftover **planetesimals** from which the planets were made nearly 4.6 billion years ago. Most are in the "main belt," which is a doughnut-like volume of space between Mars and Jupiter (about 2.1 to 3.2 astronomical units [AU] from the Sun; one AU is equal to the mean distance between Earth and the Sun). The Trojans are two groups of asteroids around 60 degrees ahead of (and behind) Jupiter in its **orbit** (5.2 AU from the Sun). Asteroids range in location from within Earth's orbit to the outer solar system, where the distinction between asteroids and comets blurs.

Some asteroids orbit at a solar distance where their year is matched to Jupiter's year. For example, the Hilda asteroids circle the Sun three times for every two revolutions of Jupiter. Other Jupiter-asteroid relationships are unstable, so asteroids are missing from those locations. For example, gaps occur in the main belt where asteroids orbit the Sun twice and three times each **Jovian** year. These gaps are called Kirkwood gaps. Any asteroids originally formed in such locations have been kicked out of the asteroid belt by Jupiter's strong gravitational forces, so no asteroids remain there.

Many asteroids are members of groups with very similar orbital shapes, tilts, and solar distances. These so-called families were formed when asteroids smashed into each other at interasteroidal velocities of 5 kilometers per

planetesimals objects in the early solar system that were the size of large asteroids or small moons, large enough to begin to gravitationally influence each other

orbit the circular or elliptical path of an object around a much larger object, governed by the gravitational field of the larger object

Jovian relating to the planet Jupiter

Eros is a 34-kilometer-long, Earth-approaching asteroid.

minerals crystalline arrangements of atoms and molecules of specified proportions that make up rocks

ultraviolet the portion of the electromagnetic spectrum just beyond (having shorter wavelengths than) violet

infrared portion of the electromagnetic spectrum with wavelengths slightly longer than visible light

wavelength the distance from crest to crest on a wave at an instant in time

carbonaceous meteorites the rarest kind of meteorites, they contain a high percentage of carbon and carbon-rich compounds

radar a technique for detecting distant objects by emitting a pulse of radio-wavelength radiation and then recording echoes of the pulse off the distant objects

second (3 miles per second). Fragments from such explosive disruptions became separate asteroids.

Asteroid Sizes, Shapes, and Compositions

Ceres, the first asteroid to be discovered (on January 1, 1801), remains the largest asteroid found to date; it is about 1,000 kilometers (620 miles) in diameter. Dozens of asteroids range from 200 to 300 kilometers (124 to 186 miles) in diameter, thousands are the size of a small city, and hundreds of billions are house-sized. Indeed, asteroids grade into the rocks that occasionally burn through our atmosphere as fireballs and the even smaller grains of sand that produce meteors ("shooting stars") in a clear, dark sky. Collected remnants are called meteorites. All are debris from the cratering and catastrophic disruptions of inter-asteroidal collisions.

Asteroids are small and distant, so even in telescopes they are only faint points of light gradually moving against the backdrop of the stars. Astronomers use telescopes to measure asteroid motions, brightnesses, and the spectral colors of sunlight reflected from their surfaces. Asteroid brightnesses change every few hours as they spin, first brightening when they are broadside to us and fading when end-on. From these data, astronomers infer that most asteroids have irregular, nonspherical shapes and spin every few minutes (for some very small asteroids) to less often than once a month.

Different **minerals** reflect sunlight (at **ultraviolet**, visible, and **infrared wavelengths**) in different ways. So the spectra of asteroids enable astronomers to infer what they are made of. Many are made of primitive materials, such as rocky minerals and flecks of metal, from which it is believed the planets were made. Such is the case with the ordinary chondrites, the most common meteorites in museums. Most asteroids are exceedingly dark in color, and are apparently rich in carbon and other black compounds, including the uncommon **carbonaceous meteorites**. Such fragile, C-type materials are abundant in space but often disintegrate when passing through Earth's atmosphere. C-type asteroids may even contain water ice deep below their surfaces.

While most asteroids survived fairly unchanged from the earliest epochs of solar system history, others were heated and melted. The metal flecks sank to form iron cores (like nickel-iron meteorites), while lighter rocks floated upwards and flowed out across their surfaces, like lavas do on Earth. Vesta, one of the largest asteroids, appears to be covered with lava; certain lava-like meteorites probably came from Vesta. Metallic asteroids are rare but are readily recognized by Earth-based **radar** observations because metal reflects radar pulses well.

New techniques in astronomy, such as radar delay–Doppler mapping and adaptive optics (which unblurs the twinkling of visible light induced by Earth's atmosphere), have revealed a variety of asteroid shapes and configurations. One asteroid, named Antiope, is a double body: Two separate bodies, each 80 kilometers (50 miles) across and separated by 160 kilometers (100 miles), orbit about each other every sixteen hours. Other asteroids have satellites (e.g., moonlets) and still others have very odd shapes (e.g., dumbbells).

The surface of the asteroid Eros looks similar to a desert on Earth. In reality, the environment of an asteroid is highly dissimilar to Earth's, with low gravity, no atmosphere, and a rotation period of a little more than five-and-one-quarter hours.

Spacecraft Studies of Asteroids

The best (though most expensive) way to study an asteroid, of course, is to send a spacecraft. Three main-belt asteroids—Gaspra, Ida, and Mathilde—were visited in the 1990s by spacecraft en route to other targets. But even during the few minutes available for close-up observations during such high-speed encounters, scientists obtained images a hundred times sharper than the best possible images from Earth.

The most thorough study of an asteroid was of Eros by the Near Earth Asteroid Rendezvous spacecraft (which was renamed NEAR Shoemaker, after American astronomer Eugene Shoemaker, who first thought of the enterprise). Eros is a 34-kilometer-long (21-mile-long), Earth-approaching asteroid. NEAR Shoemaker orbited Eros until February 12, 2001, when it

The Near Earth Asteroid Rendezvous (NEAR) spacecraft, renamed NEAR Shoemaker after scientist Eugene Shoemaker, was the first of NASA's Discovery Program spacecrafts, providing small-scale, low-cost planetary missions.

chondrite meteorite a type of meteorite that contains spherical clumps of loosely consolidated minerals

was landed on the asteroid's surface. Its instruments were designed specifically for asteroid studies. It revealed Eros to be an oddly shaped, heavily cratered object, with ridges and grooves, and covered by a million boulders, each larger than a house. Eros is made of minerals much like the ordinary **chondrite meteorites**.

Near Earth Asteroids

A few asteroids escape from the main belt through Kirkwood gaps and move around the Sun on elongated orbits that can cross the orbits of Mars and Earth. If an asteroid comes within 0.3 AU of Earth, it is called a near Earth asteroid (NEA). More than half of the estimated 1,000 NEAs larger than 1 kilometer (0.6 mile) in diameter have been discovered. Orbits of NEAs are not stable, and within a few million years they collide with the Sun, crash into a planet, or are ejected from the solar system.

stratosphere a middle portion of Earth's atmosphere above the tropopause (the highest place where convection and "weather" occurs)

The Threat of Impacts. If a 2-kilometer (1.2-mile) NEA struck Earth, it would explode as 100,000 megatons of TNT, more than the world's nuclear weapons arsenal. It would contaminate the **stratosphere** with so much Sun-darkening dust that humans would lose an entire growing season worldwide, resulting in mass starvation and threatening civilization as we know it. Such a collision happens about once every million years, so there is one chance in 10,000 of one occurring during the twenty-first century. A 10- or 15-kilometer (6- or 9-mile) asteroid, like the one that caused the extinction of the dinosaurs 65 million years ago, hits every 50 or 100 million years with a force of 100 million megatons.

impact winter the period following a large asteroidal or cometary impact when the Sun is dimmed by stratospheric dust and the climate becomes cold worldwide

Though the chances of dying by asteroid impact are similar to the chances of dying in an air crash, society has done little to address the impact hazard. Modest telescopic searches for threatening objects are underway in several countries. Given months to a few years warning, ground zero could be evacuated and food could be saved to endure an **impact winter**. If given many years, or decades, of warning, high-tech space missions could be launched in an attempt to study and then divert the oncoming body. SEE ALSO ASTEROID MINING (VOLUME 4); IMPACTS (VOLUME 4); CLOSE EN- COUNTERS (VOLUME 2); GALILEI, GALILEO (VOLUME 2); METEORITES (VOLUME 2); PLANETESIMALS (VOLUME 2); SHOEMAKER, EUGENE (VOLUME 2); SMALL BOD- IES (VOLUME 2).

Clark R. Chapman

Bibliography

Chapman, Clark R. "Asteroids." In *The New Solar System*, 4th ed., ed. J. Kelly Beatty, Carolyn Collins Petersen, and Andrew Chaikin. New York and Cambridge, UK: Sky Publishing Corp. and Cambridge University Press, 1999.

Gehrels, Tom, ed. *Hazards Due to Comets and Asteroids.* Tucson: University of Ari- zona Press, 1994.

Veverka, Joseph, Mark Robinson, and Pete Thomas. "NEAR at Eros: Imaging and Spectral Results." *Science* 289 (2000):2088–2097.

Yeomans, Donald K. "Small Bodies of the Solar System." *Nature* 404 (2000):829–832.

Internet Resources

Arnett, Bill. "Asteroids." <http://www.seds.org/nineplanets/nineplanets/asteroids .html>.

Near-Earth Object Program. NASA Jet Propulsion Laboratory, California Institute of Technology. <http://neo.jpl.nasa.gov/>.

Astrobiology *See Astrobiology (Volume 4).*

Astronomer

An astronomer is an individual who studies the universe primarily using tele- scopes. Astronomers rely on both observations of celestial objects, includ- ing planets, stars, and galaxies, and physical theories to better understand how these objects formed and work. Although professional astronomers con- duct most astronomy research today, amateur sky watchers continue to play a key role.

Astronomy has been practiced since the beginning of recorded history. Many ancient civilizations employed people with some knowledge of the night sky and the motions of the Sun and Moon, although in many cases the identities of these ancient astronomers have long since been lost. At that time the work of astronomers had both practical importance, in the form of keeping track of days, seasons, and years, as well as religious implications. Astronomers did not emerge as true scientists until the Renaissance, when new observations and theories by astronomers such as Nicholas Copernicus (1473–1543) of Poland and Galileo Galilei (1564–1642) of Italy challenged the beliefs of the church. Since then astronomers have gradually emerged as scientists in the same class as physicists and chemists, employed primar- ily by universities and government research institutions.

Two Types of Astronomers

In the early twenty-first century, astronomers can be grouped into two dif- ferent types, observational and theoretical. Observational astronomers use telescopes, on Earth and in space, to study objects ranging from planets and moons to distant galaxies. They analyze images, spectra, and other data in an effort to gain new knowledge about the objects under examination.

Astronomers use many different tools, including telescopes, satellites, computers, and radio telescope dishes, to gather and study information on the universe.

Theoretical astronomers, on the other hand, may never venture near a telescope. They work with computers, or even just pencil and paper, to develop models and theories to explain astronomical phenomena. In many respects observational astronomers are closer to the classical image of an astronomer, whereas theoretical astronomers are more strongly rooted in the worlds of physics and mathematics. The two groups do work closely together: Observational astronomers provide data to help theoretical astronomers develop and refine models, and in turn seek observational evidence for the theoreticians' work.

The Difference Between Astronomy and Astrology

Astronomers are often confused with astrologers, although the two are very different. Astrologers attempt to divine information about the future through the locations of the Sun and planets in the sky. Astrology is opposed by nearly all astronomers, who not only reject the notion that the positions of celestial objects govern the future but also note that many of the data and definitions used by astrologers are inaccurate. Astronomy and astrology, however, were once more closely tied together: In medieval times, many astronomers relied on astrology as a primary means of making a living.

supernova an explosion ending the life of a massive star

Not all astronomers are paid to do their work. There are a large number of amateur astronomers who pursue astronomy as a hobby rather than as a full-time job. They play a useful role in astronomical research, because they can observe the full sky far better than professional astronomers, who focus on small regions of the sky at a particular time. Amateur astronomers have made many asteroid, comet, and **supernova** discoveries. Automated sky surveys by professional astronomers, though, have began to make more of the discoveries that were once made almost exclusively by amateur astronomers. SEE ALSO ASTRONOMY, HISTORY OF (VOLUME 2); ASTRONOMY, KINDS OF (VOLUME 2); CAREERS IN ASTRONOMY (VOLUME 2).

Jeff Foust

Bibliography

Goldsmith, Donald. *The Astronomers.* New York: St. Martin's Press, 1991.

Internet Resources

A New Universe to Explore: Careers in Astronomy. American Astronomical Society. <http://www.aas.org/~education/career.html>.

Odenwald, Sten. "Ask an Astronomer." <http://itss.raytheon.com/cafe/qadir/qanda .html>.

Astronomy, History of

In ancient times, people watched the sky and used its changing patterns throughout the year to regulate their planting and hunting. The Sun seemed to move against the background of stars. A few bright objects (Mercury, Venus, Mars, Jupiter, and Saturn) wandered against the same background. Greek philosopher Aristotle (384–322 B.C.E.) tried to make sense of all this by proposing a system of the universe with Earth in the center (known as a geocentric system). Revolving around Earth were the Sun, the five known planets, and the Moon. This system satisfied the Greek desire for uniformity with its perfectly circular orbits as well as everyone's common sense of watching sunrise and sunset.

Greek astronomer and mathematician Ptolemy refined Aristotle's theory in 140 C.E. by adding more circles to obtain better predictions. For over a thousand years people used his scheme to predict the motions of the planets. Polish astronomer Nicholas Copernicus (1473–1543) was dissatisfied with its increasingly inaccurate predictions. He looked for a method that would be both accurate and mathematically simpler in structure. Although he did not achieve great accuracy, he was able to produce a beautiful scheme with the Sun in the center of the universe (known as a heliocentric system). His system improved on Ptolemy's plan by determining with fair accuracy the relative distance of all the planets from the Sun. However, it still used circles. The plan became a matter of religious controversy because some people did not want to displace humankind from the important spot as the center of the universe.

In 1609, German astronomer Johannes Kepler (1571–1630) showed with careful mathematical calculations that the orbits were not circles but ellipses. (An ellipse is a mathematically determined oval.) Also in 1609, Italian mathematician and astronomer Galileo Galilei (1564–1642) first used a telescope to observe celestial objects. He discovered moons orbiting Jupiter, phases of Venus, sunspots, and features on the Moon that made it seem more like a planet. None of these discoveries proved that the Copernican heliocentric theory was correct, but they offered evidence that Aristotle was wrong. For example, the phases of Venus indicated that Venus orbited the Sun (but did not prove that Earth did also). The discovery of sunspots and lunar surface features proved that the Sun and Moon were not perfect unblemished spheres. Galileo also did experiments to explore gravity and motion. English physicist and mathematician Isaac Newton (1642–1727) articulated the laws of gravity and motion. He also used a prism to split light into its component colors (spectroscopy).

In 1860 Italian astronomer Angelo Secchi (1818–1878) first classified stellar spectra. In the twentieth century, astronomers used spectra to find temperatures and line-of-sight motions of stars and galaxies. Stellar temperature and distance, when combined with the theory of how stars are powered by **fusion**, provide the basis for the current theory of stellar evolution. American astronomer Edwin P. Hubble (1889–1953) discovered that galaxies are moving away from each other as the universe expands. These motions of galaxies and changes of their component stars and gas over time indicate the evolution of the universe. It has thus become clear that although we can map our location with respect to the galaxies, we live in the midst of an expanding universe for which no center can be measured. SEE ALSO AGE OF THE UNIVERSE (VOLUME 2); CASSINI, GIOVANNI (VOLUME 2); COPERNICUS, NICHOLAS (VOLUME 2); COSMOLOGY (VOLUME 2); EINSTEIN, ALBERT (VOLUME 2); EXPLORATION PROGRAMS (VOLUME 2); GALAXIES (VOLUME 2); GALILEI, GALILEO (VOLUME 2); GRAVITY (VOLUME 2); HERSCHEL FAMILY (VOLUME 2); HUBBLE CONSTANT (VOLUME 2); HUBBLE, EDWIN P. (VOLUME 2); HUYGENS, CHRISTIAAN (VOLUME 2); KEPLER, JOHANNES (VOLUME 2); KUIPER, GERARD PETER (VOLUME 2); NEWTON, ISAAC (VOLUME 2); PLANETARY EXPLORATION, FUTURE OF (VOLUME 2); SAGAN, CARL (VOLUME 2); SHAPLEY, HARLOW (VOLUME 2); SHOEMAKER, EUGENE (VOLUME 2); STARS (VOLUME 2); TOMBAUGH, CLYDE (VOLUME 2).

Mary Kay Hemenway

Bibliography

Kuhn, Thomas S. *The Copernican Revolution: Planetary Astronomy in the Development of Western Thought.* Cambridge, MA: Harvard University Press, 1957.

Nicolson, Iain. *Unfolding Our Universe.* Cambridge, UK: Cambridge University Press, 1999.

Astronomy, Kinds of

Astronomers study light, and almost everything we know about the universe has been figured out through the study of light gathered by telescopes on Earth, in Earth's atmosphere, and in space. This light comes in many different **wavelengths** (including visible colors), the sum of which comprises what is known as the electromagnetic spectrum. Unfortunately, Earth's atmosphere blocks almost all wavelengths in the electromagnetic spectrum. Only the visible and radio "windows" are accessible from the ground, and they thus have the longest observational "history." These early restrictions on the observational astronomer also gave rise to classifying "kinds" of astronomy based on their respective electromagnetic portion, such as the term "radio astronomy."

Over the past few decades, parts of the **infrared** and submillimeter have become accessible to astronomers from the ground, but the telescopes needed for such studies have to be placed in high-altitude locations (greater than 3,050 meters [10,000 feet]) or at the South Pole where water absorption is minimal. Other options have included balloon experiments, airborne telescopes, and short-lived rocket experiments.

Presently, the field of astronomy is enriched immensely by the accessibility of several high-caliber airborne telescopes (e.g., Kuiper Airborne Observatory [KAO], Stratospheric Observatory For Infrared Astronomy [SOFIA]) and space telescopes, all of which are opening up other, previously blocked windows of the electromagnetic spectrum (such as gamma ray,

fusion releasing nuclear energy by combining lighter elements such as hydrogen into heavier elements

wavelength the distance from crest to crest on a wave at an instant in time

infrared portion of the electromagnetic spectrum with waves slightly longer than visible light

X ray a form of high-energy radiation just beyond the ultraviolet portion of the spectrum

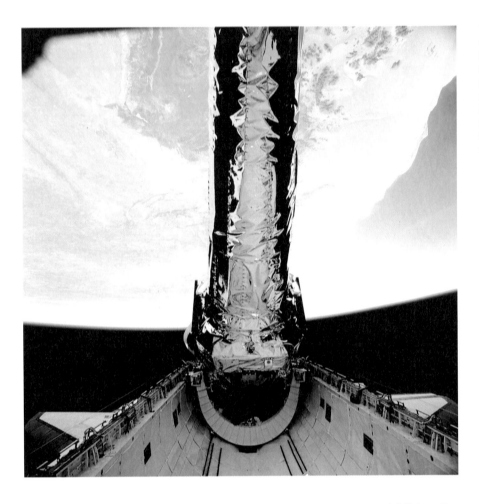

The Chandra X-Ray Observatory, pictured just prior to release from space shuttle Columbia's payload bay, has detected new classes of black holes and is giving astronomers new information about exploding stars.

X ray, **ultraviolet**, far infrared, millimeter, and microwave). Additionally, modern astronomers often need to piece together information from different parts of the electromagnetic spectrum to build up a picture of the physics/chemistry of their object(s) of interest. The table on page 12 summarizes some of the links between wavelength, objects/physics of interest, and current/planned observing platforms. It provides a flavor of how the field of astronomy today varies across wavelength, and hence, by the energy of the object sampled.

The field of astronomy is also quite vast in terms of the physical nature, location, and frequency of object types to study. The field can be broken down into four categories:

1. Solar and **extrasolar planets** and planet formation, star formation, and the **interstellar** medium;
2. Stars (including the Sun) and stellar evolution;
3. Galaxies (including the Milky Way) and stellar systems (clusters, superclusters, large scale structure, **dark matter**); and
4. Cosmology and fundamental physics.

The Study of Planets, Star Formation, and the Interstellar Medium

One of the most important developments in the first category over the past few years has been the detection of several planets orbiting other stars along

ultraviolet the portion of the electromagnetic spectrum just beyond (having shorter wavelengths than) violet

extrasolar planets planets orbiting stars other than the Sun

interstellar between the stars

dark matter matter that interacts with ordinary matter by gravity but does not emit electromagnetic radiation; its composition is unknown

Approximate Wavelengths (m)	Wavelengths Other Units	Photon Energies Greater Than	Frequency	Name for Spectral Brand	Produced by Temperatures in Region of (K)	Examples of Astrophysical Objects of Interest	Examples of Present/ Planned Telescopes to Use for Observations
10^{-13} 10^{-12} 10^{-11}		80.6MeV 80.6MeV 0.8MeV		Gamma-ray	10^8	Cosmic rays, gamma-ray bursters, nuclear processes	**Space only:** CGRO (1991–2000), INTEGRAL (2002–), GLAST (2005–)
10^{-10} 10^{-9}	1Å, 0.1nm 10Å, 1nm	80.6keV 8.06keV		Hard X-ray	10^7	Accretion disks in binaries, black holes, hot gas in galaxy clusters, Seyfert galaxies	**Space only:** ROSAT (1990–1999), ASCA (1993–), Chandra (1999–), XMM (2000–)
10^{-8}	100Å, 10nm	0.806keV		Soft X-ray	10^6	Supernovae remnants, neutron stars, X-ray stars, superbubbles	
10^{-7}	1000Å, 100nm	80eV		XUV/EUV Far UV	10^5	White dwarfs, flare stars, O stars, plasmas	**Space only:** EUVE (1992–), FUSE (1999–)
2×10^{-7}	200nm			Ultraviolet	10^5	Hot/young stars, Orion-like star nurseries, interstellar gas, helium from the big bang, solar corona, Ly alpha forest sources	**Space only:** HST (1990–), Astro-½ (1990, 1995), SOHO (1996–)
4×10^{-7}	400nm			Violet / Visible	10^4	B stars, spiral galaxies, nebulae, Cepheids, QSOs	**Ground:** Keck, Gemini (1999–), Magellan (1999–), Subaru (1999–) VLT (1999–), MMT (2000–),
7×10^{-7}	700nm			Red	10^4	K, M stars, globular clusters, galaxy mass	**Space:** HST
$8{-}50 \times 10^{-7}$	0.8-5µm			Near-infrared		Circumstellar dust shells comets, asteroids, high z galaxies, brown dwarfs	**Ground:** CHFT, CTIO, IRTF, Keck Magellan, Subaru, UKIRT, VLT **Space:** ISO (1995–98), SIRTF (2002–)
$5{-}30 \times 10^{-6}$	5–30µm			Mid–infrared	10^3	Cool interstellar dust, PAHs, organic molecules, planetary nebulae, molecular hydrogen	**Ground:** IR optimized telescopes: IRTF, UKIRT, Gemini **Airborne:** SOFIA (2005–) **Space:** ISO (1995–1998), SIRTF (2002–)
$3{-}20 \times 10^{-5}$	30–200µm			Far-infrared		Ultraluminous/starburst galaxies, debris disks, Kuiper Belt Objects	**Airborne:** SOFIA **Space:** ISO, SIRTF
$3.5{-}10 \times 10^{-4}$	350mm–1mm			Sub-millimeter		High z galaxies/proto-galaxies; molecular clouds; interstellar dust	**Ground:** HHT, JCMT, SMA (1999–) **Space:** SWAS (1998–), FIRST (2008–)
10^{-3}	1 mm	300,000MHz, 300GHz		Millimeter	100	Molecules in dark dense interstellar clouds (CO)	**Gound:** IRAM, ALMA
10^{-2}	1cm	30,000MHz, 30GHz		Microwave	10	Cosmic microwave background	**Space:** COBE (1989–), MAP (2001–)
10^{-1}	10cm	3000MHz, 3GHz		Microwave	1	Galaxy studies, Hydrogen clouds (21cm), masers	
1	1m	300MHz				Quasars, radio galaxies, hot gasses in nebulae	**Ground:** Arecibo, VLA, VLBA, MERLIN **Space:** VSOP (1997–)
10	10m	30MHz		Radio	<1	Synchroton radiation (electronics spiraling in magnetic fields) from supernovae remnants, magnetic lobes of radio galaxies	
10^2	100m	3MHz					
10^3	1 km			Long wave		No data yet. We could explore cosmic ray origins, pulsars, super-novae remnants, and look for coherent emission.	No missions planned, space only due to opaqueness of Earth's ionosphere. Lunar telescope(s) perhaps.
10^4	10km and greater		<30kHz	Very long wave/very low frequency			

SOURCE: Different "kinds" of astronomy separated by wavelength. Adapted and expanded from J. K. Davies, *Astronomy from Space*, 1997, Table 1.1, p.2.

with the detections, through deep infrared sky surveys, of substellar objects (**brown dwarfs**), whose spectral characteristics have been found to be similar to that of giant planets. Additionally, through superb Hubble Space Telescope (HST) imaging with its infrared camera and through infrared instruments on large ground-based telescopes, astronomers have started to directly observe the protostellar disks out of which planets are forming.

Astronomers have learned that the formation of stars and protostellar disks start in the interstellar medium, the vast "vacuum" of gas and dust between the stars, but astronomers are only just learning what the structure of the interstellar medium really is and how it affects and is affected by stellar birth (dust-enshrouded stars) and death (planetary nebulae and **supernovae**). Another step forward is to understand star formation in other galaxies, for astronomers readily see active star formation in the arms of spiral galaxies and in the collisions of galaxies.

The Study of Stars and Stellar Evolution

The study of stars and their evolution is perhaps one of the oldest subfields of astronomy, and has benefited greatly from observational evidence dating back over hundreds of years. This is the core of astronomy because stars are truly the fundamental blocks of the universe, creating and destroying chemical elements, acting as light posts in galaxies, and giving insights into understanding mysterious phenomena, such as **black holes** and gamma-ray bursts. Understanding such exotic and high-energy events is critical to the advancement of astronomy and fundamental physics, where such "events" occur in conditions impossible to create on Earth. Astronomers are even continuing to learn new things about the nearest star, the Sun, through, for example, recent amazing images (e.g., solar storm activity) from the Solar and Heliospheric Observatory (SOHO) satellite.

The Study of Galaxies and Stellar Systems

Just as stars are the building blocks of galaxies, galaxies are the building blocks of the universe. The study of their types, sizes, distribution, and interactions with neighbors is essential to understanding the nature and future of the universe. The study of the earliest galaxies (galaxy "seeds") is the main motivating factor behind building larger ground-based telescopes and more sensitive infrared space telescopes, such as the Space InfraRed Telescope Facility (SIRTF) and the **Next Generation Space Telescope** (NGST). Astronomers know from the deepest HST images that the early universe was composed of many irregular, active, star-formation-rich galaxies. Astronomers do not know, however, how such a chaotic early universe evolved to what is seen in our local group, whose component galaxies are quite different.

Among the many mysteries in the universe is the dark matter in galaxies and clusters. We know little about its amount (speculated to be roughly 10 to 100 times greater than the observed mass), structure, location, and makeup, despite evidence from beautiful HST pictures of **gravitational lenses**, and observations of hot gases in galaxy clusters measured by sensitive X-ray telescopes (e.g., German Röntgensatellit (ROSAT), Japanese Advanced Satellite for Cosmology and Astrophysics (ASCA), American Chandra).

brown dwarfs star-like objects less massive than 0.08 times the mass of the Sun, which cannot undergo thermonuclear process to generate their own luminosity

supernova an explosion ending the life of a massive star

black holes objects so massive for their size that their gravitational pull prevents everything, even light, from escaping

Next Generation Space Telescope the telescope scheduled to be launched in 2009 that will replace the Hubble Space Telescope

gravitational lenses two or more images of a distant object formed by the bending of light around an intervening massive object

quasars luminous objects that appear starlike but are highly redshifted and radiate more energy than an entire ordinary galaxy; likely powered by black holes in the centers of distant galaxies

Another very active field is the study of elusive **quasars**, observed out to a distance when the universe was less than 10 percent of its present age. Recent far infrared and X-ray data have revealed a large population of these objects, indicating that many of them might be heavily obscured by dust and therefore not seen by earlier visible light surveys. Astronomers know very little about the power mechanisms of these objects, and this field is a very active area for today's radio, X-ray, and gamma-ray astronomers.

The Study of Cosmology and Fundamental Physics

The area of cosmology and fundamental physics is perhaps the most elusive and yet also the most important field in astronomy because it encompasses the other three categories. Cosmology literally means "the study of the beginning of the universe." Cosmologists, however, strive to answer questions not only about the universe's origin but also about its evolution, contents, and future.

cosmic microwave background ubiquitous, diffuse, uniform, thermal radiation created during the earliest hot phases of the universe

It is now widely believed that the universe started with a "big bang," with the most conclusive evidence being precise measurements of variations in the big bang signature 2.7K **cosmic microwave background** by the Cosmic Background Explorer satellite in 1997. Other recent advances in this subfield have come through all-sky infrared surveys, which have mapped out the distribution of galaxies across the sky; additional observational evidence that has led to more accurate estimates of the rate of expansion of the universe and its deceleration parameter; and increased computing power for numerical simulations that attempt to solve the ever-present **many-bodied problem**.

many-bodied problem in celestial mechanics, the problem of finding solutions to the equations for more than two orbiting bodies

Astronomers can comprehend the universe only through what they can see (limited by the sensitivities of the instruments used), what they can infer from observational data and numerical simulations, and what is supported by theory. As time has progressed, so too has the toolkit of the astronomer, from easier access to satellites, large ground-based telescopes, arrays of telescopes around the world working as one, increased computing power, and more sensitive cameras and **spectrometers**. As long as there is a way to improve detection techniques and strategies, astronomers will never run out of new discoveries or rediscoveries among the many "kinds" of astronomy. SEE ALSO HUBBLE SPACE TELESCOPE (VOLUME 2); OBSERVATORIES, GROUND (VOLUME 2); OBSERVATORIES, SPACE-BASED (VOLUME 2).

spectrometer an instrument with a scale for measuring the wavelength of light

Kimberly Ann Ennico

Bibliography

Davies, John K. *Astronomy from Space: The Design and Operation of Orbiting Observatories.* Chichester, UK: Praxis Publishing, 1997.

Henbest, Nigel, and Michael Marten. *The New Astronomy,* 2nd ed. Cambridge, UK: Cambridge University Press, 1996.

Maran, Stephen P., ed. *The Astronomy and Astrophysics Encyclopedia.* New York: Van Nostrand Reinhold, 1992.

Internet Resources

The Hubble Space Telescope. Space Telescope Science Institute. <http://www.stsci.edu/hst/>.

The Solar and Heliospheric Observatory. European Space Agency/National Aeronautics and Space Administration. <http://sohowww.nascom.nasa.gov/>.

Black Holes

Black holes are objects for which the gravitational attraction is so strong that nothing, not even light, can escape from it. They exist in the universe in large numbers.

Albert Einstein's theory of **general relativity** explains the properties of black holes. ✱ The material inside a black hole is concentrated into a singularity: a single point of infinitely high density where space and time are infinitely distorted. Distant objects can escape from a black hole's gravitational pull, but objects inside the so-called **event horizon** inevitably fall toward the center (such objects would have to move faster than light to escape, which is impossible according to the laws of physics). The size of the event horizon and the distortions of the space and time surrounding it are determined by the mass and spin (rate of rotation) of the black hole. Space and time distortions cause unusual effects; for example, a clock falling into a black hole will be perceived by a distant observer to become redder and to run slower.

Two types of black holes are found in the universe: stellar-mass black holes and supermassive black holes. They are characterized by different masses and formation mechanisms.

A stellar-mass black hole forms when a heavy star collapses under its own weight in a supernova explosion. This happens after the nuclear fuel,

general relativity a scientific theory first described by Albert Einstein showing the relationship between gravity and acceleration

✱ Einstein was a renowned theoretical physicist, whose theory of special relativity produced what is arguably the most well-known equation in science: $E=mc^2$.

event horizon the imaginary spherical shell surrounding a black hole that marks the boundary where no light or any other information can escape

NOAO

HST

Through the Hubble Space Telescope, scientists were able to observe a massive black hole hidden at the center of a giant galaxy.

X rays a form of high-energy radiation just beyond the ultraviolet portion of the electromagnetic spectrum

which makes the star shine for millions of years, is exhausted. The resulting black hole is a little heavier than the Sun and has an event horizon a few miles across (for comparison, to turn Earth into a black hole it would have to be squeezed into the size of a marble). The existence of such black holes has been inferred in cases where the black hole pulls gas of a companion star that orbits around it. The gas heats up as it falls towards the black hole and then produces **X rays** that can be observed with Earth-orbiting satellites.

Supermassive black holes are found in the centers of galaxies that contain billions of stars. They may exist in most galaxies and probably formed at the same time as the galaxies themselves. They are millions or billions times as heavy as the Sun, as determined from the motions of stars and gas surrounding them. Spectacular activity can occur when gas falls onto the black hole (as observed in a few percent of all galaxies). Material is ejected in jets that emit radio waves, and the heated gas produces X-ray emission. Observations of such X rays may soon provide insight into the spin of black holes.

space-time in relativity, the four-dimensional space through which objects move and in which events happen

There are enough black holes in the universe that there should occasionally be collisions between them. Such violent events send ripples through the **space-time** fabric of the universe. Scientists are hoping to soon detect such "gravitational waves" for the first time.

English physicist Steven Hawking showed in 1974 that every black hole spontaneously and continuously loses a tiny fraction of its mass because of radiation. This Hawking radiation, however, is negligible for the known black holes in the universe and will not be detectable in the foreseeable future. SEE ALSO EINSTEIN, ALBERT (VOLUME 2); GRAVITY (VOLUME 2); STARS (VOLUME 2); SUPERNOVA (VOLUME 2).

Roeland P. van der Marel

Bibliography

Begelman, Mitchell, and Martin Rees. *Gravity's Fatal Attraction: Black Holes in the Universe.* New York: Scientific American Library, 1996.

Couper, Heather, and Nigel Henbest. *Black Holes.* New York: DK Publishing, 1996.

Thorne, Kip S. *Black Holes and Time Warps: Einstein's Outrageous Legacy.* New York: W.W. Norton & Company, 1995.

Big Bang name given by astronomers to the event marking the beginning of the universe, when all matter and energy came into being

Careers in Astronomy

During just the last few years of the twentieth century, astronomers began to find planets orbiting other stars. They also made detailed measurements of the remnant radiation left over from the **Big Bang** and identified the first appearance of structure in the universe, the structure that eventually led to the formation of stars and galaxies. Some astronomers even found tantalizing evidence that suggests that the expansion of the universe may be speeding up, perhaps because of previously unrecognized properties of space itself. These important findings and many others lead to a particularly exciting time to consider a career in astronomy or space science.

When considering such a career it is important to realize that nowadays most astronomy is *not* classical astronomy—observing or photographing astronomical objects. Instead it involves the use of physics, mathematics, or geology to understand these objects. Many "astronomy" departments at col-

leges and universities are, in fact, called departments of astrophysics or planetary science. A significant fraction of the Ph.D. candidates in astrophysics hold an undergraduate degree in physics or mathematics. Many Ph.D. candidates in physics departments choose thesis topics that involve astrophysics, because some of the most interesting topics in modern physics are topics in this field. In this article, "astronomy" should be understood to encompass astrophysics or one of the other fields mentioned above.

The open dome of the United Kingdom Infrared Telescope. Internships at universities and national observatories are invaluable for individuals studying for a career in astronomy.

Education and Training

Students interested in a career in astronomy must be prepared to work hard for a number of years during their training. The average Ph.D. takes approximately seven years to earn, and a Ph.D. is required for those interested in doing research. During their education, students pursuing doctoral degrees in astronomy take approximately twenty physics courses and a similar number of courses in mathematics. Those who enjoy science and problem solving will enjoy much of this, although it is challenging work. As early as possible, it is important for students to gain research experience. Most scientists find research much more interesting than class work. Nowadays many national observatories and universities offer research experience for undergraduates, an opportunity that should be taken advantage of when one is still in college. This kind of "internship" is the best way for students to determine whether they will really enjoy a career in astronomy.

Those aiming to become faculty members will probably hold one or more two- to three-year postdoctoral positions before they can hope to earn a tenure-track appointment. Be forewarned that less than half of those who seek a tenured faculty position actually earn one. Nevertheless, the problem-solving and analytical skills learned during training for an astronomy Ph.D. are good preparation for a number of possible jobs, not just as a professor of astronomy. Indeed, people with doctorates in astronomy have applied their problem-solving and computer skills in a variety of jobs, including computer consulting and business. Those who prepare for a career in astronomy with a flexible attitude about the job that they will eventually take are less likely to be disappointed than those who have very fixed career goals.

Where Astronomers Work

The largest employers of astronomers are colleges, universities, and the government. Large government employers included the National Aeronautics and Space Administration (NASA) and the national observatories (such as the National Optical Astronomy Observatories, with branches in Arizona, New Mexico, and Chile; the National Radio Astronomy Observatory; and the Space Telescope Science Institute). With the proliferation of national facilities and the communications network provided by the Internet, it is possible to do first-rate astronomy work at many universities and colleges, even some of the smaller ones. Many of these schools emphasize the quality of their teachers, and those interested in being hired by such a school should acquire good communications skills as well as scientific and technical training.

More and more astronomy is being done using observations made from space missions, and NASA is playing an increasingly large role in astronomy. Highlights of planned NASA missions include: the exploration of Saturn and Mars; the use of large telescopes to observe the **infrared** from space and from a 747 aircraft; and the development of a number of **interferometers** that will search for planets orbiting other stars and will eventually produce images of those planets. These kinds of missions are always done in large teams, so good teamwork and communications skills as well as good scientific and technical skills are desirable when working for NASA. For those interested in engineering—in building and testing instruments—there are also many interesting opportunities working on space missions such as these. As the equipment used in astronomy becomes more complex, the field will require more and more people skilled in engineering, interferometry, and similar techniques. It is possible that there could be a shortage of people with these skills who want to work in astronomy and space science.

Another aspect of astronomy is closely related to mathematics and computer science. Much of the theoretical work in astronomy now involves sophisticated modeling using most powerful computers. Students interested in computer-based analysis might consider applying their skills to astronomy. Theoretical astronomers build computer models of the Sun and stars, of supernovas, of high-temperature explosions that produce **X rays**, of in-

infrared portion of the electromagnetic spectrum with wavelengths slightly longer than visible light

interferometers devices that use two or more telescopes to observe the same object at the same time in the same wavelength to increase angular resolution

X rays a form of high-energy radiation just beyond the ultraviolet portion of the electromagnetic spectrum

teracting galaxies, and even of the formation of the first structure in the entire universe. Many current and planned investigations, such as the Sloan Digital Sky Survey, will observe and record hundreds of millions of objects. Astronomers are just beginning to plan for a National Virtual Observatory, which would develop new ways to analyze the large data sets that will soon be gathered.

Career Options for Those with Bachelor's and Master's Degrees

Career options for those with a bachelor's or a master's degree in astronomy are more limited than individuals who have a doctoral degree. All of the NASA missions and some of the observatories described in this entry hire research assistants or data assistants to help with operations and data analysis. People in such positions may work on very interesting science but they do not usually have the opportunity to choose their own projects or areas of investigation.

Because astronomy is such a popular subject, there are a number of opportunities for presenting astronomy to the public. Planetariums hire astronomers to work as educators, and most NASA missions hire people to provide educational services and public outreach. The standards for these positions are highly variable, and they often do not require a Ph.D.

One of the most important jobs available to someone with training in astronomy is teaching physics (and sometimes astronomy) at the high school level. Considerably less than half the people who teach physics in high school have been trained in the field, and this is one of the contributing factors to the poor science knowledge of American students. Anyone who pursues the astronomy studies described above, even through the first year or two of graduate school, would have more physics background than the typical high school physics teacher, and this could provide the background needed for teaching.

Opportunities for Women and Minorities

It is important to note that opportunities for women in astronomy have been increasing and may continue to do so. Approximately 25 percent of the Ph.Ds now granted in astronomy go to women. This is twice the percentage of Ph.Ds granted to women in physics. Studies also show, however, that women continue to drop out at higher rates than men at each career step. One reason for this at the beginning of a career is that it is harder for women to find role models, mentoring, and encouragement in a field where most of the professors are male. It is therefore very important for women interested in a career in astronomy to make contact with a woman already in the field and ask for some guidance. Organizations such as the American Astronomical Society can provide further information.

The number of minority students in astronomy is currently very small. Out of the total of 150 astronomy Ph.Ds that are awarded in United States annually, only a very small percentage of these are received by African-American and Hispanic students. Students with a minority background who are interested in the exciting field of astronomy would also profit by finding a mentor. In fact, most successful scientists, no matter what their background, took advantage of guidance or mentoring from someone in the field during their training. SEE ALSO ASTRONOMER (VOLUME 2); ASTRONOMY, HISTORY OF (VOLUME 2); ASTRONOMY, KINDS OF (VOLUME 2); GALILEI, GALILEO (VOLUME 2); HUBBLE SPACE TELESCOPE (VOLUME 2); OBSERVATORIES, GROUND-BASED (VOLUME 2); OBSERVATORIES, SPACE-BASED (VOLUME 2).

Douglas Duncan

Bibliography

Committee on Science, Engineering and Public Policy. *Careers in Science and Engineering: A Student Planning Guide to Grad School and Beyond.* Washington, DC: National Academy Press, 1996.

Internet Resources

American Astronomical Society. <http://www.aas.org>.

A New Universe to Explore: Careers in Astronomy. American Astronomical Society. http://www.aas.org/%7Eeducation/career.html>.

Careers in Space Science

Considering possible career options in space science can be as full of variety and inviting choices as selecting from an elegant buffet or wandering through the stalls of an exotic overseas bazaar. Space science now encompasses practically all areas of science, and space research draws on an even broader collection of skills and specialties beyond the pure sciences.

One way to think about careers in space science is to notice that there are two main areas. First, there are specialties in which scientists place their tools in space so that they can see aspects of nature that cannot be examined from the ground. These approaches, which might be called "science *from* space," include research in astronomy and research that looks back at Earth from space. In the second category are researchers who take advantage of unique aspects of operating in an orbital laboratory or on another planet to do experiments or exploration that would not be possible otherwise. They conduct "science *in* space" by making space, itself, their laboratory.

Science from Space

Astronomers build or use telescopes that are launched into space to study the universe by measuring the **infrared**, X-ray, and gamma-ray light that cannot be detected below Earth's atmosphere. They also make observations in the **visible spectrum** but with much greater clarity than astronomers often can from the ground. These space astronomy studies examine the Sun, nearby stars, more distant galaxies, and even objects at the very edges of the universe.

Science from space can involve looking in as well as looking out. Information about our own planet, Earth, can often be obtained best by getting a genuinely global view from space. Such research includes studies in climatology, atmospheric science, **meteorology**, geology and geophysics, ecology, and oceanography, just to name a few. Looking at Earth from space is a good example of how space science can span a full range of goals. Those goals can include exploring very basic questions about how nature operates, gathering information that has important and immediate value to help so-

infrared portion of the electromagnetic spectrum with wavelengths slightly longer than visible light

visible spectrum the part of the electromagnetic spectrum with wavelengths between 400 nanometers and 700 nanometers; the part of the electromagnetic spectrum to which human eyes are sensitive

meteorology the study of atmospheric phenomena or weather

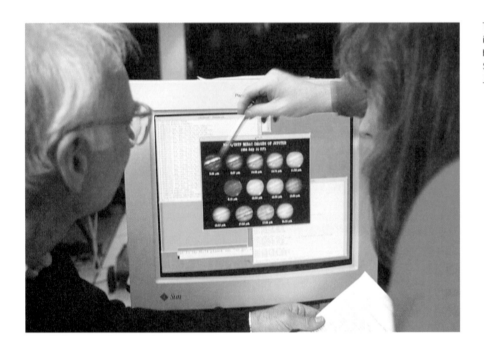

Two astronomers review images of the collision between Comet Shoemaker-Levy 9 and Jupiter.

ciety deal with natural hazards, or the management of natural resources, agriculture, forestry, or environmental problems.

Science in Space

Science *in* space takes advantage of being in the immediate presence of the objects of study, or of some unique properties of spaceflight, such as the existence of very low gravity (so-called microgravity) or a nearly perfect **vacuum**. In the former case, space scientists study the properties of the space environment itself, including the hot gas that flows out from the Sun to fill the solar system and the high-energy atomic particles that create **radiation belts** around many planets. Those belts can pose a hazard to astronauts or to any robotic spacecraft that fly through them. Planetary scientists explore other objects in the own solar system. They use orbiting telescopes, spacecraft that are sent to orbit other planets, robots that land and move around on the surface of another planet, and spacecraft that retrieve samples of planetary material and bring them back to Earth for analysis. These scientists also study the natural satellites of other planets as well as asteroids and comets.

The microgravity environment of spaceflight creates opportunities for in-space laboratories that span a wide range of scientific topics. These include biomedical studies of how weightlessness affects human performance and how to minimize those effects on astronauts. It also includes more basic studies in biology that examine the role of gravity in the development and functionality of plants and animals. Researchers in the physical sciences find space laboratories to be equally useful because they offer a unique setting for experiments in materials science, studies of combustion and the behavior of fluids, and a number of other areas of both basic and applied physics and chemistry.

Many areas of space science are now reaching across the traditional specialties of science and emerging as new multidisciplinary fields. Two

vacuum a space where air and all other molecules and atoms of matter have been removed

radiation belts two wide bands of charged particles trapped in a planet's magnetic field

notable examples are studies of global change and the field of astrobiology. In studying global change, scientists combine expertise from many Earth science specialties (e.g., oceanography, hydrology, atmospheric science, ecology) and use the vantage point of space to monitor how Earth is changing and to predict and understand the consequences of those changes. Astrobiology is a relatively new field in which researchers seek to understand how life formed in the universe and how it has evolved. Astrobiologists also want to discover whether there was or is now life beyond Earth, and to learn from those studies about the possibilities for life beyond Earth in the future. As a result of the breadth of such profound scientific questions, astrobiologists draw heavily on expertise in biology, chemistry, astronomy, and planetary science.

In all fields of space science, those who conduct research may find themselves involved in many phases of the work. That is, they may help design the experiments, the instruments, and even the spacecraft that carry them. They help build and test the instruments to be launched into space and then help operate them, and they are often involved in analyzing and interpreting the measurements that are returned to Earth. In a small number of cases a few lucky researchers get to go with their experiments into space as scientist astronauts.

Careers Outside Pure Science

Space science offers career opportunities that extend beyond pure scientific fields. Space science depends not only on scientists but also equally heavily on engineers, mathematicians, information technology experts, and other technical specialists who help make the research possible. In fact these members of a space research team often outnumber the scientists on the project. They work side by side with the scientists to build and operate experiments and to prepare data or samples that are returned from the experiments for analysis.

Finally, where do people in space science work? That is also a question with many answers. Most space scientists are on university staffs and faculties, but they also reside in industry laboratories, government laboratories such as the National Aeronautics and Space Administration's field centers, and in laboratories of nonprofit organizations. Regardless of where they work, people who earn their living in space science have a common bond. They share in the excitement and fascination that comes from pursuing some of the most challenging questions in contemporary science, and they know that they are helping to open new frontiers in exploration and to bring the benefits of science back to Earth. SEE ALSO ASTROBIOLOGY (VOLUME 4); ENVIRONMENTAL CHANGES (VOLUME 4); LIFE IN THE UNIVERSE, SEARCH FOR (VOLUME 2); MICROGRAVITY (VOLUME 2).

Joseph K. Alexander

Internet Resources

A New Universe to Explore: Careers in Astronomy. American Astronomical Society. <http://www.aas.org/%7Eeducation/career.html>.

For Kids Only: NASA Earth Science Enterprise. "NASA Career Expo." <http://kids.earth.nasa.gov/archive/career/index.html>.

NASA Spacelink. "Careers." <http://spacelink.nasa.gov/Instructional.Materials/Curriculum.Support/Careers/>.

Cassini, Giovanni Domenico

Italian Astronomer
1625–1712

Born in Perinaldo, Italy, Giovanni ✳ Domenico Cassini (1625–1712) was an astronomer best known for his discoveries connected with the planet Saturn. At the age of twenty-five, Cassini was named chair of astronomy at the University of Bologna and held that position for nineteen years. He determined the rotation rates of Jupiter in 1665, of Mars in 1666, and of Venus (erroneously) in 1667. In 1668, Cassini computed tables that predicted the motion of Jupiter's four known moons. This led directly to Danish astronomer Ole (or Olaus) Römer's determination of the speed of light in 1675.

In 1669, King Louis XIV of France invited Cassini to Paris to direct the city's observatory. At the Paris Observatory, Cassini, now using Jean Dominique as his first name, continued his astronomical observations, at times using the extremely long "aerial telescopes" developed by Dutch astronomer Christiaan Huygens.

In Paris, Cassini discovered the second satellite of Saturn, Iapetus, in 1671 and correctly explained its brightness variations. He found another satellite of Saturn, Rhea, in 1672. In 1675 Cassini observed a band on Saturn and found that its ring had a division, now named the Cassini Division. Cassini discovered two more of Saturn's satellites, Tethys and Dione, in 1684.

Among his other projects, Cassini used innovative methods to make the best measure—at the time—of the astronomical unit (the average distance between Earth and the Sun). Cassini also studied **atmospheric refraction** and conducted a **geodetic survey**.

Cassini is the namesake of a joint program of the National Aeronautics and Space Administration, the European Space Agency, and the Italian Space Agency to study the Saturn system beginning in 2004. SEE ALSO HUYGENS, CHRISTIAAN (VOLUME 2); JUPITER (VOLUME 2); SATURN (VOLUME 2); SMALL BODIES (VOLUME 2).

Stephen J. Edberg

✳ Giovanni Domenico Cassini is sometimes known as "Gian Cassini."

atmospheric refraction the bending of sunlight or other light caused by the varying optical density of the atmosphere

geodetic survey determination of the exact position of points on Earth's surface and measurement of the size and shape of Earth and of Earth's gravitational and magnetic fields

Bibliography

Abetti, Giorgio. *The History of Astronomy*, trans. Betty Burr Abetti. New York: Henry Schuman, 1952.

Beatty, J. Kelly. "A 'Comet Crash' in 1690?" *Sky and Telescope* 93, no. 4 (1997):111.

Berry, Arthur. *A Short History of Astronomy* (1898). New York: Dover Publications, Inc., 1961.

Bishop, R., ed. *Observer's Handbook, 2000.* Toronto: Royal Astronomical Society of Canada, 1999.

Close Encounters

Most asteroids follow fairly regular paths in **orbits** between Mars and Jupiter. A small fraction, about one in a thousand, have evolved from their regular orbits by slow gravitational effects of the planets, mainly Jupiter, to

orbits the circular or elliptical paths of objects around a much larger object, governed by the gravitational field of the larger object

Cosmic impacts can cause major damage to Earth's surface and ecosystem. This meteorite crater is located near Winslow, Arizona.

elliptical having an oval shape

astronomical units one AU is the average distance between Earth and the Sun (152 million kilometers [93 million miles])

travel in more **elliptical** orbits that cross the paths of other planets, including Earth. The first of these discovered was Eros, found in 1898, which crosses the orbit of Mars but not Earth. The first space mission dedicated primarily to visiting an asteroid was the Near Earth Asteroid Rendezvous (NEAR) mission (later renamed NEAR Shoemaker in honor of American astronomer Eugene Shoemaker), which orbited Eros for a year in 2000–2001, before touching down on its surface on February 12, 2001.

Even in 1694, when Edmund Halley discovered that the orbit of the comet that bears his name crosses Earth's orbit, he suggested the possibility of a collision with Earth by a comet, and he rightly suggested that such an event would have a catastrophic effect on Earth and its inhabitants. In 1932, two more asteroids were discovered, named Amor and Apollo, which pass close enough to Earth to suggest the possibility of eventual collision with the planet.

Today scientists refer to asteroids that can come closer than 1.3 **astronomical units** (AU) to the Sun (0.3 AU to Earth) as near Earth asteroids (NEAs), or collectively along with comets that come that close, near Earth objects (NEOs). By January 2002, 1,682 NEAs had been discovered, 572 of which were estimated to be 1 kilometer (0.6 mile) or larger in diameter. Scientists estimate that the total number of NEAs larger than 1 kilometer in diameter is about 1,000, so somewhat more than half of them had been found by January 2002. The largest asteroid in an orbit actually crossing Earth's orbit is around 10 kilometers (6 miles) in diameter. Scientists do not believe that there are any undiscovered objects larger than 4 or 5 kilometers (2.5 or 3 miles) in diameter.

The Frequency and Energy of Impacts

Given that these cosmic bullets are flying around Earth, the expected frequency of impacts on Earth can be estimated. Any one NEA has a likelihood of hitting Earth in about 500 million years. Since there are about 1,000 NEAs larger than 1 kilometer, one impact about every 500,000 years can be expected. The energy of such an impact can also be estimated. A piece of rock 1 kilometer in diameter, traveling at 20 kilometers per second (12.4 miles per second) on impact, should release an energy equivalent to almost 100,000 megatons of TNT, or about the total energy of all the nuclear weapons in the world. Such a blast should make a crater nearly 20 kilometers (12.4 miles) in diameter.

Past Collisions

Evidence abounds of past collisions, on Earth as well as on the surfaces of almost all other solid-surface bodies in the solar system. Impact craters up to hundreds of kilometers across are clearly visible on the face of the Moon and have been found and counted on Mercury, Venus, Mars, planetary satellites, and even the asteroids themselves.

In 1980, the father and son team of Louis and Walter Alvarez, along with two other colleagues, offered a revolutionary explanation for the extinction of the dinosaurs, as well as most other species inhabiting Earth at that time (65 million years ago). They found the rare element iridium in the thin clay layer that caps the rocks of the Cretaceous era. The element was present in concentrations far too high for a terrestrial explanation, but just about right for the debris left from a cosmic impact by an asteroid or comet about 10 kilometers (6 miles) in diameter. This hypothesis, which was first met with widespread skepticism, has gained strength with many subsequent supporting discoveries, including the identification of the "smoking gun"—the remains of the crater at the tip of the Yucatan Peninsula in Mexico. Known as the Chicxulub Crater, it is buried in sediments and invisible from the surface, except for a ring of sinkholes outlining the original rim, approximately 200 kilometers (125 miles) in diameter. Impact cratering is now recognized as an important geological process, which can even affect the evolution of life on Earth.

Potential Effects of a Collision

The world received a "wake-up call" in July 1994 when the pieces of the comet Shoemaker-Levy 9 slammed into the planet Jupiter, leaving giant dark spots in the clouds, easily visible from Earth through a small telescope. Some of the spots were as large as the entire Earth. Based on these observations and computer models of the expected effects of a cosmic impact on Earth, it is estimated that an asteroid 1 to 2 kilometers (0.6 to 1.2 miles) in diameter would form an impact crater more than 20 kilometers (12.4 miles) in diameter. In addition, it would throw enough dust into the upper atmosphere to block out the Sun for about a year, producing a global "impact winter."

Such a climatic catastrophe could lead to global crop failures and the starvation of perhaps a quarter of the world's population. The individual numbers boggle the mind: more than a billion deaths, but only once in 500,000 years. Yet the quotient is quite understandable: an average of some thousands of deaths per year, or in the same range as the death toll from commercial airline accidents, floods, earthquakes, volcanic eruptions, and

Scientists estimate that an asteroid 1 to 2 kilometers in diameter hitting Earth's surface could create a crater about 20 kilometers in diameter. Such an impact could cause an "impact winter," with catastrophic results.

✳ The term "Spaceguard Survey" is borrowed from Arthur C. Clarke's science fiction novel, *Rendezvous with Rama* (1973), detailing the story of a huge and mysterious object that appears in the solar system.

other such disasters that are taken very seriously. Because the frequency of occurrence is so low—indeed there has never been a catastrophic asteroid impact in recorded history—humans have paid less attention to this risk than to the others mentioned. But the consequences are as terrible as the intervals are long, so the importance is about the same as the other natural hazards, with one significant difference. This hazard alone (with the possible exception of a very massive volcanic eruption) has the potential to end human civilization globally.

Preparations for and Responses to Potential Collisions

What can, or should, be done? As a first step, it makes sense to simply look and find all the asteroids and Earth-approaching comets out there and see if one has our name on it. Beginning in the late 1990s, several governments and agencies embarked on what has been loosely called the Spaceguard Survey. ✳ The goal of this survey is to find at least 90 percent of all NEAs larger than 1 kilometer (0.6 mile) in diameter, the lower size limit for objects that could cause a global catastrophe. By the year 2001, the project was about half complete, and it is likely to be finished by 2010. With continued effort, ever-smaller asteroids can be found and cataloged, providing assurance that nothing is coming Earth's way in the foreseeable future (i.e., about the next fifty years).

But if we find that something is coming, what can we do? With many years warning, only a small push of a few centimeters per second would divert an asteroid from a collision course to a near miss. Even without knowing quite how to do it, it is easy to estimate that the energy needed is within the range available from nuclear weapons, and the rocket technology to deliver a bomb to an asteroid is available. Whether such a system should be developed in advance of any specific threat is a more difficult question and one that will need to be carefully addressed by both scientists and policymakers. SEE ALSO ASTEROIDS (VOLUME 2); COMETS (VOLUME 2); IMPACTS (VOLUME 4).

Alan Harris

Bibliography

Lewis, John S. *Rain of Iron and Ice: The Very Real Threat of Comet and Asteroid Bombardment.* Reading, MA: Addison-Wesley Publishing Co., 1996.

Steel, Duncan. *Target Earth.* Pleasantville, NY: Reader's Digest Association, 2000.

Internet Resources

Asteroid and Comet Impact Hazards. Ames Space Research Division, National Aeronautics and Space Administration. <http://impact.arc.nasa.gov/index.html>.

Near-Earth Object Program. NASA Jet Propulsion Laboratory, California Institute of Technology. <http://neo.jpl.nasa.gov/>.

Tumbling Stone. The Spaceguard Foundation and NEO Dynamic Site. <http://spaceguard.ias.rm.cnr.it/tumblingstone/>.

Comets

A bright comet is a spectacular astronomical event. Throughout history, comets have left a strong impression on those who have witnessed their appearances. The name comes from the Greek *kometes*, meaning "the long-haired one." Ancient Greeks thought comets to be atmospheric phenomena, part of the "imperfect" changeable Earth, not of the "perfect" immutable heavens. Today we know they are "icy conglomerates," as proposed in 1950 by Fred Whipple—that is, chunks of ice and dust left over from the formation of the solar system some 4.6 billion years ago.

Comets are among the most primitive bodies in the solar system. Because of their orbits and small sizes, comets have undergone relatively little processing, unlike larger bodies, such as the Moon and Earth, which have been modified considerably since they formed. The chemical composition of comets contains a wealth of information about their origin and evolution as well as the origin and evolution of the solar system. Hence, comets are often called cosmic fossils.

When a comet is far from the Sun, it is an inert icy body. As it approaches the Sun, heat causes ices in the nucleus to **sublimate**, creating a cloud of gas and dust known as the coma. Sunlight and **solar wind** will push the coma gas and dust away from the Sun creating two tails. The dust tail is generally curved and appears yellowish because the dust particles are scattering sunlight. The gas (or ion) tail is generally straight and it appears blue

sublimate to pass directly from a solid phase to a gas phase

solar wind a continuous, but varying, stream of charged particles (mostly electrons and protons) generated by the Sun; it establishes and affects the interplanetary magnetic field; it also deforms the magnetic field about Earth and sends particles streaming toward Earth at its poles

because its light is dominated by emission from carbon monoxide ions. The appearance of comets in photographs can give the erroneous impression that they streak through the night sky like a **meteor** or a shooting star. In fact, comets move slowly from night to night with respect to the stars and can sometimes be visible for many weeks, as was the case with comet Hale-Bopp in 1997 and with comet Halley during its 1985–1986 appearance.

Comet Halley is not the brightest comet, but it is the most famous, mainly because it is the brightest of the predictable comets. It was named after Edmund Halley, an eighteenth-century British astronomer who was the first to calculate the orbits of comets. Comet Halley's orbit has an average period of seventy-six years. Its closest approach to the Sun (perihelion) is between the orbits of Venus and Mercury (0.59 **astronomical units**), and its aphelion (farthest distance from the Sun) is at 35 AU, beyond Neptune's orbit. The orbit has an inclination of 162 degrees with respect to the **ecliptic**. This means that comet Halley orbits the Sun clockwise when seen from the north, whereas Earth orbits the Sun counterclockwise.

The study of comets is a very active field of science. In 1986 a flotilla of spacecraft were used to study comet Halley. In the first decade of the twenty-first century, several spacecraft are scheduled to be launched to encounter and study a number of comets. In addition to space-based studies, ground-based observations of comets have yielded a wealth of information.

The Comet's Nucleus

All of the activity in a comet originates in its nucleus, which is composed of roughly equal amounts of ices and dust. Water ice is the most abundant of the ices, comprising about 80 percent of the total. So far, only the nuclei of comets Halley and Borrelly have been imaged in detail. Comet Halley turned out to be larger, darker, and less spherical than expected by most astronomers. The images of comet Borrelly's nucleus obtained in September 2001 by NASA's Deep Space 1 spacecraft show considerable similarity with those of comet Halley. Halley's nucleus is peanut-shaped, approximately 18 kilometers (11 miles) long and 8 kilometers (5 miles) wide. The reflectivity (or albedo) is 4 percent, which is as dark as coal. The size, albedo, and approximate shape of several other cometary nuclei have been determined. Comet Halley's nucleus seems to be typical among comets with relatively short orbital periods, and there are much larger nuclei such as that of comet Hale-Bopp. So far, the cometary nuclei studied in detail appear to have most of their surface covered by an inert mantle or crust. The active (exposed ice) fraction of their surface is small; in comet Halley, this fraction is somewhere between 15 and 30 percent.

The development of a crust can suppress the activity of cometary nuclei and give them an asteroidal appearance. The best example to date is comet Wilson-Harrington, which was discovered in 1949 and was lost until it was rediscovered as an inert object and given the asteroid number 4015. The behavior of this object has added credence to the long-held expectation that some Earth-crossing asteroids are extinct or dormant comet nuclei.

The Composition of Comets

The composition of cometary nuclei is primarily inferred from studies of the coma components, namely gas, plasma (ions), and dust. So far, twenty-

Edmund Halley was an eighteenth-century British astronomer who first calculated the orbit of comets. The most famous comet, Halley, is named after him.

meteor the physical manifestation of a meteoroid interacting with Earth's atmosphere; this includes visible light and radio frequency generation, and an ionized trail from which radar signals can be reflected

astronomical units one AU is the average distance between Earth and the Sun (152 million kilometers [93 million miles])

ecliptic the plane of Earth's orbit

Comet Hale-Bopp (seen in the center) was visible from Earth for several weeks in 1997. Comets travel at a relatively slow rate although they give the appearance of streaking through the sky.

four different molecules have been identified in comets, ten of which were discovered in comet Hale-Bopp. The molecules observed in comets and their relative abundances are very similar to those observed in dense **interstellar** molecular cloud cores, which is the environment where star formation occurs. Thus, it appears that comets underwent little processing in the **solar nebula** and they preserve a good record of its original composition.

interstellar between the stars

solar nebula the cloud of gas and dust out of which the solar system formed

Information on the composition of cometary dust particles was scarce before 1986. Studies of the dust in comet Halley and other comets confirmed that some of the grains are silicates, more specifically crystalline olivine $(Mg, Fe)_2 SiO_4$ and pyroxene $(Mg,Fe,Ca) SiO_3$. Another major component of the dust in comet Halley was organic dust. These small solid particles were discovered by the visiting spacecraft and were called "CHON" because they were composed almost exclusively of the elements carbon, hydrogen, oxygen, and nitrogen.

The Origins of Comets

Dutch astronomer Jan Hendrik Oort noted in 1950 that the source of new comets was a shell located between 20,000 and 100,000 AU from the Sun. The existence of the Oort cloud is now widely accepted. Astronomers believe that comets in the Oort cloud formed near Uranus and Neptune and were gravitationally scattered by these two planets into their current location. In addition to the Oort cloud, there is another reservoir that was proposed in 1951 by Dutch-born American astronomer Gerard Peter Kuiper as a ring of icy bodies beyond Pluto's orbit. This Kuiper belt is considered to be the main source of Jupiter-family comets, which are those with low-inclination and short-period orbits. SEE ALSO CLOSE ENCOUNTERS (VOLUME 2); COMET CAPTURE (VOLUME 4); IMPACTS (VOLUME 4); KUIPER BELT (VOLUME 2); KUIPER, GERARD PETER (VOLUME 2); OORT CLOUD (VOLUME 2).

Humberto Campins

COMET-PLANET COLLISIONS

Many comets are in Earth-crossing orbits and collisions do occur between comets and planets. A spectacular example of such a collision was the impact of comet Shoemaker-Levy 9 with Jupiter in July 1994. It is now well established that an impact with an asteroid or comet created a large crater at the edge of the Yucatan Peninsula 65 million years ago. Known as the Chicxulub impact, this event almost certainly caused the extinction of the dinosaurs. Efforts are underway to study the population of potential hazards from both comets and asteroids in sufficient detail to predict and prevent future large impacts.

Bibliography

Oort, Jan H. "The Structure of the Cloud of Comets Surrounding the Solar System and a Hypothesis Concerning Its Origin." *Bulletin of the Astronomical Institute of the Netherlands* 11 (1950):91–110.

Whipple, Fred L. "A Comet Model I: The Acceleration of Comet Encke." *Astrophysical Journal* 111 (1950):375–394.

Copernicus, Nicholas

Polish Astronomer
1473–1543

Nicholas Copernicus was the first to argue the theory that the Sun, not Earth, is the center of the solar system.

Nicholas Copernicus was a Polish astronomer who changed humankind's view of the universe. Greek astronomers, particularly Ptolemy, had argued that Earth was the center of the universe with the Sun, Moon, planets, and stars orbiting around it. This geocentric (Earth-centered) model, however, could not easily explain retrograde motion, the apparent backwards movement that planets exhibit at some points in their paths across the sky. Ptolemy and others had proposed a complicated system of superimposed circles to explain retrograde motion under the geocentric model. Copernicus realized that if all the planets, including Earth, orbited the Sun, then retrograde motion resulted from the changing of perspective as Earth and the other planets moved in their orbits.

Copernicus published his heliocentric (Sun-centered) theory in the book *De revolutionibus orbium coelesticum* (On the revolutions of the celestial orbs). The Catholic Church, however, had accepted the geocentric model as an accurate description of the universe, and anyone arguing against this model faced severe repercussions. At the time, Copernicus was gravely ill, so he asked Andreas Osiander to oversee the book's publication. Osiander, concerned about the Church's reaction, wrote an unsigned preface to the book stating that the model was simply a mathematical tool, not a true depiction of the universe. Copernicus received the first copy of his book on his deathbed and never read the preface. The telescopic discoveries of Italian mathematician and astronomer Galileo Galilei (1564–1642) and the mathematical description of planetary orbits by German astronomer Johannes Kepler (1571–1630) led to the acceptance of Copernicus's heliocentric model. SEE ALSO ASTRONOMY, HISTORY OF (VOLUME 2); GALILEI, GALILEO (VOLUME 2); KEPLER, JOHANNES (VOLUME 2).

Nadine G. Barlow

Bibliography

Andronik, Catherine M. *Copernicus: Founder of Modern Astronomy.* Berkeley Heights, NJ: Enslow Publishers, 2002.

Gingerich, Owen. *The Eye of Heaven: Ptolemy, Copernicus, Kepler.* New York: American Institute of Physics, 1993.

Cosmic Rays

Cosmic rays are, in fact, not rays, but high energy subatomic particles of cosmic origin that continually bombard Earth. The measurements scientists make of them, both on the ground and from probes in space, are the only direct measurements that are made of matter originating outside the solar system.

Among the cosmic rays are electrons, protons, and the complete nuclei of all the elements. Their energies range from below the rest mass of an electron, easily attainable in terrestrial accelerators, up to energies 10^{11} times the rest mass of a proton. Matter with such energies is moving at speeds so close to the speed of light that there is an enormous **relativistic time dilation**, so that in its proper frame only 10^{-11} of the time has elapsed that an observer on Earth would have measured. An early verification of German-born American physicist Albert Einstein's **special theory of relativity** came from explaining how unstable **mesons** produced by cosmic rays impinging on the upper atmosphere (whose lifetime was less than the time it took for them to reach the detectors on Earth) managed to survive without decaying. According to special relativity, these high energy mesons would not have had enough time in their own reference frame to decay.

Although cosmic rays have been known for more than a century, neither their precise origin nor their source of energy is known. Austrian-born American physicist Viktor Hess demonstrated their cosmic origin in 1912, using balloon flights to show that the penetrating, ubiquitous, ionizing radiation increases in intensity with altitude. It was not until the 1930s, with increased understanding of nuclear physics, that the "radiation" was recognized to be charged particles.

The low energy particles measured—below 10^{10} **eV**—are dominated by the effects of our environment in the solar system and the unpredictability of space weather. Incoming galactic cosmic rays are scattered on magnetic irregularities in the solar wind, resulting in "solar modulation" of the galactic cosmic ray spectrum. At low energies, many of the particles themselves originate in solar flares, or are accelerated by shocks in the solar wind.

At mid-energies, 10^{10} to 10^{15} eV, the particles measured are galactic, show a smooth **power law energy spectrum**, and show a composition of nuclei roughly consistent with **supernovae ejecta**, modified by their subsequent diffusion through the galaxy. Bulk acceleration in the supernova blast wave, and diffusive acceleration in shocks in the remnant can probably account for particles up to 10^{14} eV. They diffuse throughout the **interstellar medium**, but remain trapped within the galaxy for several million years by the magnetic field and scattering by **magnetohydrodynamic waves**.

Particles have been detected with energies up to about 10^{21} eV. There is no generally accepted mechanism for accelerating them above about 10^{15} eV. One speculation is that collapsing **superstrings** could produce particles with the **grand unified theory** (GUT) energy of 10^{25} eV; the particles then decay and lose energy. Above 10^{19} eV neither the spectrum nor the composition are well-known because the events are rare and the detection methods indirect.

In 1938 French physicist Pierre Auger discovered extensive air showers. When a single high energy particle impinges on the atmosphere, it generates a cascade that can contain 10^9 particles. Information about the primary

relativistic time dilation effect predicted by the theory of relativity that causes clocks on objects in strong gravitational fields or moving near the speed of light to run slower when viewed by a stationary observer

special theory of relativity the fundamental idea of Einstein's theories, which demonstrated that measurements of certain physical quantities such as mass, length, and time depended on the relative motion of the object and observer

mesons any of a family of subatomic particle that have masses between electrons and protons and that respond to the strong nuclear force; produced in the upper atmosphere by cosmic rays

eV an electron volt is the energy gained by an electron when moved across a potential of one volt. Ordinary molecules, such as air, have an energy of about 3×10^{-2} eV

power law energy spectrum spectrum in which the distribution of energies appears to follow a power law

supernovae ejecta the mix of gas enriched by heavy metals that is launched into space by a supernova explosion

interstellar medium the gas and dust found in the space between the stars

magnetohydrodynamic waves a low frequency oscillation in a plasma in the presence of a magnetic field

superstrings supersymmetric strings are tiny, one dimensional objects that are about 10^{-33} cm long, in a 10-dimensional spacetime. Their different vibration modes and shapes account for the elementary particles we see in out 4-dimensional spacetime

grand unified theory states that, at a high enough energy level (about 10^{25} eV), the electromagnetic force, strong force, and weak force all merge into a single force

muons the decay product of the mesons produced by cosmic rays; muons are about 100 times more massive than electrons but are still considered leptons that do not respond to the strong nuclear force

Čerenkov light light emitted by a charged particle moving through a medium, such as air or water, at a velocity greater than the phase velocity of light in that medium; usually a faint, eerie, bluish, optical glow

anisotropy a quantity that is different when measured in different directions or along different axes

cosmic microwave background ubiquitous, diffuse, uniform, thermal radiation created during the earliest hot phases of the universe

nucleus can be deduced from the lateral distribution of the **muons** and electrons that reach the ground, and from the pulse of **Čerenkov light** emitted as the shower descends through the atmosphere. If the spectrum, composition, and **anisotropy** above 5×10^{19} eV, where there should be a cutoff in the spectrum because of interactions on the 2.7°K **cosmic microwave background** photons, can be measured, and these are consistent, these cosmic rays will identify sites where some of the most exotic and energetic events in the universe occur.

Cosmic rays represent a significant component in the energy balance of our galaxy. The energy density in cosmic rays in the galactic disk is comparable to that found in starlight and in the galactic magnetic field, and therefore must play an important, if so far poorly understood, role in the cycle of star formation. By maintaining a residual ionization in the cores of dense molecular clouds, star formation is inhibited because the magnetic field cannot diffuse out. On the other hand, cosmic rays streaming along the magnetic field in the diffuse interstellar medium could provoke cloud condensation through MHD instabilities. SEE ALSO GALAXIES (VOLUME 2); SOLAR PARTICLE RADIATION (VOLUME 2); SOLAR WIND (VOLUME 2); SPACE ENVIRONMENT, NATURE OF THE (VOLUME 2); STARS (VOLUME 2); SUN (VOLUME 2); SUPERNOVA (VOLUME 2); WEATHER, SPACE (VOLUME 2).

Susan Ames

Bibliography

Friedlander, Michael W. *A Thin Cosmic Rain: Particles from Outer Space.* Cambridge, MA: Harvard University Press, 2000.

Gaisser, Thomas K. *Cosmic Rays and Particle Physics.* Cambridge, UK: Cambridge University Press, 1990.

Sokolsky, Pierre. *Introduction to Ultrahigh Energy Cosmic Ray Physics.* Boston, MA: Addison-Wesley Publishing Company, Inc. 1989.

Cosmology

Cosmology is the study of the origin and evolution of the universe. In the last half of the twentieth century, astronomers made enormous progress in understanding cosmology. The discovery that the universe apparently began at a specific point in time and has continued to evolve ever since is one of the most revolutionary discoveries in science.

The History of the Universe: In the Beginning

The universe began in what astronomers dubbed the "Big Bang"—an initial event, after which the universe began to expand. Current estimates place the Big Bang at about 13 to 15×10^9 years ago. During the first seconds after the Big Bang, the universe was extremely hot and dense. The physics needed to understand the universe in these early stages is very speculative because it is impossible to recreate these conditions in an experiment today to check the predictions of the theory. Before 10^{-44} seconds after the Big Bang, the four fundamental forces of nature—gravity, the electromagnetic force, and the strong and weak nuclear forces—were unified into a single force. At 10^{-44} seconds, gravity separated from the others; at 10^{-34} seconds, the strong force became separated; and at 10^{-11} seconds, the weak force separated from the electromagnetic force.

During this period the universe began a sudden burst of exponential expansion—faster than the speed of light. This expansion is called "inflation" and explains why the universe we observe is so uniform. Temperatures were so hot (10^{27} K) before inflation that the familiar particles that make up atoms today (**protons** and **neutrons**) were not stable—the universe was a hot soup of quarks (particles that are hypothesized to make up baryons), leptons (**electrons** and neutrinos), photons, and other exotic particles.

The History of the Universe: Formation of the Elements, Stars, and Galaxies, and the Cosmic Microwave Background

As the universe expanded after inflation it continued to cool. For the first three minutes conditions everywhere were similar to those at the center of stars today, and **fusion** of protons into deuterium, helium, and lithium took place. Most of the helium we see today in stars is believed to have been produced during these early minutes. The universe was an extremely opaque plasma, and photons dominated the mass density and dynamical evolution of the universe. When the universe cooled sufficiently to allow the free electrons to recombine with the hydrogen and helium nuclei, suddenly the opacity dropped, and the photons were free to stream through space unimpeded. These photons are seen today as the cosmic microwave background, a bath of light that is seen in all directions today. The experimental detection of the cosmic microwave background was one of the great triumphs of the Big Bang theory. Recombination and the subsequent production of the cosmic microwave background occurred about 180,000 years after the Big Bang.

At this point the matter distribution of the universe was still fairly uniform, with only small density fluctuations from place to place. As the universe expanded, the slightly overdense regions began to collapse. Sheets and filaments in the gas formed, which drained into dense clumps where star formation began. Eventually, these protogalactic fragments merged and galaxies and **quasars** formed. The universe began to look like it does today.

The Future of the Universe: Einstein's Biggest Blunder or Most Amazing Prediction?

Cosmologists predict the future of the universe as well as study its past. Whether the universe will expand forever or eventually slow down, turn around, and recollapse depends on how fast the galaxies are moving apart today and how much gravity there is to counter the expansion—quantities that in principle can be measured.

German-born American physicist Albert Einstein (1879–1955) described the modern theory of gravity, general relativity. He used the idea that space could be curved to reformulate English physicist and mathematician Isaac Newton's (1642–1727) theory of gravity. In general relativity, the mass of an object curves the space around it, and parallel lines no longer go on forever without intersecting. In many textbooks the curvature

As the universe expands, overdense regions in space collapse, and the resulting protogalactic fragments merge to form galaxies.

protons positively charged subatomic particles

neutrons subatomic particles with no electrical charge

electrons negatively charged subatomic particles

fusion releasing nuclear energy by combining lighter elements such as hydrogen into heavier elements

quasars luminous objects that appear star-like but are highly redshifted and radiate more energy than an entire ordinary galaxy; likely powered by black holes in the centers of distant galaxies

BLACK HOLES AND HAWKING RADIATION

Hawking radiation: A particle-antiparticle pair can be created spontaneously near a black hole. If one member of the pair falls into the event horizon, its partner can escape, carrying away energy, called Hawking radiation, from the black hole. This radiation is named after English theoretical physicist Stephen Hawking, who first hypothesized that this could be an important way that black holes evaporate.

A noted cosmologist, Hawking is working on the basic laws that govern the universe. This creative visionary was born in January 1942, and now studies a variety of issues related to the Big Bang and black holes. His seminal work, *A Brief History of Time* (1988), was a popular best-seller. Despite his disability because of an incurable disease, amyotrophic lateral sclerosis (ALS), the wheelchair-bound Hawking continues active research into theoretical physics, mixed with a fast-paced agenda of travel and public lectures. Communicating via computer system and a speech synthesizer, Hawking is still active in his quest to decipher the nature of space and time.

neutron stars the dense cores of matter composed almost entirely of neutrons that remain after a supernova explosion has ended the lives of massive stars

of space is represented by a sphere or a saddle shape—but in reality, space is three-dimensional, and the "curvature" is not in a particular direction. Einstein wrote down what are called "field equations" that described how the curvature of space can be calculated from mass and energy. When he solved the equations he realized that even if the universe is infinite, isotropic (the same in all directions), and homogeneous (the same density everywhere), it would not be static. Depending on the geometry, it would expand or contract. American astronomer Edwin P. Hubble (1889–1953) had not yet discovered that the universe expands, so in 1917 Einstein added a "parameter" lambda, called the cosmological constant, to the field equations. Later, when Hubble showed that the universe is expanding, and that there was no need to add a cosmological constant to the field equations, Einstein called the cosmological constant "the biggest blunder of my life."

Were Einstein alive to day, he would be amazed to learn about recent observations that suggest that the cosmological constant is not zero and that the expansion is accelerating. In this case, the curvature of space is not so easily related to the dynamical evolution of the universe. At the beginning of the twenty-first century, theorists had not come up with a theory for the origin of a non-zero lambda that has testable predictions. Certainly, more observations are called for to confirm or refute this result.

Nonetheless, the conditions in the universe in the distant future can be described, given the physics that is understood today. If the universe is *closed*, then the Hubble expansion will eventually stop, and the universe will then collapse. If the density of the universe is, for the sake of argument, about twice the critical density for closing the universe, then the expansion stops about 50 billion years after the Big Bang. At about 85 billion years after the Big Bang, the density of the universe will again be about what it is today. At this point, the nearby galaxies will appear to move toward us, more distant galaxies will be standing still, and the very distant galaxies will be moving away. Eventually, the galaxies will all touch, and the universe will continue to contract and heat. Soon the stars will be cooler than the universe as a whole, so radiation will not be able to flow out of them, and they will explode. As a result, 100 billion years after the Big Bang will come the big crunch. At this point the universe may become a black hole—or it may bounce, and cycle again.

If the universe is *open* or *flat*, the Hubble expansion goes on forever. Physical processes that take such a long time that they are irrelevant in to-day's universe will eventually have time to occur. After 1 trillion (10^{12}) years, star formation will have used up all the available gas, and no new stars will form. Stellar remnants such as white dwarfs, **neutron stars**, and black holes will remain. After 10^{18} years, galaxies will evaporate—their stars will disperse into space. After 10^{40} years, protons and neutrons will decay into positrons and electrons. After that, only black holes will exist. The black holes will eventually evaporate by Hawking radiation. At 10^{100} years after the Big Bang, all of the black holes, even the supermassive ones in quasars, will be gone. The universe will be very black and cold indeed.

Conclusion

The questions asked by cosmologists are some of the most simple and yet most profound questions intelligent creatures can ask. What is the origin of this beautiful and complex universe we live in, and what is its ultimate fate? Amazing progress was made over the last hundred years in cosmology, but clearly many important parts of the story are yet to be discovered. SEE ALSO AGE OF THE UNIVERSE (VOLUME 2); EINSTEIN, ALBERT (VOLUME 2); GALAXIES (VOLUME 2); HUBBLE CONSTANT (VOLUME 2); HUBBLE, EDWIN P. (VOLUME 2); SHAPLEY, HARLOW (VOLUME 2); WHAT IS SPACE? (VOLUME 2).

Jill Bechtold

Bibliography

Guth, Alan H., and Alan P. Lightman. *The Inflationary Universe: The Quest for a New Theory of Cosmic Origins.* Reading, MA: Addison-Wesley Publishing, 1998.

Hogan, Craig J., and Martin Rees. *The Little Book of the Big Bang: A Cosmic Primer.* New York: Copernicus Books, 1998.

Livio, Mario, and Allan Sandage. *The Accelerating Universe: Infinite Expansion, the Cosmological Constant, and the Beauty of the Cosmos.* New York: John Wiley & Sons, 2000.

Rees, Martin J. *Just Six Numbers: The Deep Forces that Shape the Universe.* New York: Basic Books, 2001.

Earth

Imagine that you are describing planet Earth to someone who has never seen it. How would you describe its appearance? What would you say about it? What things about Earth are typical of all planets? What things are unique?

To describe Earth, you might say that it is the third planet from the Sun in this solar system, and that it is 12,756 kilometers (7,909 miles) in diameter. Someone else might say that Earth is a fragile-looking blue, brown, and white sphere. A third person might say that Earth is the only planet in our system, as far as we know, with life. All of these descriptions are true; they are very different, however, from the descriptions of Earth that someone living in the 1950s or earlier would have given. Before we began to travel into space and to send spacecraft to observe other planets, we did not realize how different, or how similar, our planet was from other planets. And we were so busy examining the details and small regional differences of our world that we did not think about the planet as a whole.

Planet Earth

Our knowledge of Earth has been fundamentally changed by the knowledge we have gained about the other planets in the solar system. We have come to realize that, in some ways, Earth is very similar to its nearest neighbors in space. Like all of the other planets in our system, Earth orbits around our star (the Sun). It is the largest of the inner planets, just slightly larger than Venus; and it experiences seasons (as does Mars) due to the tilt of its rotation axis.

Like many other planets in our system, Earth has a natural satellite. We call our single satellite the "Moon" and have used that term to describe all of the other moons in our system, although Earth and its Moon are unusually closer in size than is common. One of the ways in which Earth is similar to its nearest neighbors is that all of the **"rocky" planets** have been

"rocky" planets nickname given to inner or solid-surface planets of the solar system, including Mercury, Venus, Mars, Earth, and the Moon

Faults, such as the San Andreas Fault in California, are indications that Earth is still geologically active.

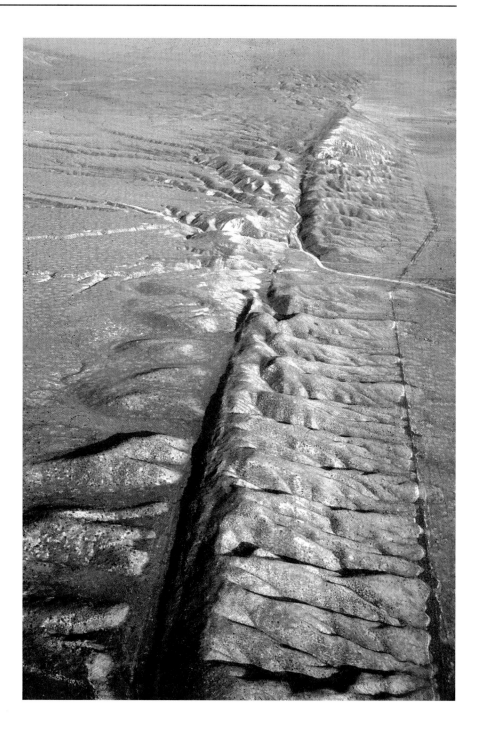

tectonism process of deformation in a planetary surface as a result of geological forces acting on the crust; includes faulting, folding, uplift, and down-warping of the surface and crust

affected by four fundamental geological processes: volcanism, **tectonism**, erosion, and impact cratering.

The surface of our planet is a battleground between the processes of volcanism and tectonism that create landforms and the process of erosion that attempts to wear away these landforms. Geologically speaking, Earth is a "water-damaged" planet, because water is the dominant agent of erosion on the surface of our world. On planets with little or no atmosphere, erosion of the surface may occur due to other processes, such as impact cratering. On the rocky planets the dominant mechanism of erosion may differ, and the styles or details of the volcanic or tectonic landscape may differ, but the fundamental geological processes remain the same.

Of the four fundamental processes, the one that may be unexpected is impact cratering. In fact, prior to our exploration of the Moon, impact cratering was not considered important to Earth. Those few impact craters identified on Earth were treated as curiosities. Now, after studying the other planets, we realize that impact cratering is an important and continuing process on all planets, including Earth. Impacts from **meteorites**, comets, and occasionally large asteroids have occurred throughout the history of Earth and have been erased by Earth's dynamic and continuing geology. The formation of an impact crater can significantly affect the geology, atmosphere, and even the biology of our world. For example, scientists believe that an impact that occurred about 65 million years ago on the margin of the Yucatan Peninsula was a possible cause of the extinction of the dinosaurs and many other species.

A Uniquely Different Planet

Although Earth is in some ways a typical rocky planet, several of its most interesting features appear to be unique. For example, a global map of Earth with the ocean water removed shows a very different planet from our neighboring rocky planets. The patterns made by continents, oceans, aligned volcanoes, and linear mountains are the result of the process geologists call plate tectonics.

We know from the study of earthquake waves moving through Earth that our planet is made up of three main layers: the crust, mantle, and core.

WHAT IS VOLCANISM?

Volcanism is a geological term used to describe the complete range of volcanic eruptions, volcanic landforms, and volcanic materials. Volcanism is driven by the internal heat of a planet and provides evidence of the way in which heat is released from that planet. The type and abundance of volcanoes on the surface of a planet can provide evidence about the level of geologic activity of the planet.

meteorite any part of a meteoroid that survives passage through Earth's atmosphere

WHAT IS TECTONISM?

Tectonism is a geological term used to describe major structural features and the processes that create them, including compressional or tensional movements on a planetary surface that produce faults, mountains, ridges, or scarps. Tectonic or structural movements are driven by the internal heat of a planet, and those movements on Earth produce earthquakes. Faults, ridges, or mountains on a planetary surface imply that the planet was or is still geologically active.

convection the movement of heated fluid caused by a variation in density

basalt a dark, volcanic rock with abundant iron and magnesium and relatively low silica; common on all of the terrestrial planets

The upper layer of Earth (consisting of the crust and the upper mantle) is broken into rigid plates that move and interact in various ways. Where plates are moving together or one plate is moving beneath another, mountains such as the Himalayas or explosive volcanoes such as the Cascades are formed. Where plates are moving apart, such as along the mid-oceanic ridges, new crust is formed by the slow eruption of lava. Where two plates slide along each other, such as the San Andreas Fault zone in California, major earthquakes occur. The movement of the plates is caused by the **convection** of the mantle beneath them; that convection is driven by the planet's internal heat, derived from radioactive decay of certain elements. Similarly, rotation and convection in the fluid metallic outer core is responsible for Earth's uniquely strong magnetic field. Plate tectonics can be thought of as a giant recycling mechanism for Earth's crust.

The concept of plate tectonics is a relatively new idea, and it is central to our understanding of Earth's dynamic geology. Nevertheless, planetary geologists have found no clear evidence of past or present Earth-style plate tectonics on any of the other rocky planets; Earth seems to be unique in this regard.

Earth is also unique in that no other planet in the solar system currently has the proper temperature and atmospheric pressure to maintain liquid water on its surface. Water exists on Earth as gas (water vapor), liquid, and solid (ice), and all three forms are stable at Earth's surface temperature and pressure. Water may be the single most important criteria for life as it has developed on Earth. And the presence of life, in turn, has changed and affected the composition of the atmosphere and the surface of Earth. For example, the rock type limestone would not be possible without marine life, and limestone formation may have significantly altered the distribution of carbon dioxide on Earth.

Mars and Venus also have atmospheres, but they are primarily composed of carbon dioxide. Earth's atmosphere is approximately 76 percent nitrogen and 20 percent oxygen with traces of water vapor, carbon dioxide, and ozone. Although water is not a major component by percent, it is a very important part of Earth's atmosphere. Earth's surface water and atmosphere are linked to form a single system. Water evaporates from the oceans, moves through the atmosphere as vapor or cloud droplets, precipitates onto the surface as rain or snow, and returns to the oceans by way of rivers. Clouds cover approximately 50 percent of Earth's surface at any one time, and they play an important role in maintaining the balance of atmospheric and surface temperatures on our planet.

Our atmosphere and water work together to form a general category of rocks on Earth that is not known to exist on any neighboring planets. On Earth's surface, sedimentary rocks, such as quartz-rich sandstone or marine limestone, are very common; they cover approximately 70 percent of the surface of our planet in a very thin veneer. Although Mars may surprise us, initial studies of our nearest neighbors indicate that the volcanic rock **basalt** is the basic building block of planetary crust (including most of Earth's subsurface crust) and the most common rock type on the surface of the other rocky planets. Once again, Earth is unique. And as we explore other planets around other suns, typical Earth sandstone might be as exotic and rare as gold. SEE ALSO CLOSE ENCOUNTERS (VOLUME 2); EARTH—WHY LEAVE?

(VOLUME 4); MARS (VOLUME 2); MOON (VOLUME 2); NASA (VOLUME 3); SO-LAR WIND (VOLUME 2).

Jayne Aubele

Bibliography

Cloud, Preston. *Oasis in Space: Earth History from the Beginning.* New York: W. W. Norton and Company, 1988.

Hamblin, W. Kenneth, and Eric H. Christiansen. *Exploring the Planets.* New York: Macmillan Publishing Company, 1990.

Harris, Stephen L. *Agents of Chaos.* Missoula, MT: Mountain Press Publishing Company, 1990.

Moore, Patrick, and Garry Hunt, eds. *Atlas of the Solar System.* New York: Crescent Books, 1990.

Einstein, Albert

German-born, Swiss-educated Physicist
1879–1955

Albert Einstein was a scientist who revolutionized physics in the early twentieth century with his theories of relativity. Born in Ulm, Germany, in 1879, Einstein was interested in science from an early age. While he performed well in school, he disliked the academic environment and left at the age of fifteen. He took an entrance exam for the Swiss Federal Institute of Technology (ETH) in Zurich but failed; only after completing secondary school was he able to gain entrance to ETH, where he graduated in 1900. Unable to find a teaching position, Einstein accepted a job in the Swiss patent office in 1902.

During his time as a patent clerk Einstein made some of his most important discoveries. In 1905 he published three papers, which brought him recognition in the scientific community. In one he described the physics of Brownian motion, the random motion of particles in a gas of liquid. In another paper he used the new field of quantum mechanics to explain the photoelectric effect, where metals give off **electrons** when exposed to certain types of light. Einstein published his third, and arguably most famous, paper in 1905, which outlined what later became known as the special theory of relativity. This theory showed how the laws of physics worked near the speed of light. The paper also included the famous equation $E=mc^2$, explaining how energy was equal to the mass of an object times the speed of light squared.

These papers allowed Einstein to exchange his patent clerk job for university positions in Zurich and Prague before going to Berlin as director of the Kaiser Wilhelm Institute of Physics. Shortly thereafter he published the general theory of relativity, which describes how gravity warps space and time. This theory was confirmed in 1919 when astronomers measured the positions of stars near the Sun during a solar eclipse and found that they had shifted by the amount predicted if the Sun's gravity had warped the light.

The acceptance of Einstein's general theory turned him into an international celebrity. During the 1920s he toured the world, giving lectures.

Albert Einstein developed the famous equation $E=mc^2$.

electrons negatively charged subatomic particles

In 1922 he won the Nobel Prize for physics, although it was officially awarded for his work studying the photoelectric effect, not relativity. In 1932 he accepted a part-time position at Princeton University in Princeton, New Jersey, and planned to split his time between Germany and the United States. But when the Nazis took power in Germany one month after he arrived at Princeton, Einstein decided to stay in the United States.

Einstein spent the rest of his scientific career in an unsuccessful pursuit of a theory that would explain all the fundamental forces of nature. He also took a greater role outside of physics. In 1939 he cowrote a letter to President Franklin Roosevelt, urging him to investigate the possibility of developing an atomic bomb and warning him that Germany was likely doing the same. After the war he urged world leaders to give up nuclear weapons to preserve peace. In ill health for several years, he died in Princeton in 1955. SEE ALSO AGE OF THE UNIVERSE (VOLUME 2); ASTRONOMY, HISTORY OF (VOLUME 2); BLACK HOLES (VOLUME 2); COSMIC RAYS (VOLUME 2); COSMOLOGY (VOLUME 2); GRAVITY (VOLUME 2); WORMHOLES (VOLUME 4); ZERO-POINT ENERGY (VOLUME 4).

Jeff Foust

Bibliography

Brian, Denis. *Einstein: A Life*. New York: John Wiley & Sons, 1996.

Clark, Ronald W. *Einstein: The Life and Times*. New York: Avon Books, 1972.

Internet Resources

"Albert Einstein." University of St. Andrews, School of Mathematics and Physics. <http://www-groups.dcs.st-and.ac.uk/~history/Mathematicians/Einstein.html>.

Weisstein, Eric. "Einstein, Albert (1879–1955)." <http://www.treasure-troves.com/bios/Einstein.html>.

Exploration Programs

Prior to missions to the Moon and the planets in the solar system our knowledge of what lay beyond Earth was minimal. Five millennia of astronomical observation had produced an incomplete picture of the solar system. Although the Moon and planets were neighbors, there was only so much that could be learned from even the best telescopes. Only by sending spacecraft and astronauts on programs of exploration could we examine our neighbors in space more closely.

The first objective for both the United States and the Soviet Union was reaching the Moon. In September 1959 the Soviet probe Luna 2 struck the Moon. Three weeks later, Luna 3 sent back the first grainy images of the Moon's farside. For the United States, the Ranger project of the 1960s marked the first effort to launch probes toward the Moon. A variety of difficulties plagued the first several Ranger missions, and it was not until Ranger 7, in July 1964, that the program achieved complete success. Two more Ranger spacecraft were launched, including Ranger 8, which took 7,300 images before crash-landing in the Sea of Tranquility, where the Apollo 11 astronauts would land four and a half years later.

The Ranger program was the first of three intermediate steps leading to Apollo. Next came the Lunar Orbiter program, which photographed

Lunar surface experiments provided much data for scientists to evaluate, including several hundred kilograms of samples. The United States' Apollo astronauts often collected surface information themselves, while the Soviets used unpiloted Luna probes to gather their data.

potential Apollo landing sites. Altogether, five Lunar Orbiter spacecraft were launched from 1966 to 1967. By the end of the fourth mission, Lunar Orbiter probes had surveyed 99 percent of the front and 80 percent of the backside of the Moon. While Lunar Orbiters snapped photographs overhead, the Soviets and Americans perfected soft landing techniques. In February 1966, a 100-kilogram Soviet probe, shaped like a beach ball, touched down on the Moon and returned the first images of the lunar surface.

The Americans countered the Soviet success with a program called Surveyor. Once on the surface, the tripod-shaped Surveyors evaluated the lunar soil and environment. Surveyor 1 made a successful soft landing in three centimeters of dust in the Ocean of Storms in June 1966. Surveyors 3, 5, 6 and 7 landed at different sites and carried out experiments on the surface, including analyzing the chemical composition of the lunar soil. All told,

flyby flight path that takes the spacecraft close enough to a planet to obtain good observations; the spacecraft then continues on a path away from the planet but may make multiple passes

Surveyors acquired almost 90,000 images from five landing sites. The success of the Ranger and Surveyor programs and that of the five Lunar Orbiters gave the National Aeronautics and Space Administration (NASA) the confidence that humans could go the Moon.

In July 1969, Apollo 11 became the first mission to land humans on the Moon when Neil Armstrong and Edwin "Buzz" Aldrin piloted their lunar module "Eagle" to the Sea of Tranquility. Four months later, Apollo 12 landed at the site where Surveyor 3 had touched down in the Sea of Storms. Four more Apollo missions visited the Moon through December 1972. By the end of the program, Apollo astronauts had returned nearly 380 kilograms of samples from the Moon. Besides the samples, data from lunar-orbital experiments and information from lunar surface experiments were returned. Over the same time period, the Soviet Union retrieved several hundred grams of lunar material using Luna probes.

Beyond the Moon

As plans were getting under way to explore the Moon, NASA also focused on the rest of the solar system. The Mariner series of missions were designed to be the first U.S. spacecraft to reach other planets. Mariner 2 became Earth's first interplanetary success. After a flawless launch, the Mariner 2 spacecraft encountered Venus at a range of 35,000 kilometers in December 1962. As it flew by, Mariner 2 scanned the planet and revealed that Venus has an extremely hot surface. Mariner 2 also measured the solar wind, a constant stream of charged particles flowing outward from the Sun.

In July 1965, Mariner 4 provided the first close look at Mars. The twenty-two fuzzy images returned by Mariner 4 revealed a planet pocked with craters. Four years later, Mariner 6 and 7 provided 200 more images of the Red Planet. Late in 1971, Mariner 9 went into orbit around Mars and a new era of Mars exploration dawned. Previous missions had been **flybys**, but Mariner 9 became the first artificial satellite of Mars. Upon arrival, a dust storm obscured the entire planet, but after the dust cleared, Mariner 9 revealed a place of incredible diversity that included volcanoes and a canyon stretching 4,800 kilometers. More surprisingly, Mariner 9 radioed back images of ancient riverbeds carved in the landscape.

Mariner 9 was followed in 1976 by Viking 1 and 2, each consisting of a lander and an orbiter. Each orbiter-lander pair entered Mars orbit; then the landers separated and descended to the planet's surface. The Viking 1 lander touched down on the western slope of Chryse Planitia ("Plains of Gold") on July 20, 1976, the seventh anniversary of the Apollo 11 Moon landing. Within an hour of landing, the first photos of Mars' surface were radioed back to Earth. Besides taking photographs, both Viking 1 and 2 landers conducted biology experiments to look for signs of life. These experiments discovered unusual chemical activity in the Martian soil, but provided no clear evidence for the presence of microorganisms. However, both landers provided a wealth of data about the Martian surface, and the Viking 1 and 2 orbiters took thousands of images from above.

While the United States focused much attention on Mars throughout the 1960s and 1970s, the Soviet Union flew a series of missions to Venus. Venera 4 became the first mission to place a probe into the Venusian atmosphere in June 1967. In June 1975, probes released by Venera 9 and 10

transmitted the first black and white images of Venus' surface. Other missions followed, including Venera 15 and 16, which produced **radar** images of the Venusian surface. Venus was also a target of NASA's Mariner 10 mission in 1973 to 1974, which used a "gravity assist" to send the spacecraft on to Mercury. Gravity assist techniques were to play a crucial role in NASA's next phase of planetary exploration—journeys to Jupiter and beyond.

Pioneer 10 and 11 were the first spacecraft to venture beyond the asteroid belt into the realm of the outer planets. Pioneer 11 safely passed through the asteroid belt and passed 42,000 kilometers (26,098 miles) below Jupiter's south pole in December 1974, exactly a year after Pioneer 10's closest approach. Using Jupiter's immense gravity like a slingshot, Pioneer 11 encountered Saturn in September 1979. After passing Saturn, Pioneer 11 plunged into deep space, carrying a plaque similar to that aboard Pioneer 10 in the hope that intelligent life would someday find it.

NASA mission designers recognized that the giant outer planets—Jupiter, Saturn, Uranus and Neptune—would soon align in such a way that a single spacecraft might be able to use gravity assists to hop from one planet

radar a technique for detecting distant objects by emitting a pulse of radio-wavelength radiation and then recording echoes of the pulse off the distant objects

The Viking 2 lander looks out over Mars' Utopia Plain in 1976. Both the Viking 1 and 2 missions took thousands of photos of the planet and conducted biological experiments to search for signs of life, uncovering unusual chemical activity but no clear proof of microorganisms.

to the next. Taking advantage of this alignment, NASA approved the Voyager Project. Voyager 1 made its closest approach to Jupiter in March 1979, and Voyager 2 came within 570,000 kilometers (354,182 miles) of Jupiter in July 1979. Voyager 1 and 2 flybys of Saturn occurred nine months apart, with the closest approaches occurring in November 1980 and August 1981. Voyager 1 then headed out of the orbital plane of the planets. However, Voyager 2 continued onward for two more planetary encounters, coming within 81,500 (50,642 miles) kilometers of Uranus' cloud tops in January 1986, and making a flyby of Neptune in August 1989. Both Voyagers continue to operate and are approaching interstellar space.

While the Voyager missions were highly successful, the pace of planetary exploration slowed in the 1980s. One of the few new missions was Magellan, which went into orbit around Venus in August 1990. Over the next four years Magellan used radar to map 99 percent of the Venusian surface. After concluding its radar mapping, Magellan made global maps of Venus's gravity field. Flight controllers also tested a new maneuvering technique called aerobraking, which uses a planet's atmosphere to slow a spacecraft.

NASA managers also followed up the initial **reconnaissance** of Jupiter with the Galileo mission. En route to Jupiter, Galileo flew by two asteroids—Gaspra and Ida—the first such visits by any spacecraft. Galileo arrived at Jupiter in December 1995, and dropped an instrumented probe into the giant planet's atmosphere. Since then, Galileo has made dozens of orbits of Jupiter, usually flying close to one of its four major moons. Among its discoveries, Galileo uncovered strong evidence that Jupiter's moon Europa has a saltwater ocean beneath its surface.

reconnaissance a survey or preliminary exploration of a region of interest

In the mid-1980s, the European Space Agency launched its first deep space mission, part of an ambitious international mission to Comet Halley. The plan was to send an armada of five spacecraft—two Soviet (Vega 1 and 2), two Japanese (Sakigake and Suisei) and one European (Giotto)—towards the comet in 1986. A series of images sent back by Giotto revealed the comet nucleus to be a dark, peanut-shaped body, about 15 kilometers long. NASA did not send a mission to Comet Halley for budgetary reasons. Instead, NASA planed to return to Mars after a seventeen-year pause. In September 1992, the United States launched Mars Observer, but the mission ended in failure when contact was lost with the spacecraft as it approached Mars.

A New Strategy: Faster, Better, Cheaper

The loss of Mars Observer prompted NASA to rethink its strategy for planetary exploration. The few large and expensive missions that characterized the preceding fifteen years were replaced by greater numbers of focused, cheaper missions, a strategy described as "faster, better, cheaper." Ironically, it was a joint project between the U.S. military and NASA, called Clementine, which underscored the potential of this concept. Clementine was launched in January 1994 and mapped the lunar surface, providing preliminary evidence of ice at the Moon's poles.

In the early 1990s, NASA established the Discovery program to select low-cost solar system exploration missions with focused science goals. The first Discovery mission was the Near Earth Asteroid Rendezvous (NEAR) mission. NEAR entered orbit around the asteroid Eros in February 2000, beginning a yearlong encounter. The car-sized spacecraft gathered ten times more data during its orbit than originally planned, and completed all the mission's science goals before becoming the first spacecraft to land on the surface of an asteroid.

Mars Pathfinder, the second Discovery class mission, landed on Mars on July 4, 1997, assisted by airbags to cushion the impact. The landing site, known as Ares Vallis, was chosen because scientists believed it was the site of an ancient catastrophic flood. Onboard Pathfinder was a six-wheeled rover named Sojourner. From landing until the last transmission in September 1997, Mars Pathfinder returned more than 16,500 images from the lander and 550 images from the rover, in addition to chemical analyses of rocks and soil plus data on winds and other weather phenomena.

Coinciding with Pathfinder's mission was the arrival of the Mars Global Surveyor (MGS) spacecraft in orbit around the Red Planet in September 1997. Although not a Discovery mission, MGS applied the aerobraking skills

magnetometer an instrument used to measure the strength and direction of a magnetic field

minerals crystalline arrangements of atoms and molecules of specified proportions that make up rocks

ion propulsion a propulsion system that uses charged particles accelerated by electric fields to provide thrust

pioneered by Magellan. After a year and a half trimming its orbit, MGS began its prime mapping mission in March 1999. From orbit, MGS took pictures of gullies and debris features that suggest there may be current sources of liquid water at or near the planet's surface. In addition, **magnetometer** readings showed that the planet's magnetic field is localized in particular areas of the crust. MGS completed its primary mission in January 2001, but continues to operate in an extended mission phase.

Since Mars Pathfinder, more Discovery class missions have been launched, including Lunar Prospector, Stardust, and Genesis. The small spin-stabilized Lunar Prospector spacecraft lifted off in January 1998 and spent almost two years measuring the Moon's magnetic and gravitational fields and looking for natural resources, such as **minerals** and gases, which could be used to sustain a human lunar base or manufacture fuel. Mission scientists believe that Lunar Prospector detected between 10 to 300 million tons of water ice scattered inside craters at the lunar poles.

Stardust, the fourth Discovery mission, was launched in February 1999 and is slated to fly through the cloud of dust that surrounds Comet Wild–2, and bring cometary samples back to Earth in January 2006. Stardust will be the first mission to return extraterrestrial material from outside the orbit of the Moon. Launched in August 2001, the Genesis spacecraft is headed toward an orbit around Lagrangian 1 (L1), a point between Earth and the Sun where the gravity of both bodies is balanced. Once it has arrived, Genesis will begin collecting particles of solar wind that imbed themselves in specially designed high purity wafers. After two years, the sample collectors will be returned to Earth.

In addition to the Discovery program, the U.S. space agency has embarked on a series of New Millennium missions to test advanced technologies. The first New Millennium mission was Deep Space 1, which validated **ion propulsion** and tested other new technologies, such as autonomous optical navigation, several microelectronics experiments, and software to plan and execute onboard activities with only general direction from the ground.

While the trend in planetary exploration has been toward cheaper, smaller missions, the joint American/European Cassini mission to Saturn represents the opposite approach. Launched in October 1997, Cassini is the most ambitious effort in planetary space exploration ever mounted and involves sending a sophisticated robotic spacecraft to orbit the ringed planet and study the Saturnian system over a four-year period. Onboard Cassini is a scientific probe called Huygens that will parachute through the atmosphere to the surface of Saturn's largest moon, Titan. Cassini will enter Saturn orbit in July 2004, and the Huygens probe will descend to the surface of Titan in November of that year. Building on the spectacular success of exploration programs over the past forty years, future missions are planned to the Moon, Mars, Mercury, Jupiter's moons and beyond. SEE ALSO APOLLO (VOLUME 3); APOLLO LUNAR LANDING SITES (VOLUME 3); ASTROBIOLOGY (VOLUME 4); LIFE IN THE UNIVERSE, SEARCH FOR (VOLUME 2); PLANETARY EXPLORATION (VOLUME 1); PLANETARY EXPLORATION, FUTURE OF (VOLUME 2); ROBOTIC EXPLORATION OF SPACE (VOLUME 2).

John F. Kross

Bibliography

Compton, William D. *Where No Man Has Gone Before: A History of Apollo Lunar Exploration Missions.* Washington, DC: NASA Historical Series (NASA SP-4214), 1989.

Dewaard, E. John, and Nancy Dewaard. *History of NASA: America's Voyage to the Stars.* New York: Exeter Books, 1984.

Washburn, Mark. *Distant Encounters. The Exploration of Jupiter and Saturn.* New York: Harcourt Brace Jovanovich, 1983.

Yenne, Bill. *The Encyclopedia of US Spacecraft.* New York: Exeter Books, 1988.

Internet Resources

Deep Space 1. Jet Propulsion Laboratory. <http://nmp.jpl.nasa.gov/ds1/>.

Discovery Program. NASA Goddard Space Flight Center. <http://nssdc.gsfc.nasa.gov/planetary/discovery.html>.

Galileo Program. Jet Propulsion Laboratory/NASA Current Mission Series. <http://www.jpl.nasa.gov/missions/current/galileo.html>.

Genesis Program. Jet Propulsion Laboratory/NASA Current Mission Series. <http://www.jpl.nasa.gov/missions/current/genesis.html>.

Lunar Prospector. NASA Ames Research Center. <http://lunar.arc.nasa.gov/>.

Mars Pathfinder Program. Jet Propulsion Laboratory/NASA History Series. <http://www.jpl.nasa.gov/missions/past/marspathfinder.html>.

Near Earth Asteroid Rendezvous (NEAR). The John Hopkins University Applied Physics Laboratory. <http://near.jhuapl.edu/>

Stardust Program. Jet Propulsion Laboratory/NASA Current Mission Series. <http://www.jpl.nasa.gov/missions/current/stardust.html>.

Viking Program. Jet Propulsion Laboratory/NASA History Series. <http://www.jpl.nasa.gov/missions/past/viking.html>.

Voyager Program. Jet Propulsion Laboratory/NASA History Series. <http://www.jpl.nasa.gov/missions/current/voyager.html>.

Extrasolar Planets

The question of whether or not other planetary systems similar to our own exist has intrigued astronomers and the general public alike for centuries. It was only in the in the 1990s that astronomers began to discover direct evidence for planets outside the solar system.

Method of Detection

The planets of the solar system are visible because of the sunlight that they reflect. Unfortunately, planets that orbit other stars are far too faint relative to their stars for current astronomical telescopes to observe them directly as faint points of light next to their much brighter stars. Instead, astronomers use a variety of techniques to indirectly infer the presence of these extrasolar planets.

The method by which all of the known extrasolar planets have been discovered is the radial velocity (or Doppler) technique. The mass of the orbiting planet causes the central star to be pulled around in an orbit. Astronomers detect the resulting small, periodic shifts in the apparent speed of the star. By measuring the shape of the resulting Doppler curve over time, they are able to deduce a lower limit on the mass of the planet and estimate the separation of the planet from the star. This planet-star distance is typically expressed in astronomical units (AUs); one AU is the average distance from Earth to the Sun.

Extrasolar Discoveries

The first detection of a planet orbiting another Sun-like star was accomplished in 1995 using the radial velocity technique. This planet, orbiting the star 51 Pegasi, was found by two Swiss astronomers, Michel Mayor and Didier Queloz, of the Geneva Observatory. A pair of American astronomers, Geoffrey Marcy and Paul Butler, soon followed with the announcement of several other new planets. Each of these teams has since expanded into large groups that are now surveying thousands of Sun-like stars in search of new worlds.

These radial velocity surveys are able to detect only massive planets, that is, planets which have masses similar to Jupiter. ✴ Less-massive planets produce a correspondingly smaller tug on their parent star and are thus more difficult to detect. Recently, observers of radial velocity have increased the sensitivity of their measurements, and announced the discovery of several planets with masses similar to that of Saturn (one-third that of Jupiter's mass). Future improvements in the technique should allow for the detection of planets with even lower masses. Unfortunately, stars themselves are somewhat variable, and the flutter that results from their intrinsic variability implies that small, rocky worlds similar to Earth (a mere 1/300th of Jupiter's mass) will not be detectable by this method.

By early 2002, more than seventy extrasolar planets have been discovered. Based on the total number of stars in the current surveys, this implies that at least 7 percent of Sun-like stars have at least one planet. This estimate, however, is only a lower limit: The surveys have not been in operation long enough to see planets at large distances from their stars. Our own Jupiter takes nearly twelve years to circle the Sun. Planetary systems similar to our own—that is, those with massive planets in large, circular orbits far from the central star—would not yet have been detected. One of the chief goals of the radial velocity surveys between 2002 and 2012 is to search for such systems.

After astronomers find a planet orbiting a given star, they continue to monitor that star in the hope of detecting additional planets. In 1999, Butler and colleagues announced the first detection of a multiple-planet system, in orbit about the Sun-like star Upsilon Andromedae. Recently, six other stars have been demonstrated to harbor multiple planets, and many other stars show hints that they too possess multiple planets.

Hot Jupiters

The first distinct subclass of extrasolar planets to emerge from radial-velocity surveys consists of the so-called Hot Jupiters (or 51-Peg-type planets). These planets have masses similar to that of Jupiter, but they are located 100 times closer to their stars than Jupiter is from the Sun. The existence of such planets challenges conventional theories of planet formation. Gas giants such as Jupiter presumably form at large distances from their stars, where the environment is sufficiently cool for a core of ice and rock to **coagulate** and nucleate the formation of the planet. If this theory is correct, then the Hot Jupiters must have undergone a migration from the site of their formation to their current location. The cause of this migration mechanism and the details of why it did not operate in the solar system are the subjects of intensive theoretical investigation.

✴ **Jupiter is the largest planet in the solar system.**

coagulate to cause to come together into a coherent mass

Due to the proximity to the parent star, there is a reasonable chance—one in ten—that the orbit of a Hot Jupiter is tilted at just the right angle so that the planet will be observed, with each orbit, to pass in front of the disk of star. The resulting dimming of the light from the star, called a transit, was first observed in 1999 for the Sun-like star HD209458. These observations proved that the radial velocity variations, by which the planet had been initially detected, truly were due to an orbiting planet (and not some form of undiagnosed variability in the star). Moreover, for the first time, astronomers were to estimate both the physical size and mass of a planet and thus calculate its density. Their conclusion was that this planet was indeed a gas giant, similar in mass and size to Jupiter. Later, astronomers further scrutinized this star with the Hubble Space Telescope. By observing how light of different colors is filtered by the outer reaches of the planet, they detected its atmosphere, the first such detection for a planet outside the solar system.

The presence of this extrasolar planet, 150 light-years from Earth, was discovered in 1999 when astronomers detected its subtle gravitational pull on the yellow, Sun-like star HD209458.

Astronomical Missions

The successes of the techniques described above, and the realization that these ground-based methods will not allow for the detection of small, rocky worlds similar to Earth, have inspired several astronomical satellites.

In 2009 the National Aeronautics and Space Administration (NASA) plans to launch the Space Interferometry Mission (SIM). Using SIM, astronomers will perform very precise **astrometry** to detect the **reflex motion** of stars due to orbiting planets. They hope to survey hundreds of stars for large **terrestrial planets** (greater than 5 Earth masses), orbiting at distances of several AU from their stars.

In the decade after SIM, NASA will launch the Terrestrial Planet Finder, with the objective of enabling astronomers to detect extrasolar planets that are true analogs of Earth and to study the atmospheres of those planets. In particular, NASA plans to search for atmospheric components, such as ozone, which may be attributable to life. SEE ALSO HUBBLE SPACE TELESCOPE (VOLUME 2); JUPITER (VOLUME 2); STARS (VOLUME 2); SUN (VOLUME 2).

David Charbonneau

Bibliography

Doyle, Laurance R., Hans-Jorg Deeg, and Timothy M. Brown. "Searching for Shadows of Other Earths." *Scientific American* 283, no 3 (2000):58.

Marcy, Geoffrey, and Paul Butler. "Hunting Planets Beyond." *Astronomy* 28, no. 3 (2000):42.

astrometry the measurement of the positions of stars on the sky

reflex motion the orbital motion of one body, such as a star, in reaction to the gravitational tug of a second orbiting body, such as a planet

terrestrial planets small rocky planets with high density orbiting close to the Sun; Mercury, Venus, Earth, and Mars

dark matter matter that interacts with ordinary matter by gravity but does not emit electromagnetic radiation; its composition is unknown

light years one light year is the distance that light in a vacuum would travel in one year, or about 9.5 trillion kilometers (5.9 trillion miles)

Galaxies

Galaxies are collections of stars, gas, and dust, combined with some unknown form of **dark matter**, all bound together by gravity. The visible parts come in a variety of sizes, ranging from a few thousand **light years** with a billion stars, to 100,000 light-years with a trillion stars. Our own Milky Way galaxy contains about 200 billion stars.

Types of Galaxies

The invisible parts of galaxies are known to exist only because of their influence on the motions of the visible parts. Stars and gas rotate around galaxy centers too fast to be gravitationally bound by their own mass, so dark matter has to be present to hold it together. Scientists do not yet know the size of the dark matter halos of galaxies; they might extend over ten times the extent of the visible galaxy. What we see in our telescopes as a giant galaxy of stars may be likened to the glowing hearth in the center of a big dark house.

Imagine viewing a galaxy through a small telescope, as pioneering astronomers William and Caroline Herschel and Charles Messier did in the late eighteenth century. You would see mostly a dull yellow color from countless stars similar to the Sun, all blurred together by the shimmering Earth atmosphere. This light comes from stars that formed when the universe was only a tenth of its present age, several billion years before Earth existed.

American astronomer Edwin P. Hubble used a larger telescope starting in the 1920s and saw a wide variety of galaxy shapes. He classified them into

Spiral galaxies such as NGC 4414, a dusty spiral galaxy, form from fast spinning hydrogen gas. Central regions of the galaxy contain older yellow and red stars, with younger blue stars on the outer spiral arms, which are replete with clouds of interstellar dust.

elliptical, with a smooth texture; disk-like with spirals; and everything else, which he called irregular.

Elliptical Galaxies. Elliptical galaxies are three-dimensional objects that range from spheres to elongated spheroids like footballs. Some may have developed from slowly rotating hydrogen clouds that formed stars in their first billion years. Others may have formed from the merger of two or more smaller galaxies. Most ellipticals have very little gas left that can form new stars, although in some there is a small amount of star formation within gas acquired during recent mergers with other galaxies.

Spiral Galaxies. Spiral galaxies, which include the Milky Way, formed from faster-spinning clouds of hydrogen gas. Theoretical models suggest they got this spin by interacting with neighboring galaxies early in the universe. The center of a spiral galaxy is a three-dimensional bulge of old stars, surrounded by a spinning disk flattened to a pancake shape.

Hubble classified spiral galaxies according to the tightness of the spirals that wind around the center, and the relative size of the disk and bulge. Galaxies with big bulges tend to have more tightly wrapped spirals; they are designated type Sa. Galaxies with progressively smaller bulges and more open arms are designated Sb and Sc. Barred galaxies are similar but have long central barlike patterns of stars; they are designated SBa, SBb, and Sbc, while intermediate bar strengths are designated SAB.

Type Sa galaxies rotate at a nearly constant speed of some 300 kilometers per second (186 miles per second) from the edge of the bulge to the far outer disk. Sc galaxies have a rotation speed that increases more gradually from center to edge, to typically 150 kilometers per second (93 miles per second). The rotation rate and the star formation rate depend only on the average density. Sa galaxies, which are high density, converted their gas into stars so quickly that they have very little gas left for star formation today. Sc galaxies have more gas left over and still form an average of a few new stars each year. Some galaxies have extremely concentrated gas near their centers, sometimes in a ring. Here the star formation rate may be higher, so these galaxies are called starbursts.

The pinwheel structures of spiral galaxies result from a concentration of stars and gas in wavelike patterns that are driven by gravity and rotation. Bright stars form in the concentrated gas, highlighting spiral arms with a bluish color. Theoretical models and computer simulations match the observed spiral properties. Some galaxies have two long symmetric arms that give them a "grand design." These arms are waves of compression and **rarefaction** that ripple through a disk and organize the stars and gas into the spiral shape. These galaxies change shape slowly, on a timescale of perhaps ten rotations, which is a few billion years. Other galaxies have more chaotic, patchy arms that look like fleece on a sheep; these are called flocculent galaxies. The patchy arms are regions of star formation with no concentration of old stars. Computer simulations suggest that each flocculent arm lasts only about 100 million years.

Irregular Galaxies. Irregular galaxies are the most common type. They are typically less than one-tenth the mass of the Milky Way and have irregular shapes because their small sizes make it difficult for spiral patterns to develop. They also have large reservoirs of gas, leading to new star formation. The varied ages of current stars indicate that their past star formation rates were highly nonuniform. The dynamical processes affecting irregulars are not easily understood. Their low densities and small sizes may make them susceptible to environmental effects such as collisions with larger galaxies or intergalactic gas clouds. Some irregulars are found in the debris of interacting galaxies and may have formed there.

Some small galaxies have elliptical shapes, contain very little gas, and do not see any new star formation. It is not clear how they formed. The internal structures of irregulars and dwarf ellipticals are quite different, as are their locations inside clusters of galaxies (the irregulars tend to be in the outer parts). Thus it is not likely that irregulars simply evolve into dwarf ellipticals as they age.

rarefaction decreased pressure and density in a material caused by the passage of a sound wave

The spiral arms and dust clouds of the Whirlpool galaxy are the birthplace of large, luminous stars. Galaxies are composed of billions to trillions of stars, as well as gas, dust, and dark matter, all bound together by gravity.

Active Galaxies, Black Holes, and Quasars

In the 1960s, Dutch astronomer Maarten Schmidt made spectroscopic observations of an object that appeared to be a star but emitted strong radio radiation, which is uncharacteristic of stars. He found that the normal **spectral lines** emitted by atoms were shifted to much longer **wavelengths** than they have on Earth. He proposed that this redshift was the result of rapid motion away from Earth, caused by the cosmological expansion of the universe discovered in the 1920s by Hubble. The velocity was so large that the object had to be very far away. Such objects were dubbed quasi-stellar objects, now called **quasars**. Several thousand have been found.

With the Hubble Space Telescope, astronomers have recently discovered that many quasars are the bright centers of galaxies, some of which are interacting. They are so far away that their spatial extents cannot be resolved through the shimmering atmosphere. Other galaxies also show the unusually strong radio and infrared emissions seen in quasars; these are called active galaxies.

spectral lines the unique pattern of radiation at discrete wavelengths that many materials produce

wavelength the distance from crest to crest on a wave at an instant in time

quasars luminous objects that appear starlike but are highly redshifted and radiate more energy than an entire ordinary galaxy; likely powered by black holes in the centers of distant galaxies

Energy sources for active galaxies are likely to be black holes, the mass of which can be millions to billions of times greater than that of the Sun.

black holes objects so massive for their size that their gravitational pull prevents everything, even light, from escaping

X rays a form of high-energy radiation just beyond the ultraviolet portion of the electromagnetic spectrum

radio lobes two regions of radio emission above and below the planes of active galaxies; thought to originate from powerful jets being emitted from the accretion disk surrounding the massive black hole at the center of active galaxies

The energy sources for quasars and active galaxies are most likely **black holes** with masses of a billion suns. Observers sometimes note that black holes are surrounded by rapidly spinning disks of gas. Theory predicts that these disks accrete onto the holes because of friction. Friction also heats up the disk so much that it emits **X rays**. Near the black hole, magnetic and hydrodynamic processes can accelerate some of the gas in the perpendicular direction, forming jets of matter that race far out into intergalactic space at nearly the speed of light. Nearby galaxies, including the Milky Way, have black holes in their centers too, but they tend to be only one thousand to one million times as massive as the Sun.

Active spiral galaxies are called Seyferts, named after American astronomer Carl Seyfert. Their spectral lines differ depending on their orientation, and so are divided into types I and II. The lines tend to be broader, indicating more rapid motions, if Seyfert galaxies are viewed nearly face-on. Active elliptical galaxies are called BL Lac objects (blazars) if their jets are viewed end-on; they look very different, having giant **radio lobes**, if their jets are viewed from the side. These radio lobes can extend for hundreds of millions of light-years from the galaxy centers.

Galaxy Interactions

Galaxies generally formed in groups and clusters, so most galaxies have neighbors. The Milky Way is in a small group with another large spiral galaxy (Andromeda, or Messier 31), a smaller spiral (Messier 33), two prominent irregulars (the Large and Small Magellanic Clouds), and two dozen tiny galaxies. In contrast, the spiral galaxy Messier 100 is in a very large cluster, Virgo, which has at least 1,000 galaxies. With so many neighbors, galaxies regularly pass by each other and sometimes merge together, leading to violent gas compression and star formation. In dense cluster centers, galaxies merge into giant ellipticals that can be 10 to 100 times as massive as the Milky Way. There is a higher proportion of elliptical and fast-rotating spiral galaxies in dense clusters than in small groups. Presumably the dense environments of clusters led to the formation of denser galaxies.

The Milky Way Galaxy

In the 1700s the philosophers Thomas Wright, Immanuel Kant, and Johann Heinrich Lambert speculated that our galaxy has a flattened shape that makes the bright band of stars called the Milky Way. Because English physicist and mathematician Isaac Newton (1642–1727) showed that objects with mass will attract each other by gravity, they supposed that our galaxy disk must be spinning in order to avoid collapse. In the early 1800s William Herschel counted stars in different directions. The extent of the Milky Way seemed to be about the same in all directions, so the Sun appeared to be near the center.

In the 1900s American astronomer Harlow Shapley studied the distribution of **globular clusters** in our galaxy. Globular clusters are dense clusters of stars with masses of around 100,000 Suns. These stars are mostly lower in mass than the Sun and formed when the Milky Way was young. Other galaxies have globular clusters too. The Milky Way has about 100 globular clusters, whereas giant elliptical galaxies are surrounded by thousands of globulars.

Shapley's observations led to an unexpected result because he saw that the clusters appear mostly in one part of the sky, in a spherical distribution around some distant point. He inferred that the Sun is near the edge of the Milky Way—not near its center as Herschel had thought. Shapley estimated the distance to clusters using variable stars. Stars that have finished converting hydrogen into helium in their cores change their internal structures as the helium begins to ignite. For a short time, they become unstable and oscillate, changing their size and brightness periodically; they are then known as variable stars. American astronomer Henrietta Leavitt (1868–1921) discovered that less massive, intrinsically fainter stars vary their light faster than higher mass, intrinsically brighter stars. This discovery was very important because it enabled astronomers to determine the distance to a star based on its period and apparent brightness. Much of what we know today about the size and age of the universe comes from observations of variable stars.

Shapley applied Leavitt's law to the variable stars in globular clusters. He estimated that the Milky Way was more than 100,000 light years across, several times the previously accepted value. He made an understandable mistake in doing this because no one realized at the time that there are two different types of variable stars with different period-brightness relations: the so-called RR Lyrae stars in globular clusters are fainter for a given period than the younger **Cepheid variables**.

The Discovery of Galaxies

In the 1920s astronomers could not agree on the size of the Milky Way or on the existence of other galaxies beyond. Several lines of conflicting evidence emerged. Shapley noted that nebulous objects tended to be everywhere except in the Milky Way plane. He reasoned that there should be no special arrangement around our disk if the objects were all far from it, so

---

WHEN GALAXIES COLLIDE

Collisions between galaxies can form spectacular distortions and bursts of star formation. Sometimes bridges of gas and stars get pulled out between two galaxies. In head-on collisions, one galaxy can penetrate another and form a ring. Interactions can create bars in galaxy centers and initiate spiral waves that make grand design structure. Close encounters can also strip gas from disks, which then streams through the cluster and interacts with other gas to make X rays.

globular clusters roughly spherical collections of hundreds of thousands of old stars found in galactic haloes

Cepheid variables a class of variable stars whose luminosity is related to their period; periods can range from a few hours to about 100 days—the longer the period, the brighter the star

this peculiar distribution made him think they were close. Actually the objects are distant galaxies, and dust in the Milky Way obscures them. The distance uncertainty was finally settled in the 1930s when Hubble discovered a Cepheid variable star in the Andromeda galaxy. He showed from the period-brightness relationship that Andromeda is far outside our own galaxy.

Galaxy investigations will continue to be exciting in the coming decades, as new space observatories, such as the **Next Generation Space Telescope**, and new ground-based observatories with flexible mirrors that compensate for the shimmering atmosphere, probe the most distant regions of the universe. Scientists will see galaxies in the process of formation by observing light that left them when the universe was young. We should also see quasars and other peculiar objects with much greater clarity, leading to some understanding of the formation of **nuclear black holes**. SEE ALSO AGE OF THE UNIVERSE (VOLUME 2); BLACK HOLES (VOLUME 2); COSMOLOGY (VOLUME 2); GRAVITY (VOLUME 2); HERSCHEL FAMILY (VOLUME 2); HUBBLE CONSTANT (VOLUME 2); HUBBLE, EDWIN P. (VOLUME 2); HUBBLE SPACE TELESCOPE (VOLUME 2); SHAPLEY, HARLOW (VOLUME 2).

Debra Meloy Elmegreen and Bruce G. Elmegreen

Next Generation Space Telescope the telescope scheduled to be launched in 2009 that will replace the Hubble Space Telescope

nuclear black holes black holes that are in the centers of galaxies; they range in mass from a thousand to a billion times the mass of the Sun

Bibliography

Berendzen, Richard, Richard Hart, and Daniel. Seeley. *Man Discovers the Galaxies.* New York: Columbia University Press, 1984.

Bothun, Gregory. "Beyond the Hubble Sequence: What Physical Processes Shape Galaxies." *Sky and Telescope* 99, no. 5 (2000):36–43.

Elmegreen, Debra Meloy. *Galaxies and Galactic Structure.* Upper Saddle River, NJ: Prentice Hall, 1998.

Elmegreen, Debra Meloy, and Bruce G. Elmegreen. "What Puts the Spiral in Spiral Galaxies?" *Astronomy* Vol. 21, No. 9 (1993):34–39.

Ferris, Timothy *Coming of Age in the Milky Way.* New York: Anchor Books, 1989.

Sandage, Allan. *The Hubble Atlas of Galaxies.* Washington, DC: Carnegie Institution of Washington, 1961.

Sawyer, Kathy. "Unveiling the Universe." *National Geographic* 196 [supplement] (1999):8–41.

Galilei, Galileo

Italian Astronomer, Mathematician, and Physicist
1564–1642

Galileo Galilei (commonly known as Galileo) was a founder of modern physics and modern astronomy. He was born in Pisa, Italy, in 1564, and was a professor from 1592 through 1610 at Padua, which was part of the Venetian Republic. While in Pisa, he noticed a chandelier swinging in the cathedral and developed the physical law that shows that pendulums of the same length swing in the same time interval. Using a pendulum for timing, he experimentally worked out how objects accelerate while falling. In these experiments, he rolled objects down an inclined plane; the traditional story that he dropped weights from the Leaning Tower of Pisa was a myth.

In 1609 Galileo heard of a device that existed that could magnify distant objects. Using his experimental abilities, he ground lenses and assembled a telescope. He demonstrated its possibilities for aiding commerce by showing Venetian nobles that they could see ships approaching farther out than ever before. Starting that same year, Galileo also turned his telescope toward the sky. He subsequently discovered that the Moon had mountains and craters on it, that Jupiter had moons orbiting it, and that Venus went through a complete set of phases. These observations indicated that Greek philosopher Aristotle's (384–322 B.C.E.) view of the universe as unchanging and perfect was not true, and Galileo endorsed Polish astronomer Nicholas Copernicus's (1473–1543) idea that the Sun instead of Earth is the center of the solar system. Galileo's book *Sidereus nuncius* (The starry messenger; 1610) brought his discoveries to a wide audience.

Soon Galileo discovered sunspots, showing that the Sun is not a perfect body. But a controversy with a Jesuit astronomer over who discovered sunspots set the Roman Catholic Church against him. In 1616 the Church's Inquisition warned him against holding or defending Copernicus's ideas. To get his agreement, they showed him instruments of torture.

Galileo was relatively quiet until his book *Dialogo sopra i due massimi sistemi del mondo* (Dialogue on the two great world systems) was published in 1632. It was written in his native Italian instead of the scholarly Latin, to spread his discussion widely. The Inquisition then convicted him of teaching Copernicanism and sentenced him to house arrest. But even under those conditions, and the blindness that came on, he continued his scientific work. He died in Florence in 1642. In 1992 Pope John Paul II agreed that Galileo was correct to endorse Copernicanism, though Galileo was not pardoned. SEE ALSO ASTRONOMY, HISTORY OF (VOLUME 2); COPERNICUS, NICHOLAS (VOLUME 2); JUPITER (VOLUME 2); MOON (VOLUME 2); RELIGION (VOLUME 4); SATURN (VOLUME 2); VENUS (VOLUME 2).

Jay Pasachoff

Galileo Galilei faced an inquisition from the Roman Catholic Church in 1616 for endorsing Nicholas Copernicus's theory that the Sun, not Earth, was the center of the solar system.

Bibliography

Machamer, Peter, ed. *The Cambridge Companion to Galileo.* Cambridge, UK: Cambridge University Press, 1998.

MacLachlan, James H. *Galileo Galilei: First Physicist.* New York: Oxford University Press, 1987.

Reston, James, Jr. *Galileo: A Life.* New York: HarperCollins, 1994.

Government Space Programs

While the United States leads the world in space initiatives and exploration, it is not the only country with active interests off the planet. Rivaling the achievements of the National Aeronautics and Space Administration (NASA) in space exploration is Russia, which inherited the Soviet Union's space assets and cherished space history. Although economic uncertainties undermine the stability and future of the Russian space program, at the end of 2001 it remained the only country, other than the United States, which could launch people into orbit.

The Russian Focus on Space Stations

As a major partner in the International Space Station program, Russia is responsible for sending Progress unpiloted cargo ships and Soyuz capsules to the outpost. The Soyuz spacecraft is a small, three-person vessel that serves

The Soyuz-Frigate rocket is prepared for launch in July 2000. It carries the Cluster satellite of the joint ESA-NASA mission.

as an emergency escape system for the station crew. Russian cosmonauts are scheduled to be part of every space station crew, and the commander's post is to alternate between a Russian cosmonaut and an American astronaut. Rosviakosmos, the Russian Aviation and Space Agency, works closely with the prime Russian aerospace contractor, the Korolev Rocket & Space Corporation Energia, which is also known as RKK Energia.

Russian companies built the station's base block, called Zarya, under a subcontract with the Boeing Company. Russia built and paid for the station's service module, named Zvezda, which serves as the living quarters for the station's crew and as the early command and control center. Russia has plans to build two research modules and docking compartments for the space station. Energia entered into a commercial agreement with the U.S. company Spacehab to develop one of the modules, which also could be used as temporary living quarters for visitors.

Russia had its own space station until 2001, when ground controllers shepherded the Mir space station through a fiery demise in Earth's atmosphere and burial at sea. Attempts to commercialize Mir failed and the Russian government ran out of funds to operate the space station. Most of the limited Russian government funding for space is earmarked for the International Space Station program.

Although Russian government funding for its space program is less than what is spent by the United States and many other countries, Russia has been remarkably resourceful in coming up with ways to finance and launch space hardware. For example, to the consternation of NASA and the other partners in the International Space Station program, Russia earned about $20 million by flying the first space tourist, American Dennis Tito, to the station in April of 2001. That amount of money would not even pay for a shuttle launch in America, but in Russia $20 million is enough to pay for several Soyuz and Progress flights to the station.

The Cooperative Efforts of Europe

Europe has been active in space for decades, working independently and with both the Americans and Russians long before the former Cold War foes began working together. While individual European countries maintain national space programs, most space initiatives are a combined effort managed through the fifteen-nation European Space Agency (ESA), which was founded in 1975. The countries that belong to ESA are Austria, Belgium, Denmark, Finland, France, Germany, Ireland, Italy, the Netherlands, Norway, Portugal, Spain, Sweden, Switzerland, and the United Kingdom.

Europe operates four space centers: the European Space Research and Technology Centre in the Netherlands; a control center in Germany; a hub for collecting and distributing information from Earth observation satellites in Italy; and the European Astronaut Centre in Germany, home to ESA's sixteen-member astronaut corps. Europe has its own launch system, the Ariane family of rockets, and a dedicated launch site in Kourou, French Guiana. Ariane rockets are sold commercially through Arianespace, which was formed in 1980 to market Ariane launch services worldwide. ESA's program includes both robotic and human space initiatives. ESA developed the Spacelab equipment that flew more than two dozen missions on NASA's space shuttles, and the agency sent astronauts to live on the Russian space station Mir.

ESA is a prime partner in the International Space Station program. Its contributions include the Columbus space laboratory, slated for launch in 2004, and a robotic arm for the Russian segments of the station. ESA also has plans for an automated cargo ferry for the station. Europe is a partner with NASA in the Hubble Space Telescope, the Ulysses solar probe, several Earth observation satellite systems, and several space-based observatories, including the Solar and Heliospheric Observatory, which is studying the Sun. ESA built the Huygens probe, which is en route to Saturn aboard the Cassini spacecraft. Huygens is to parachute through the hazy atmosphere of Titan, the largest moon of Saturn. Other planetary research projects include the Mars Express mission and the Rosetta comet probe. Space technology initiatives include the Artemis telecommunications satellites, the Galileo navigation satellites, and the SMART-1 spacecraft, the purpose of which is to demonstrate the use of solar-electric propulsion on a mission to the Moon.

Japan's Wide-Ranging Space Program

Japanese efforts to develop and market its own space launch system have been marred by difficulties, but the country has been a dedicated and stable space partner for the United States and Europe. Japan's space efforts are coordinated by the National Space Development Agency of Japan (NASDA), but several institutes, including the Institute of Space and Astronautical Sci-

China's space program launched the second of a series of unpiloted test flights that the government intends to culminate in a piloted space voyage early in the twenty-first century.

vacuum a space where air and all other molecules and atoms of matter have been removed

ence and the Science and Technology Agency, are also involved in space programs. Japan has a small astronaut corps that trains at the Johnson Space Center in Houston, Texas, alongside NASA astronauts.

Since the launch of its first satellite in February 1970, Japan has pursued advanced space technology. On July 4, 1998, Japan became only the third country in history to launch a probe to another planet, sending the Nozomi probe for a 2004 encounter with Mars. For the International Space Station, the Japanese are building a science laboratory called Kibo that includes an exposed back porch and a small robotic crane to operate experiments in the **vacuum** of space. In addition, Japan is developing a cargo transfer vehicle to ferry supplies to the station. Japan also is working on a system to land a spacecraft on the Moon. The project, which is targeted for launch in 2003, is called Selene ("Moon goddess") NASDA is backing a wide range of research efforts, including the development of a next-generation reusable space plane, new communications satellites, and Earth **remote sensing** systems.

Canada's Five-Pronged Program

The Canadian Space Agency has five major interests in space: Earth remote sensing, space science, human presence in space, satellite communications, and space technologies. Canada has a small but enthusiastic astronaut corps, which trains primarily at NASA's Johnson Space Center. Canada's first Earth-observing satellite, Radarsat, was launched in November 1995 and is being used for a variety of commercial and scientific projects including agricultural research, cartography, hydrology, forestry, oceanography, ice studies, and coastal monitoring.

Along with Russia, Europe, and Japan, Canada joins the United States as a full partner in the International Space Station program. Canada is providing a $1.6-billion remote manipulator system for the space station, which includes a 17.4-meter-long (57-foot-long) robotic arm, a mobile base, and robotic fingers to handle delicate assembly tasks.

The Ambitious Chinese Program

China has an ambitious space plan, which hopes to launch its own astronauts into orbit in 2003. Russia has trained Chinese astronauts at its cosmonaut training center in Star City, outside of Moscow. China unveiled its new spacecraft in a one-day, unpiloted test flight on November 20, 1999. In January 2001 the Shenzhou ("magic vessel") flew for a second test flight that lasted for a week. China has already launched its first navigation positioning satellite, the Beidou Navigation Test Satellite–1. Several institutes have joined together to develop a pair of microsatellites to map Earth and monitor natural disasters. Chinese officials have stated that the long-range goal of China's human space program is to build a space station. China, which is not a member of the International Space Station partnership, has developed and operated an unpiloted orbital platform in space.

The nation's Long March boosters have flown more than seventy missions since their debut in 1970, although the rocket has had some significant setbacks and spectacular failures. China also has had mixed success marketing its sixteen versions of the Long March rockets, with twenty-one commercial **payloads** flown through mid-2001. The country also is developing a new system of reusable launchers, as well as liquid- and solid-fuel boosters to carry small payloads into orbit.

India's Emerging Program

Another emerging player in the world space community is India, which founded its space program in 1972 and has launched at least twenty-six satellites, nine of which have been dedicated to improving the country's communications. India is developing its own heavy-lift launcher to send communications satellites into desirable orbital slots 35,786 kilometers

remote sensing the act of observing from orbit what may be seen or sensed below Earth

payload any cargo launched aboard a rocket that is destined for space, including communications satellites or modules, supplies, equipment, and astronauts; does not include the vehicle used to move the cargo or the propellant that powers the vehicle

low Earth orbit an orbit between 300 and 800 kilometers above Earth's surface

(22,300 miles) above the planet. The country has delivered its own Earth-imaging satellites to **low Earth orbit** and flew a commercial mission in May 1999 with the Polar Satellite Launch Vehicle.

Other space initiatives include a proposal to send a robotic scientific probe to the Moon in 2005, in what would be India's first venture into deep space. This mission would also make India only the fourth nation—after the United States, Russia, and Japan—to send a spacecraft to the Moon. The lunar probe would be launched on the new heavy-lift booster under development, the Geostationary Satellite Launch Vehicle, which made its debut test flight on April 18, 2001.

Israel's Boosters and Satellites

With national security an overriding concern, the young Israeli space program is focused on remote sensing technology, launch vehicle development, and lightweight minisatellites. Israel's Shavit launcher made its debut on September 19, 1988, when it placed the Ofeq 1 engineering test satellite into a low Earth orbit.

ballistic the path of an object in unpowered flight; the path of a spacecraft after the engines have shut down

The Shavit booster is a small, three-stage, solid propellant booster based on a **ballistic** missile design. Israel Aircraft Industries, which developed the booster, is continuing to work on expanding the rocket's capabilities. An upgraded Shavit ("comet") was launched in 1995 to place the Ofeq 3 satellite into orbit. A launch failure in 1998 claimed the fourth satellite in the series, but a more advanced follow-on program, the Earth Resources Observation Satellite, has been successful.

Brazil's Developing Program

Brazil's space program is still young. Efforts to develop launch technology were stalled with the 1998 failure of its VLS-1 space booster. The accident also claimed a Brazilian research satellite. In October 1999, a joint Chinese-Brazilian Earth remote-sensing satellite was launched on a Chinese Long March booster, and the countries signed an agreement a year later to jointly develop and fly a follow-on mission. Brazil also developed a satellite to monitor the Amazon rain forest and has positions reserved in low Earth orbit for an eight-satellite communications network.

A junior partner in the International Space Station program, Brazil agreed to provide experiment platforms for use on the orbital outpost. The Technology Experiment Facility is intended to provide experiments involving long-term exposure to the space environment. Brazil also will provide a pallet that can be used to attach small payloads to the station's outer truss segments. Other equipment that Brazil has promised to supply to the space station include a research facility for optical experiments and Earth observations using the station's telescope-quality window; and an unpressurized cargo carrier that can be mounted in the shuttle's cargo bay and loaded with equipment for the space station. In exchange for these contributions, Brazil will have access to space station facilities for research and will be able to fly a Brazilian astronaut to the station to conduct experiments. SEE ALSO INTERNATIONAL COOPERATION (VOLUME 3); NASA (VOLUME 3).

Irene Brown

Bibliography

Johnson, Nicholas, and David Rodvold. *Europe and Asia in Space, 1993–1994.* Kirtland Air Force Base, NM: Kaman Sciences. Air Force Phillips Laboratory, 1995.

Simpson, J. "The Israeli Satellite Launch." *Space Policy* (May 1989):117–128.

Internet Resources

Vick, Charles, ed. *Space Policy Project.* Federation of American Scientists. <http://www.fas.org/spp/guide/index.html>.

Gravity

The term "gravity" implies to many the notion of weight. Since antiquity, objects have been observed to "fall down" to the ground, and it therefore seemed obvious to associate gravity with Earth itself. Earth pulls all material bodies downward, but some appear to fall faster. For example, a rock and a feather fall to the ground at appreciably differing rates, and the logical conclusion of such great intellects as Greek philosopher Aristotle (384–322 B.C.E.) was that heavier objects fall faster than lighter ones. In fact, many erroneously believe this today, but it is found not to be true when tested in a controlled experimental manner. Air resistance is the confusing culprit and, when removed or minimized, all bodies are observed to hit the ground in the same amount of time when dropped from the same height.

Newton's Law of Universal Gravitation

In 1687, English physicist and mathematician Isaac Newton examined the laws of motion and universal gravitation in a classic text, *The Principia*, making it possible to explain and predict the motions of the planets and their newly discovered moons. Gravity is not just a property of Earth but of any matter in the universe. The essence of Newton's law of universal gravitation is demonstrated by imagining a "point mass," which is a certain amount of matter concentrated into a space of virtually zero volume. Now, suppose there is another point mass located some distance away from the first mass. According to Newton, these two masses mutually attract one another along the straight line drawn directly between them. In other words, the first mass feels a "pull" towards the second mass and the second mass feels an equal amount of "pull" towards the first. Of course, the universe contains far more than just these two isolated masses. The gravitational interaction is between any given mass and any other mass. A particular mass has a total gravitational force acting upon it that is the **vector sum** of all the attractions from every other mass paired with it. Every other mass will attract the mass in question independently, as if the others are not present. Intervening matter does not block gravity.

vector sum sum of two vector quantities taking both size and direction into consideration

The more massive and closer neighbors to our imaginary test mass will exert a larger gravitational force on it than less massive, more distant objects. The force between the test mass and any other point mass is directly proportional to the product of these masses and inversely proportional to the square of the distance between them. Expressing the statement in the form of an algebraic equation yields:

$$F_{grav} = G \times \frac{m_1 \times m_2}{d^2}$$

F_{grav} is the gravitational force existing between point masses m_1 and m_2, and d is the distance between the two masses. G is a constant making the units consistent. Its value was unknown to Newton and was later experimentally determined.

Real objects are not point masses but occupy a volume of space and have an infinite variety of shapes. Newton's law applies here by assuming that any object is composed of many particles, each of which is a close approximation to the ideal point mass previously described. Since gravitation is a very weak force compared to electrical or nuclear interactions, small objects that are normally encountered are not held together by self-gravitation. Instead, the electrically based chemical and molecular bonds are responsible. Nonetheless, the object behaves gravitationally like a collection of point particles each pulling independently on any other separate object's collection of point particles.

Fortunately, most large celestial bodies, such as planets and stars, are nearly spherical in shape, have mass that is symmetrically distributed, and are fairly distant from each other compared to their diameters. Under these assumptions, we can treat each object as a point particle and use Newton's formulation. Near Earth, an object's weight is the combined attraction of every particle in it with every particle that makes up the planet. Since Earth is a rather symmetrically distributed sphere, the net attraction of all its mass points on the object is directed (more or less) toward its center, and the object accelerates or "falls" straight down when released. The attraction is mutual, as Earth accelerates "upward" towards the falling object. But Earth is very massive compared to the object, so its inertia or resistance to acceleration is much greater. Its acceleration is immeasurable and we simply observe objects falling "down" to the ground. Heavier objects accelerate downward at the same rate as lighter ones (neglecting air resistance) because of their correspondingly greater inertia.

Einstein's General Theory of Relativity

Throughout the 1800s, Newton's law of gravitation was applied with increasing precision to the observed orbits of planets and double stars. The planet Neptune was discovered in 1846 from the gravitational disturbance it created on the orbit of Uranus. Even modern space science relies on Newton's law of gravitation to determine how to send spacecraft to any place in the solar system with pinpoint accuracy. To better understand gravity's fundamental nature and account for observable departures from Newton's law, however, an entirely new approach was needed. German-born American physicist Albert Einstein provided this in 1915 with the general theory of relativity.

Rather than the "action-at-a-distance" concept inherent to Newton's formulation, Einstein reasoned that a mass literally distorts the shape of the "space" surrounding it. If a beam of light is sent through empty space, it will define a "straight-line" path and hence the shortest distance between two points. The presence of mass, however, will cause the beam to bend its direction of propagation from a straight line and therefore define a curvature to space itself.

To visualize this, imagine a stretched rubber sheet onto which a large mass is placed. This mass creates a depression in the area surrounding it

Astronauts and cosmo-
nauts prepare to share a
meal in the Zveda Service
Module on the Interna-
tional Space Station
(ISS). The absence of
gravity can turn the sim-
ple act of opening a can
into a challenge.

while the membrane is essentially "flat" farther out. The larger the mass, the larger and deeper the depression. If another smaller mass is placed on the sheet, it will "fall" into the dimple well created by the heavier object and appear to be "attracted" to it. Likewise, if friction could be eliminated, it is possible to project the lighter mass into the edge of the well at just the right speed and angle to cause it to circle the massive object indefinitely just as the planets orbit the Sun. The Sun is massive enough, Einstein calculated, to cause a measurable deviation in the direction of distant starlight passing near it. The accurate positional measurement of stars appearing near the Sun's edge was successfully made in 1919 during a total solar eclipse, and Einstein's predictions were verified. SEE ALSO EINSTEIN, ALBERT (VOLUME 2); MICROGRAVITY (VOLUME 2); NEWTON, ISAAC (VOLUME 2); ZERO GRAVITY (VOLUME 3).

Arthur H. Litka

Bibliography

Baum, Richard, and William Sheehan. *In Search of Planet Vulcan: The Ghost in New-ton's Clockwork Universe.* New York: Plenum Press, 1997.

Galileo, Galilei. Stillman Drake, trans. *Discoveries and Opinions of Galileo.* Garden City, NY: Doubleday, 1957.

Newton, Isaac. *Mathematical Principles of Natural Philosophy and His System of the World,* trans. Andrew Motte. Berkeley: University of California Press, 1934.

Thorne, Kip S. *Black Holes and Time Warps: Einstein's Outrageous Legacy.* New York: W. W. Norton & Company, 1994.

Herschel Family

William Herschel, his sister Caroline, and his son John constitute one of the most famous families in astronomy.

William Herschel (1738–1822) was born in Hanover, Germany, and moved to England in 1757 to pursue a career as a musician. However, his interests shifted and William began studying astronomy in 1766. He cataloged celestial objects in an attempt to determine the three-dimensional structure of the galaxy. William discovered over 800 double stars and showed that many of them revolve around each other. On March 13, 1781, William became the first person to discover a new planet: Uranus. He also discovered four moons: Titania and Oberon at Uranus, and Enceladus and Mimas at Saturn.

nebulae clouds of interstellar gas and/or dust

Caroline Herschel (1750–1848) joined William in England in 1772 to pursue a career as a singer. She began assisting William full-time with his astronomical observations in 1782 and also started observing on her own that year. She discovered eight comets, a record by a female astronomer until 1987. Caroline also compiled catalogs of star clusters and **nebulae**.

John (1792–1871) used the Sun's spectrum to determine its chemical composition and made long-term observations of solar phenomena. In the 1830s, he traveled to South Africa to observe southern hemisphere star clusters and nebulas and revised the nomenclature of southern stars. SEE ALSO ASTRONOMY, HISTORY OF (VOLUME 2); COMETS (VOLUME 2); SMALL BODIES (VOLUME 2); STARS (VOLUME 2); URANUS (VOLUME 2).

Nadine G. Barlow

Bibliography

Lubbock, Constance A. *The Herschel Chronicles: The Life-Story of William Herschel and His Sister Caroline Herschel.* Cambridge, UK: University Press, 1993.

Hubble Constant

Big Bang name given by astronomers to the event marking the beginning of the universe, when all matter and energy came into being

In the standard **Big Bang** model, the universe expands according to the Hubble law, a simple relation expressed as $v = H_o d$, where v is the velocity of a galaxy at a distance d, and H_o is the Hubble constant. The Hubble constant characterizes both the scale and age of the universe. A measurement of the Hubble constant, together with the ages of the oldest objects in the universe, and the average density of the universe, are all separately required to describe the universe's evolution. Measuring an accurate value of H_o was one of the motivating reasons for building the Hubble Space Telescope (HST).

The measurement of most distances in astronomy cannot be done directly because the size scales are simply too big. In general, the basis for estimating distances in astronomy is the inverse square radiation law, which states that the brightness of an object falls off in proportion to the square of its distance from us. (We all experience this effect in our own lives. A street light in the distance appears fainter than the one beside us.) Astronomers identify objects that exhibit a constant brightness (so-called "standard candles"), or those where the brightness is perhaps related to a quantity that is independent of distance (for example, period of **oscillation**, rotation rate, or color). The standard candles must then be independently calibrated (to absolute physical units) so that true distances (in meters or megaparsecs, where 1 megaparsec = 3.08×10^{22} meters) can be determined using the inverse square law.

Cepheid Variables

The most precise method for measuring distances is based on the observations of **Cepheid variables**, stars whose atmospheres pulsate regularly for periods ranging from 2 to about 100 days. Experimentally it has been established that the period of pulsation is correlated with the brightness of the star. High resolution is the key to discovering Cepheids in other galaxies—in other words, the telescope must have enough resolving power to distinguish Cepheids from other stars that contribute to the overall light of the galaxy. The resolution of the Hubble Space Telescope is about ten times better than can be generally obtained through Earth's turbulent atmosphere.

The reach of Cepheid variables as distance indicators is limited, however, even with the HST. For distances beyond 20 megaparsecs or so, brighter objects than ordinary stars are required; for example, bright supernovae or the brightnesses of entire galaxies. The absolute calibration for all of these methods is presently established using the Cepheid distance scale. A Key Project of the HST has provided Cepheid distances for a sample of galaxies useful for setting the absolute distance scale using these and other methods.

Until recently, a controversy has existed about the value of the Hubble constant, with published distances disagreeing by a factor of two. However the new Cepheid distances from the HST have provided a means of calibrating several distance methods. For the first time, to within an uncertainty of 10 percent, all of these methods are consistent with a value of the Hubble constant in the range of about 60 to 70 kilometers (37.28 to 43.5 miles) per second per megaparsec. This implies an age of the universe of between 13,000 and 15,000 million years. SEE ALSO HUBBLE, EDWIN P. (VOLUME 2); HUBBLE SPACE TELESCOPE (VOLUME 2); AGE OF THE UNIVERSE (VOLUME 2).

Wendy L. Freedman

Bibliography

Barrow, John. D. *The Origin of the Universe.* New York: Basic Books, 1994.

Ferguson, Kitty. *Measuring the Universe.* New York: Walker & Co., 1999.

Freedman, Wendy L. "The Expansion Rate of the Universe." *Scientific American* 1 (1988):92–97.

Hubble, Edwin P.

American Astronomer
1889–1953

American astronomer Edwin Powell Hubble's (1889–1953) key discovery was his finding that the universe is expanding. ✳ Hubble received undergraduate degrees in math and astronomy from the University of Chicago.

Edwin P. Hubble is best known for his determination that the more distant the galaxy, the quicker it moves away from Earth. This implied that the universe was expanding and led the way to the Big Bang theory.

galaxy a system of as many as hundreds of billions of stars that have a common gravitational attraction

nebulae clouds of interstellar gas and/or dust

Big Bang name given by astronomers to the event marking the beginning of the universe, when all matter and energy came into being

infrared portion of the electromagnetic spectrum with wavelengths slightly longer than visible light

ultraviolet the portion of the electromagnetic spectrum just beyond (having shorter wavelengths than) violet

Upon graduation, he was awarded a Rhodes scholarship to Oxford University, where he studied law. After some time as a lawyer and teacher, he returned to the University of Chicago to pursue a doctorate in astronomy. During his studies, World War I (1914–1918) began. Hubble enlisted in the army and rose to the rank of major.

After the war, Hubble worked at Mount Wilson Observatory, California, which then contained the largest telescope in the world. In the early 1920s, scientists knew about our own **galaxy**, the Milky Way, but they did not know if anything was outside of it. Some had conjectured that **nebulae**, faint cloudy features in the night sky, were actually "island universes" or other galaxies. Hubble measured the distance to some of these nebulas and found that they indeed lay far outside the Milky Way. In further studies, he showed that these nebulas are actually other galaxies, and he went on to classify them.

After this work, he made the most remarkable discovery of his career. He found that the more distant the galaxy, the faster it is moving away from Earth. This relationship implies that the universe is expanding, and this knowledge led to the formation of the **Big Bang** theory describing the formation of the universe. The constant that describes the relationship between galaxy speed and distance is called the Hubble constant. SEE ALSO ASTRONOMY, HISTORY OF (VOLUME 2); ASTRONOMY, KINDS OF (VOLUME 2); GALAXIES (VOLUME 2); HUBBLE CONSTANT (VOLUME 2); HUBBLE SPACE TELESCOPE (VOLUME 2).

Derek L. Schutt

Bibliography

Christianson, Gale E. *Edwin Hubble: Mariner of the Nebulae.* New York: Farrar, Straus, Giroux, 1995.

Sharov, Alexander S., and Igor D. Novikov. *Edwin Hubble: The Discoverer of the Big Bang Universe*, trans. Vitalie Kisin. Cambridge, UK: Cambridge University Press, 1993.

Hubble Space Telescope

The National Aeronautics and Space Administration's Hubble Space Telescope (HST) is the first major **infrared**-optical-**ultraviolet** telescope to be placed into orbit around Earth. The telescope is named after American astronomer Edwin P. Hubble, who found galaxies beyond the Milky Way in the 1920s, and discovered that the universe is uniformly expanding.

Located high above Earth's obscuring atmosphere, at an altitude of 580 kilometers (360 miles), the HST has provided the clearest views of the universe yet obtained in optical astronomy. Hubble's crystal-clear vision has fostered a revolution in optical astronomy. It has revealed a whole new level of detail and complexity in a variety of celestial phenomena, from nearby stars to galaxies near the limits of the observable universe. This has provided key insights into the structure and evolution of the universe across a broad scale. Its location outside of Earth's atmosphere has also provided Hubble with the ability to view astronomical objects across a wide swath of

Named after American astronomer Edwin Hubble, the Hubble Space Telescope is the first major infrared-optical-ultraviolet telescope placed into orbit around Earth. It can clearly observe objects less than one-billionth as bright as what can be seen by the human eye.

the **electromagnetic spectrum**, from ultraviolet light through visible and on to near-infrared **wavelengths**.

The heart of the telescope is the primary mirror, which is 94.5 inches (2.4 meters) in diameter. It is the smoothest optical mirror ever polished, with surface tolerance of one-millionth of an inch. It is made of fused silica glass and weighs about 670 kilograms (1,800 pounds).

Outside the blurring effects of Earth's turbulent atmosphere, the telescope can resolve astronomical objects ten times more clearly than can be seen with even larger ground-based optical telescopes. Hubble can see objects less than one-billionth as bright as what can be seen with the human eye. Hubble can detect objects as faint as thirty-first magnitude, which is comparable to the sensitivity of much larger Earth-based telescopes.

Hubble images have exceptional contrast, which allows astronomers to discern faint objects near bright objects. This enables scientists to study the environments around stars and to search for broad circumstellar disks of dust that may be forming into planets.

electromagnetic spectrum the entire range of wavelengths of electromagnetic radiation

wavelength the distance from crest to crest on a wave at an instant in time

Launch and Servicing Missions

The HST was launched by the space shuttle Discovery on April 24, 1990. Hubble initially was equipped with five science instruments: the Wide-Field Planetary Camera, the Faint Object Camera, the Faint Object **Spectrograph**, the High-Resolution Spectrograph, and the High-Speed **Photometer**. In addition, Hubble was fitted with three fine guidance sensors used for pointing the telescope and for doing precision astrometry—the measurement of small angles on the sky.

After Hubble was launched, scientists discovered that its primary mirror was misshapen because of a fabrication error. This resulted in spherical aberration: the blurring of starlight because the telescope could not bring all the light to a single focus. Using image-processing techniques to reduce the blurring in HST images, scientists were able to do significant research with Hubble until an optical repair could be developed.

In December 1993, the first HST servicing mission carried replacement instruments and supplemental optics aboard the space shuttle Endeavour to restore the telescope to full optical performance. A deployable optical device, called the Corrective Optics Space Telescope Axial Replacement (COSTAR), was installed to improve the sharpness of the first-generation instruments. The COSTAR was outfitted with pairs of small mirrors that intercepted the incoming light from the primary mirror and reconstructed the beam so that it was in crisp focus. In addition, the original Wide-Field Planetary Camera was replaced with a second camera, the Wide-Field Planetary Camera 2, which has a built-in correction for the aberration in the primary mirror.

In March 1997 the space shuttle Discovery returned to the HST for a second servicing mission. Two advanced instruments, the Near Infrared Camera and Multi-Object Spectrometer and the Space Telescope Imaging Spectrograph were installed to replace two first-generation instruments. Astronauts also replaced or enhanced several electronic subsystems and patched unexpected tears in Hubble's shiny, aluminized, thermal insulation blankets, which give the telescope its distinctive foil-wrapped appearance and protect it from the heat and cold of space.

In December 1999 a third servicing mission replaced a number of subsystems but added no new instruments. About a month before the mission a critical **gyroscope** had failed, leaving Hubble with only two operational gyros out of a total of six onboard. This had left the telescope incapable of precision pointing. The December mission restored Hubble to six fully functioning gyroscopes. The telescope's main computer was upgraded from a 1960s computer with 48 kilobytes of memory, to an Intel 486 microprocessor.

In March 2002, the next and most ambitious serving mission in the series, involving five exhausting six-hour space walks by pairs of astronauts, took place. They installed a high-efficiency camera called the advanced camera for surveys. The mission also performed "heart surgery" by replacing a complex power control unit, which required completely shutting off the telescope's electrical power. The telescope also got stubby new solar panels that increased the power enough for all of the instruments to operate simultaneously.

spectrograph an instrument that can permanently record a spectra

photometer instrument to measure intensity of light

gyroscope a spinning disk mounted so that its axis can turn freely and maintain a constant orientation in space

In 2004, the last servicing mission will install the wide-field planetary camera 3 and the cosmic origins spectrograph. Hubble will be on its own until 2010, when NASA stops the observing program and must decide whether to retrieve Hubble and install a rocket propulsion system that will put it into a safe higher orbit or let it reenter the atmosphere and largely burn up over the ocean.

How Hubble Operates

Hubble is controlled at the Goddard Space Flight Center in Greenbelt, Maryland. The Space Telescope Science Institute (STSI), located at the Johns Hopkins University in Baltimore, Maryland, directs the science mission. Space telescope research and funding engages a significant fraction of the worldwide community of professional astronomers. Astronomers compete annually for observation time on Hubble.

Observing proposals are submitted to peer review committees of astronomers. The STSI director makes the final decision and can use his or her own discretionary time for special programs. Accepted proposals must be meticulously planned and scheduled by experts at STSI to maximize the telescope's efficiency.

The space telescope is not pointed by "real-time" remote control but instead automatically carries out a series of preprogrammed commands over the course of a day. This is necessary because the telescope is in a **low Earth orbit**, which prevents any one ground station from staying directly in contact with it. Instead, controllers schedule intermittent daily linkups with the space observatory via a series of satellites in **geosynchronous orbit**.

A date "pipeline," assembled and maintained by STSI, ensures that all observations are stored on optical disk for archival research. The data are sent to research astronomers for analysis, and then made available to astronomers worldwide one year after the observation.

By the turn of the twenty-first century, Hubble had looked at over 13,000 celestial targets and stored over 6 gigabytes of data onto large optical disks. The telescope had made nearly one quarter million exposures, approximately half of these were of astronomical targets and the rest were calibration exposures.

Hubble Provides New Insights

The HST has made dramatic inroads into a broad range of astronomical frontiers. Astronomers have used Hubble to look out into the universe over distances exceeding 12 billion light-years. Because the starlight harvested from remote objects began its journey toward Earth billions of years ago, the HST looks further back into time the farther away it looks into space (as do all large telescopes). Hubble has seen back to a time when the universe was only about 5 percent of its present age.

The Hubble Deep Field. Hubble's deepest views of the universe, made with its visible and infrared cameras, are collectively called the Hubble Deep Field. These "long exposures" of the universe reveal galaxies that existed when the universe was less than 1 billion years old. The Hubble Deep Field

low Earth orbit an orbit between 300 and 800 kilometers above Earth's surface

geosynchronous orbit a specific altitude of an equatorial orbit where the time required to circle the planet matches the time it takes the planet to rotate on its axis; an object in geostationary orbit will always remain over the same geographic location on the equator of the planet it orbits

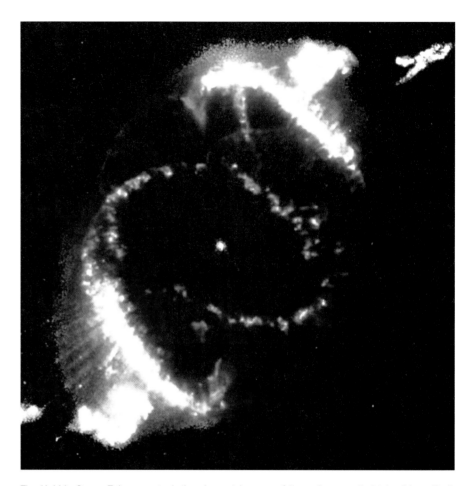

The Hubble Space Telescope took the clearest images of the universe yet obtained in optical astronomy, including this image of the Cat's Eye Nebula, revealing a new level of detail and complexity in many celestial phenomena, and providing significant insights into the structure and evolution of the universe.

also uncovered hundreds of galaxies at various stages of evolution, strung along a corridor of billions of light years. The high resolution of Hubble images enables astronomers to actually see the shapes of galaxies in the distant past and to study how they have evolved over time.

Expansion and Age of the Universe. Another key project for the HST has been to make precise distance measurements for calculating the rate of expansion of the universe. This was achieved by measuring distances to galaxies much farther out than had previously been accomplished in decades of observing.

Determining the exact value of this rate is fundamental to calculating the age of the universe. In 1998, a team of astronomers triumphantly announced that they had accurately measured the universe's expansion rate to within an accuracy of 10 percent. This brought closure to a three-decade-long debate over whether the universe is 10 or 20 billion years old. The final age appears to be between 13 and 15 billion years, but this estimate is also affected by other parameters of the universe.

The HST was also used to find out if the universe was expanding at a faster rate long ago. This was done by using Hubble to peer halfway across the universe to find ancient exploding stars called supernovae. These stars can be used to calculate vast astronomical distances because they are so bright and shine at a predictable luminosity, which is a fundamental requirement for measuring distances.

Hubble observations, as well as other observations done with ground-based telescopes, show that the universe has not decelerated. In fact, to the surprise of astronomers, the expansion of the universe is accelerating, and therefore will likely expand forever. This realization offers compelling evidence that there is a mysterious repulsive force in space, first theorized by German-born American physicist Albert Einstein (1879–1955), which is pushing the galaxies apart—in addition to the original impetus of the **Big Bang**.

This idea was bolstered in 2000 when Hubble astronomers accidentally discovered a supernova so far away, it exploded when the universe was actually decelerating. This supernova happened about 7 billion years ago, just before dark energy began accelerating the universe, like a car accelerating through a traffic light that has just turned green.

Black Holes. The HST has provided convincing evidence of the existence of supermassive black holes that are millions or even a billion times more massive than the Sun. Hubble's exquisite vision allows astronomers to zoom in on the environment around a black hole and make critical measurement of the motion of stars and gas around the hole, to precisely measure its mass. The measurements show that there is far more mass at the core of galaxies than can be accounted for by starlight. This unseen mass is locked away inside black holes.

HST observations of both quiescent and active galaxies, the latter of which pours out prodigious amounts of energy, have shown that supermassive black holes are commonly found at the hub of a **galaxy**. A Hubble census of black holes also showed that the mass of a black hole corresponds to the mass of the central bulge of a galaxy. Therefore, galaxies with large bulges have more massive black holes than galaxies with smaller bulges. This suggests that supermassive black holes may be intimately linked to a galaxy's birth and evolution.

Quasars. Hubble's keen ability to discern faint objects near bright objects allowed for definitive observations that showed the true nature of quasars, which are compact powerhouses of light that resemble stars and that reside largely at the outer reaches of the universe. HST observations conclusively showed that quasars dwell in the cores of galaxies, which means they are powered by supermassive black holes that are swallowing material at a furious rate.

Gamma-Ray Bursts. Hubble played a key role in helping astronomers resolve questions regarding the nature of mysterious gamma-ray bursts. Gamma-ray bursts are powerful blasts that come from random directions in the universe about once per day. Hubble observations found host galaxies associated with some of these blasts. This places the bursts at cosmological distances rather then being localized phenomena within our galaxy. Hubble also showed that the blasts occur among the young stars in the spiral arms of a host galaxy. This favors **neutron star** collisions or neutron star–black hole collisions as the source of the bursts.

Big Bang name given by astronomers to the event marking the beginning of the universe, when all matter and energy came into being

galaxy a system of as many as hundreds of billions of stars that have a common gravitational attraction

neutron star the dense core of matter composed almost entirely of neutrons that remains after a supernova explosion has ended the life of a massive star

Stellar Environments. The HST has unveiled a wide variety of shapes, structures, and fireworks that accompany the birth and death of stars. HST images have provided a clear look at pancake-shaped disks of dust and gas swirling around and feeding embryonic stars. Besides helping build the star, the disks are also the prerequisite for condensing planets. Hubble images also show blowtorch-like jets of hot gas streaming from deep within the disks. These jets are an "exhaust product" of star formation.

In dramatic images, HST has shown the effects of very massive young stars on their surrounding nebulae. The astronomical equivalent of a hurricane, the intense flow of visible and ultraviolet radiation from an exceptionally massive young star eats into surrounding clouds of cold hydrogen gas, laced with dust. This helps trigger a firestorm of star birth in the neighborhood around the star.

The HST has produced a dazzling array of images of colorful shells of gas blasted into space by dying stars. These intricate structures are "fossil evidence" showing that the final stages of a star's life are more complex than once thought. An aging star sheds its outer layers of gas through stellar winds. Late in a star's life, these winds become more like a gale, and consequently sculpt strikingly complex shapes as they plow into slower-moving material that was ejected earlier in the star's life.

The most dramatic star-death observation for the HST has been tracking the expanding wave of debris from the explosion of supernova 1987A. HST observations show that debris from the supernova blast is slamming into a ring of material around the dying star. The crash has allowed scientists to probe the structure around the supernova and uncover new clues about the star's final years.

Extrasolar Planets. Even Hubble's powerful vision is not adequate to see the feeble flicker of a planet near a star. Nevertheless, Hubble was still very useful for conducting the first systematic search for a special type of planet far beyond our stellar neighborhood. For ten consecutive days Hubble peered at the globular cluster 47 Tucane to capture the subtle dimming of a star due to the eclipse-like passage of a Jupiter-sized planet in front of the star. Based on extrasolar planet discoveries in our own stellar neighborhood, astronomers predicted that seventeen planets should have been discovered. However, Hubble did not find any, which means that conditions favoring planet formation may be different elsewhere in the galaxy.

Aiming at a known planet 150 light-years away, Hubble made the first-ever detection of an atmosphere around a planet. When the planet passed in front of its star, Hubble measured how starlight was filtered by skimming through the atmosphere. Hubble measures the presence of sodium in the atmosphere. These techniques could eventually lead to the discovery of oxygen in the atmospheres in inhabited terrestrial extrasolar planets. SEE ALSO ASTRONOMY, KINDS OF (VOLUME 2); EXTRASOLAR PLANETS (VOLUME 2); GYROSCOPES (VOLUME 3); HUBBLE, EDWIN P. (VOLUME 2); OBSERVATORIES, SPACE-BASED (VOLUME 2).

Ray Villard

Bibliography

Chaisson, Eric. *The Hubble Wars*. New York: HarperCollins, 1994.

Smith, Robert W. *The Space Telescope.* Cambridge, UK: Cambridge University Press, 1993.

Internet Resources

The Hubble Space Telescope. Space Telescope Science Institute. <http://hst.stsci .edu/>.

Huygens, Christiaan

Dutch Astronomer and Mathematical Physicist
1629–1695

Christiaan Huygens (1629–1695) was born in the Hague, Netherlands. He is remembered for his work in optics, astronomy, and timekeeping.

Christiaan Huygens developed and patented the pendulum time-keeping device. The pendulum system of modern grandfather clocks grew from his idea.

Huygens developed lens-shaping techniques better than those of Italian mathematician and astronomer Galileo Galilei (1564–1642) and greatly improved the telescope. This permitted the use of high magnifications, leading to Huygens's discovery of Saturn's largest moon, Titan, in 1655. Huygens was also the first to recognize a ring around Saturn, and published a thorough explanation of it in *Systema Saturnium* in 1659, resolving a long-standing mystery that began with Galileo's first observation of Saturn. This book also contained a drawing showing two dark bands on Jupiter and a dark band on Mars.

Galileo had used a pendulum for timekeeping, but Huygens invented the pendulum clock in 1656 and patented it in 1657. Huygens developed the mathematical theory of the pendulum, including a formula for its behavior. Along with his studies of the pendulum, he theorized about the motions of bodies along various curves and drew conclusions related to planetary motions governed by gravity. Oddly, though, he did not accept English physicist and mathematician Isaac Newton's explanation of gravity.

Huygens later went back to the study of optics and developed long focal length lenses used in "aerial telescopes."

Huygens is the namesake of the European Space Agency's atmospheric probe of Titan, which is being carried by the Cassini orbiter to Saturn in 2004. The Cassini Program is a joint effort of the National Aeronautics and Space Administration, European Space Agency, and the Italian Space Agency to study the Saturn system. SEE ALSO ASTRONOMY, HISTORY OF (VOLUME 2); GALILEI, GALILEO (VOLUME 2); GOVERNMENT SPACE PROGRAMS (VOLUME 2); NASA (VOLUME 3); ROBOTIC EXPLORATION OF SPACE (VOLUME 2); SATURN (VOLUME 2).

Stephen J. Edberg

Bibliography

Abetti, Giorgio. *The History of Astronomy*, trans. Berry Burr Abetti. New York: Henry Schuman, 1952.

Berry, Arthur. *A Short History of Astronomy* (1898). New York: Dover Publications, Inc., 1961.

Bishop, R., ed. *Observer's Handbook, 2000.* Toronto: Royal Astronomical Society of Canada, 1999.

Jupiter

Jupiter is the largest planet in the solar system and is easily visible in the night sky. Jupiter's mass (1.9×10^{27} kilograms [4.2×10^{27} pounds]) is nearly two and a half times the mass of the rest of the solar system's planets combined. Jupiter's volume, filled mostly with gas, is 1,316 times that of Earth. The fifth planet from the Sun, Jupiter's year is 11.86 Earth years but its day is short, only nine hours and fifty-five minutes. Jupiter resembles a small star: its composition, like the Sun's, is mostly hydrogen and helium. It emits about twice the energy that it receives from the Sun and puts out over 100 times more heat than Earth. If Jupiter had been about 50 to 100 times larger, it might have evolved into a star rather than a planet.

Historic Observations of Jupiter

Jupiter has intrigued humans since antiquity. It is named for the king of the Roman gods, and most of its twenty-eight moons are named after the god's many lovers. In 1609 and 1610, Italian mathematician and astronomer Galileo Galilei and German astronomer Simon Marius began telescopic studies of Jupiter and its system. Galileo is credited with the discovery of Jupiter's four largest moons: Io, Europa, Ganymede, and Callisto, now called the Galilean satellites in his honor. These moons had an impact on the thinking of those times. It was believed then that Earth was the center of the universe and that all the planets and moons revolved around Earth. Galileo's observations showed that the four moons revolved around Jupiter, not Earth. This discovery contributed to Galileo's doom. He was condemned by the Catholic Church, forced to recant his discovery, and only in 1992 did Pope John Paul II agree that Galileo was right to support Copernicanism.

As telescopes improved, other astronomers continued to observe Jupiter and to study its colorful bands and the long-lived storm known as the Great Red Spot. Twenty-four other smaller satellites have been discovered, from Amalthea in 1892 to Leda in 1974 to twelve new moons in 2001. Observations from Earth showed that Jupiter has a massive **magnetosphere** and that the planet emits radiation at radio **wavelengths**. From this, astronomers deduced that Jupiter is surrounded by **radiation belts**, similar to Earth's **Van Allen radiation belts**, and that the planet must have a strong magnetic field.

Spacecraft Explorations

Space missions allowed scientists to make great leaps forward in the exploration of Jupiter and its moons. The first spacecraft to fly by Jupiter were Pioneer 10 (in 1973) and Pioneer 11 (in 1974). They passed as close as 43,000 kilometers (26,660 miles) from Jupiter. Their suite of instruments made important observations of the atmosphere, magnetosphere, and space environment around the planet. In 1979 the spacecraft Voyager 1 and Voyager 2 passed close to Jupiter and its moons, making startling discoveries that included **auroras** on Jupiter, a ring system surrounding the planet, and active volcanoes on the moon Io.

In 1995, the Galileo spacecraft became the first to orbit Jupiter. It dropped a probe into the planet that survived for 57.6 minutes, until it was crushed by Jupiter's enormous pressure. The probe's instruments sent back

magnetosphere the magnetic cavity that surrounds Earth or any other planet with a magnetic field; it is formed by the interaction of the solar wind with the planet's magnetic field

wavelength the distance from crest to crest on a wave at an instant in time

radiation belts two wide bands of charged particles trapped in a planet's magnetic field

Van Allen radiation belts two belts of high energy charged particles captured from the solar wind by Earth's magnetic field

auroras atmospheric phenomena consisting of glowing bands or sheets of light in the sky caused by high-speed charged particles striking atoms in a planet's upper atmosphere

valuable information on the temperature, pressure, composition, and density of the upper atmosphere.

The Galileo probe provided scientists with their first glimpse inside the top layers of the atmosphere. One surprising discovery was that Jupiter has thunderstorms that are many times larger than those on Earth. The cause of the thunderstorms is the vertical circulation of water vapor in the top layers of Jupiter's atmosphere.

The main Galileo spacecraft has been making observations of Jupiter, its moons, and its environment since 1995, and these were slated to continue until 2002. Scientific observations continue to be made using Earth-based telescopes and the Hubble Space Telescope, which is in orbit around Earth. The combination of many sets of observations over time is extremely valuable for understanding Jupiter and its system.

The Atmosphere and Interior of Jupiter

Jupiter's atmosphere has alternating patterns of dark and light belts and zones. Within these belts and zones are gigantic storm systems such as the Great Red Spot. The locations and sizes of the belts and zones change gradually over time, and many of them can be seen through a telescope. The Great Red Spot has lasted for at least 100 years, and probably as long as 300 years. It rotates counterclockwise every six days, and this direction, plus its location in the southern hemisphere, indicates that it is a high-pressure zone. This differs from the cyclones that occur on Earth, which are low-pressure zones. The red color of the spot is something of a mystery. Several chemicals, including phosphorus, have been suggested as the cause of the red color but, on the whole, the reasons for Jupiter's different colors are not yet understood.

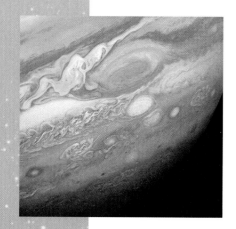

This image of Jupiter's Great Red Spot was captured by the spacecraft Voyager 1 in 1979 from a distance of 5.7 million miles (9.2 million kilometers) from the planet.

The atmosphere of Jupiter consists of about 81 percent hydrogen and 18 percent helium, with small amounts of methane, ammonia, phosphorus, water vapor, and various hydrocarbons. Observations by Galileo showed a cloud of fresh ammonia ice downstream from the Great Red Spot. Jupiter's atmosphere has strong winds, but the mechanisms that drive them are not well understood. There are at least twelve different streams of prevailing winds, and they can reach velocities of up to 150 meters per second (492 feet per second) at the equator. On Earth, winds are driven by large differences in temperature, differences that do not exist, at least not on the top part of Jupiter's atmosphere, where the temperature at the poles is about the same as that at the equator (−130°C [−202°F]).

The cloud layer, which is thought to be only about 50 kilometers (31 miles) thick, comprises only a small part of the planet. What is the interior of Jupiter like? The pressure inside Jupiter, which increases with depth, is enormous—it may reach about 100 million times the pressure on Earth's surface. Although we cannot directly observe Jupiter's interior, theory plus observations of the atmosphere and the surrounding environment suggest that below the cloud layer there is a 21,000-kilometer-thick (13,000-mile-thick) layer of hydrogen and helium. This layer gradually changes from gas to liquid as the pressure increases. Beneath this layer is a sea of liquid metallic hydrogen about 40,000 kilometers (24,800 miles) deep. Metallic hydrogen does not form on Earth, because our planet lacks the extreme pressures necessary to break up the hydrogen molecules and pack them so tightly that they break up and become electrically conductive. This electrically conductive metallic hydrogen is what drives Jupiter's strong magnetic field. Deeper still in Jupiter's interior is the core, which may be solid and rocky. It is estimated that the core is about one and a half times Earth's diameter, but ten to thirty times more massive. It is also very hot: about 30,000°C (54,000°F). This heat comes up through the layers and is detected at "hot spots" in the atmosphere, which are cloud-free holes.

Magnetic Field and Rings

Jupiter's sea of metallic hydrogen causes it to have the strongest magnetic field of any planet in the solar system. The field is inverted relative to Earth's, that is, a compass there would always point south. The region around the planet that is dominated by the magnetic field is called the magnetosphere. The stream of charged particles sent by the **solar wind** causes Jupiter's magnetosphere to be shaped like a teardrop, pointing directly away from the Sun. Inside the magnetosphere is a swarm of ions, **protons**, and **electrons**, which are called plasma. The plasma rotates along with Jupiter's magnetic field, blasting off charged particles. Some of them impact on the surfaces of the moons. On Io, volcanoes eject material into space, and the particles get caught up in Jupiter's magnetosphere. This creates a doughnut-shaped region of charged particles at about the distance from Jupiter of Io's orbit. This is called the Io plasma torus. It was first observed by the Pioneer spacecraft.

The Voyager missions showed that Jupiter is surrounded by faint rings. Unlike Saturn's rings, which are made up of icy particles, Jupiter's rings are made up of small dust particles. Two small satellites, Adrastea and Metis, are embedded within the rings. Observations by Galileo spacecraft showed that the dust comes from **meteoroids** impacting the satellites closest to Jupiter.

solar wind a continuous, but varying, stream of charged particles (mostly electrons and protons) generated by the Sun; it establishes and affects the interplanetary magnetic field; it also deforms the magnetic field about Earth and sends particles streaming toward Earth at its poles

protons positively charged subatomic particles

electrons negatively charged subatomic particles

meteoroid a piece of interplanetary material smaller than an asteroid or comet

The four largest of Jupiter's sixteen moons include, from left to right, the volcanically active Io, the icy and mottled Europa, the large and rocky Ganymede, and the heavily cratered Callisto.

The Galilean Satellites

The Galilean satellites are all different from one another. Io and Europa have greater densities than Ganymede and Callisto, suggesting that the two inner moons (Io and Europa) contain more rock, and the outer moons more water ice.

Io. Io is the most volcanically active body in the solar system. It is the only place outside Earth where eruptions of hot magma have been observed. Other planets and moons in the solar system have been volcanically active in the distant past. Io is about the same size as Earth's Moon and, had it not been for its peculiar orbit, it too would have cooled down and volcanism would have ceased. Tidal stresses are produced within Io as a result of the gravitational pull of Jupiter, Europa, and Ganymede. These stresses cause the interior of Io to heat up, leading to active volcanism. About 100 active volcanoes have been seen so far on Io, many of which were discovered from their thermal signature in **infrared** observations made by the Galileo spacecraft. Some of the active volcanoes have plumes that can reach 300 kilometers (186 miles) high. Io's surface is very young as a result of many continuous volcanic eruptions, and no impact craters have been seen. The colors of the surface—vivid reds, yellows, greens, and black—are different from those seen on other solid bodies in the solar system. These colors are a result of sulfur and silicates on the surface. Io's lavas are hotter than those seen on Earth today, reaching temperatures of 1,500°C (2,700°F). They may be similar in composition to **ultramafic lavas** on Earth, which erupted millions of years ago.

infrared portion of the electromagnetic spectrum with wavelengths slightly longer than visible light

ultramafic lavas dark, heavy lavas with a high percentage of magnesium and iron; usually found as boulders mixed in other lava rocks

Europa. Europa is particularly intriguing because of the possibility that it might harbor life. Observations by Galileo spacecraft showed that Europa's cracked surface resembles the ice floes seen in Earth's polar regions. High-resolution images show that some of the broken pieces of the ice crust have shifted away from one another, but that they fit together like a jigsaw puzzle. This suggests that the crust has been, or still is, lubricated from underneath by warm ice or liquid water. The two most basic ingredients for life are water and heat. Like Io, Europa is subject to tidal stresses because of

THE GALILEAN SATELLITES			
Name	Radius	Distance from Jupiter	Density
Io	1,821 km	421,600 km	3.53 gm/cm^{-3}
Europa	1,565 km	670,900 km	2.97 gm/cm^{-3}
Ganymede	2,634 km	1,070,000 km	1.94 gm/cm^{-3}
Callisto	2,403 km	1,883,000 km	1.85 gm/cm^{-3}

Jupiter and Ganymede's gravitational pull. While Europa has no evidence of current active silicate volcanism, the tidal stresses may cause heating of the interior, providing the other key ingredient for life. Europa's surface does show evidence of ice volcanism. There are places where material appears to have come up from underneath as slushy ice and flowed on the surface. Europa has very few impact craters, indicating that its surface is young. Slushy ice flowing over the surface probably erased many impact craters. Europa's surface composition is dominated by water, but Galileo detected other compounds, including hydrogen peroxide (H_2O_2) on the surface and a thin oxygen atmosphere. The behavior of Jupiter's magnetic field around Europa implies that there may be ions circulating globally beneath the icy surface.

Ganymede. Larger than the planets Mercury and Pluto, Ganymede was the first moon known to have a magnetic field, one of the earliest discoveries made by the Galileo mission. The field is stronger than that of Mercury. Ganymede has a core made up of metallic iron or iron sulfides. If the core is molten and moving, it would produce the strong magnetic field observed by Galileo. Ganymede's surface shows a complex geologic history. The surface is characterized by large dark areas and by bright grooved terrains. The grooves are thought to have formed when the crust separated along lines of weakness. Other images showed hillcrests and crater rims capped by ice, and old terrain cut by furrows and marked by impact craters. Observations in the ultraviolet made from the Hubble Space Telescope showed the presence of oxygen on Ganymede, and Galileo observations detected hydrogen escaping from Ganymede into space. These results indicate that Ganymede has a thin oxygen atmosphere. Astronomers believe that the atmosphere is produced when charged particles trapped in Jupiter's magnetic field come down to Ganymede's surface. The charged particles penetrate the icy surface, disrupting the water ice. The hydrogen escapes into space, whereas the heavier oxygen atoms are left behind.

Callisto. About the same size as the planet Mercury, Callisto is Jupiter's second largest moon. Its surface is heavily cratered, implying that it is extremely old, probably dating from about 4 billion years ago, which is close to the time when the solar system formed. Callisto's surface is icy and has some large impact craters and basins surrounded by concentric rings. The largest impact basin is called Valhalla, and it has a bright central region 600 kilometers (372 miles) in diameter, with rings extending to 3,000 kilometers (1,860 miles) in diameter. Galileo observations showed that Callisto has a magnetic field. Underneath its icy crust, Callisto may have a liquid ocean, which, if it is as salty as Earth's oceans, could carry enough electrical currents to produce the magnetic field. A major discovery made by the Galileo mission is that Callisto has a thin atmosphere of carbon dioxide. SEE ALSO EXPLORATION PROGRAMS (VOLUME 2); GALILEI, GALILEO (VOLUME 2); NASA (VOLUME 3); PLANETARY EXPLORATION, FUTURE OF (VOLUME 2); ROBOTIC EXPLORATION OF SPACE (VOLUME 2); SHOEMAKER, EUGENE (VOLUME 2); SMALL BODIES (VOLUME 2).

Rosaly M. C. Lopes

Bibliography

Beatty, J. Kelly, Carolyn Colins Petersen, and Andrew Chaikin, eds. *The New Solar System*, 4th ed. Cambridge, UK: Sky Publishing Corporation and Cambridge University Press, 1999.

Shirley, James. H., and Rhodes. W. Fairbridge, eds. *Encyclopedia of Planetary Sciences.* London: Chapman & Hall, 1997.

Internet Resources

NASA Jet Propulsion Laboratory, California Institute of Technology. <http://jpl.nasa .gov/>.

Kepler, Johannes

German Mathematician and Astronomer
1571–1630

Johannes Kepler was a German mathematician and astronomer who discovered three key laws that govern planetary motion. Born in Weil der Stadt, Germany, in 1571, Kepler studied astronomy and theology at the University of Tübingen before becoming an astronomy and mathematics professor in Graz, Austria, in 1594. In 1600 he accepted an invitation from Danish astronomer Tycho Brahe to become Brahe's assistant in Prague, and study the orbit of the planet Mars.

After Brahe's death in 1601, Kepler acquired Brahe's extensive astronomical records and studied them for years in an effort to prove the Copernican model of the solar system. During this time he discovered what are now known as Kepler's three laws of planetary motion. The first law, published with the second in 1609, revealed that planets do not orbit in perfect circles, as had been previously assumed, but in ellipses, with the Sun at one focus. The second law found that planets sweep out equal areas in equal periods of time. The third law, published separately in 1619, stated that the square of a planet's orbital period is proportional to the cube of the orbit's mean radius. During this time Kepler also made advances in optics and mathematics. He died after a brief illness in Regensburg, Germany, in 1630. SEE ALSO ASTRONOMY, HISTORY OF (VOLUME 2); COPERNICUS, NICHOLAS (VOLUME 2); MARS (VOLUME 2).

Jeff Foust

Johannes Kepler developed what came to be known as Kepler's three laws of planetary motion.

Bibliography

Caspar, Max. *Kepler.* New York: Dover, 1993.

Internet Resources

"Johannes Kepler." University of St. Andrews, School of Mathematics and Physics. <http://www-groups.dcs.st-and.ac.uk/~history/Mathematicians/Kepler.html>.

Kuiper Belt

Comets are some of the most spectacular objects in the night sky. About once per decade, a truly bright comet comes along and can be viewed by the unaided eye. Where do these comets come from? Where do they spend most of their lives?

Some comets orbit the Sun once every thousand years or so and can be easily viewed only when they are in the inner solar system. These are known as long period comets. Nonperiodic comets appear in the inner solar

LOCATION OF THE KUIPER BELT OF COMETS

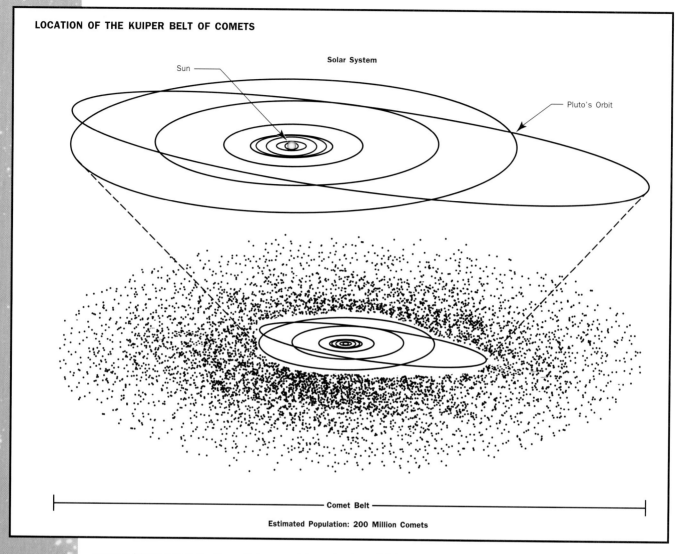

Solar System

Sun

Pluto's Orbit

Comet Belt

Estimated Population: 200 Million Comets

SOURCE: Adapted from the Space Telescope Science Institute, PR-95-26, created June 14, 1995. <http://www.boulder.swri.edu/clark/chance/09kbelt.jpg>

system only once. Some comets, however, enter the inner solar system repeatedly and predictably. These are the short period comets.

All of these comets are in orbit around the Sun but, unlike the planets, which all revolve around the Sun in the same direction and are confined to approximately the same plane as Earth's orbit (the plane of the **ecliptic**), cometary orbits show no preferred orientations. The shortest period comets (orbital periods of less than twenty years) are an exception. Comets in this group, called the Jupiter family comets (JFCs), revolve around the Sun near the plane of the ecliptic in the same direction as Earth's orbit.

Noting the different nature of the JFC orbits, astronomers sought explanations. It had been believed that all comets originated in the Oort cloud, a halo of comets at extremely large distances from the Sun. But in 1988, Martin J. Duncan, Thomas R. Quinn, and Scott Tremaine showed that it was impossible to have the random orientations of Oort cloud comets converted to the planar orientations of JFCs. They proposed that, in addition to the Oort cloud as a reservoir for comets, there must be a disk-like reser-

ecliptic the plane of Earth's orbit

voir of comets with its inner edge near Neptune. They called this disk the Kuiper belt after Dutch-born American astronomer Gerard P. Kuiper, who postulated in 1951 that the solar system could not end at Neptune because that would imply a sharp edge to the disk out of which planets formed. ✳

The objects in the Kuiper belt represent remnants from the formation of our solar system. When the planets formed 4.6 billion years ago, they formed from an agglomeration of many planetesimals, or small solid celestial bodies. Beyond Neptune, the density of planetesimals was too low and the time for them to collide and accumulate was too long for another planet to form. Thus, the planetesimals remained in the outer solar system, past Neptune's orbit. They are called Kuiper belt objects (KBOs).

The existence of the Kuiper belt was confirmed in the 1990s. In 1992 David Jewitt and Jane Luu found the first KBO, designated 1992 QB$_1$. By 2000 many surveys had been performed and a total of 345 KBOs had been found.

In 2001, some one trillion planetesimals still existed in KBO orbits from Neptune outwards. Most remained in the Kuiper belt, interacting with one another or Neptune. Some of these were just barely visible from Earth through the most sensitive telescopes; others were too faint to see. But they do exist. And, as time passes, some will be perturbed into the inner solar system, where they will become Jupiter-family comets and appear periodically. SEE ALSO COMETS (VOLUME 2); KUIPER, GERARD PETER (VOLUME 2); OORT CLOUD (VOLUME 2); ORBITS (VOLUME 2); PLANETESIMALS (VOLUME 2); SMALL BODIES (VOLUME 2).

Anita L. Cochran

✳ In 1949 Irish astronomer Kenneth Edgeworth published an analysis similar to Kuiper's, but Kuiper's work was better known; in recognition of Edgeworth's contribution, some people call the comet reservoir the Edgeworth-Kuiper belt.

Bibliography

Duncan, Martin J., Thomas R. Quinn, and Scott Tremaine. "The Origin of Short-Period Comets." *Astrophysical Journal (Letters)* 328 (1988):L69–L73.

Edgeworth, Kenneth. E. "The Origin and Evolution of the Solar System." *Monthly Notices of the Royal Astronomical Society* 109 (1949):600.

Jewitt, David, and Jane Luu. "Discovery of the Candidate Kuiper Belt Objects 1992 QB$_1$." *Nature* 362 (1993):730–732.

Kuiper, Gerard P. "On the Origin of the Solar System." In *Astrophysics: A Topical Symposium*, pp. 357–424, ed. J. A. Hynek. New York: McGraw-Hill, 1951.

Malhotra, Renu, Martin J. Duncan, and Harold Levison "Dynamics of the Kuiper Belt." In *Protostars and Planets IV*, eds. Vince Manning, Alan P. Boss, and Sara S. Russell. Tucson: University of Arizona Press, 2000.

Kuiper, Gerard Peter

Dutch-American Astronomer
1905–1973

Gerard Peter Kuiper was the father of modern planetary astronomy. His work ran the gamut from star and planetary system formation to the study of the planets themselves. He used techniques ranging from visual observations to those requiring the latest technology, including **infrared** detectors, airborne observatories, and spacecraft.

Kuiper was born in Harenkarspel, the Netherlands. While in his native country, Kuiper made important contributions to the study of binary stars,

infrared portion of the electromagnetic spectrum with wavelengths slightly longer than visible light

Gerard Peter Kuiper, astronomer and professor at Yerkes Observatory of the University of Chicago, explaining his theory that there is a disk of comets beyond Neptune's orbit.

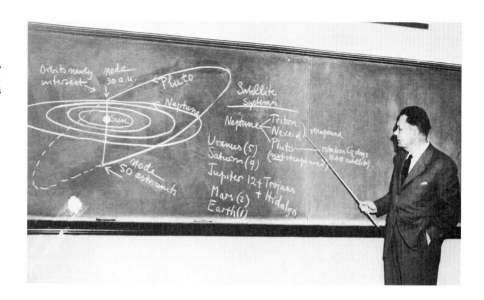

which led to work on planetary system formation after he moved to the United States.

During the winter of 1943–1944, Kuiper made **spectrographic studies** of the major planets and satellites, leading to the discovery that Saturn's largest moon, Titan, had an atmosphere containing methane. Studies of the brightnesses of the moons of Uranus and Neptune led to the discovery of additional satellites: Miranda, orbiting Uranus, in 1948; and Nereid, orbiting Neptune, in 1949.

spectrographic studies studies of the nature of matter and composition of substances by examining the light they emit

In 1951, he proposed that a disk of comet nuclei extends from the solar system's planetary zone out to as much as 1,000 times the Earth-Sun distance (the astronomical unit [AU]). This is now called the Kuiper Belt and is recognized to extend from Neptune's distance (about 30 AU) to perhaps 50 to 100 AU.

In 1960, Kuiper founded the Lunar and Planetary Laboratory at the University of Arizona. He remained active in his later years, traveling and conducting site surveys for new observatories. Kuiper died in 1973. ✷ SEE ALSO ASTRONOMY, HISTORY OF (VOLUME 2); CAREERS IN SPACE SCIENCE (VOLUME 2); COMETS (VOLUME 2); KUIPER BELT (VOLUME 2); NEPTUNE (VOLUME 2); SATURN (VOLUME 2); URANUS (VOLUME 2).

✷ A now-retired National Aeronautics and Space Administration airborne observatory that made groundbreaking infrared observations from the stratosphere was named after Kuiper.

Stephen J. Edberg

Bibliography

Cruikshank, Dale P. "Twentieth Century Astronomer." *Sky and Telescope* 47 (1974): 159–164.

Pannekoek, Anton. *A History of Astronomy.* New York: Interscience Publishers, 1961.

Internet Resources

Jewitt, David. "Kuiper Belt." Institute for Astronomy. <http://www.ifa.hawaii.edu/faculty/jewitt/kb.html>.

Life in the Universe, Search for

It is an old question, and a persistent one: Is there life elsewhere in the cosmos? Is the universe more than just an enormous collection of dead rock and glowing gas, with only one inhabited world?

The Viking 1 lander set down on the surface of Mars on July 20, 1976. It had a robotic arm that would scoop up samples of the martian soil and place them in three biology experiments onboard the lander. Although results from these experiments were ambiguous, most scientists believe that no life currently exists on the martian surface.

While speculation about life in space is an old pastime, a serious, scientific search for it is very new. Despite the impression one may get from movies and television, scientists still have not found any conclusive evidence of biology beyond Earth—not even evidence of the simplest microbes. But many scientists expect that this situation will soon change.

Part of their optimism is due to an astounding fact revealed by centuries of studying the heavens: The physics and chemistry of the farthest galaxies are the same as the physics and chemistry found on Earth. Astronomers have proven this by analyzing the light of distant objects with spectrographs. When they use these instruments to break up starlight into its constituent colors, they see the telltale "fingerprints" of atoms that are found on Earth: the ninety-two elements listed on the familiar periodical table of elements.

The light elements such as hydrogen, carbon, nitrogen, and oxygen are especially plentiful in space. These are the building blocks of all life on our planet. If the stuff of life is so commonplace, might not life itself also be widespread?

How to Find Extraterrestrial Life

There are several obvious—and a few not-so-obvious—methods used in the hunt for extraterrestrial biology.

We could simply send rockets to other worlds and look for it. Since the mid-1960s, this has been done on a limited basis. Spacecraft have landed on

WHAT IS A SPECTROGRAPH?

Spectrographs are devices used by astronomers to break up the light collected by a telescope into its various colors, or wavelengths. Usually a prism or diffraction grating is used for this purpose. The resultant "rainbow" is then recorded on film or electronically. The lines (either bright or dark) that inevitably show up in such spectra indicate the composition of the gases in the outer regions of the stars or planets being examined.

the Moon, Venus, and Mars (although only the Moon has been visited by humans), and camera-toting probes have investigated all the other solar system planets except Pluto. Of these familiar locales, only the Moon and Mars have been examined in much detail. The Moon is sterile, which given its lack of atmosphere and liquid water, is hardly surprising. Mars is less obviously dead. ✳

✳ **The continuing efforts to find Martians are described in the section "What Have We Found?"**

Another approach is to use spacecraft to gather rocks from other worlds so they can be scrutinized in the laboratory. The Moon rocks lugged back by Apollo astronauts are an example of this, and the National Aeronautics and Space Administration (NASA) hopes to eventually use robot craft to bring back small pieces of Mars.

Still another way of getting extraterrestrial evidence is to find it on Earth. When **meteorites** hit nearby worlds, they kick up bits of rock, some of which might have enough speed to escape from their planet entirely. These rocky runaways then wander around the inner solar system. Some, by chance, will hit Earth. If they are large enough to avoid being completely incinerated as they plunge through our atmosphere, they could end up in a laboratory collection. A dozen meteorites from Mars have been found to date, brought here by nature rather than NASA.

meteorite any part of a meteoroid that survives passage through a body's atmosphere

For investigating distant worlds that orbit other stars, there is no hope of sending rockets or collecting meteorite samples. Instead, astronomers can use incoming electromagnetic radiation (more commonly known as light and radio) to search for certain "signatures" of life. Making a spectrographic analysis of the light reflected by the atmosphere of a far-off planet would permit scientists to check for the presence of oxygen or methane. Either one might be a clue to the presence of bacteria or possibly more advanced biological forms. The oxygen in Earth's atmosphere is the result of billions of years' worth of exhaust gases from bacteria and plants. Much of the methane is due to the digestive activities of cows and pigs. Finding large amounts of either of these gases in the atmosphere of a distant, Earth-sized planet would suggest an inhabited world.

A final technique is to look for radio or light signals that have been deliberately sent by sophisticated beings on other planets. Hunting for artificially produced signals is known as SETI, the Search for Extraterrestrial Intelligence. Since 1960, SETI scientists have used large radio antennas (and more recently, specially outfitted conventional telescopes) to scan for signals from intelligent aliens.

Where Do We Expect to Find Life?

DNA deoxyribonucleic acid; the molecule used by all living things on Earth to transmit genetic information

All life on Earth is based on carbon chemistry and uses **DNA** as its blueprint for reproduction. Alien life might not sport DNA, but the odds are good that it would still be carbon-based. This is a sure bet because carbon has an exceptional ability to link up with other atoms into long chains, or polymers. To encourage this sort of chemical complexity, a solvent in which the atoms and molecules can easily move and meet is essential. Liquid water is the best such solvent, and therefore most researchers assume that the first step in tracking down extraterrestrial life is to find cosmic niches where liquid water is likely to both exist and persist.

Until recently, astronomers felt that liquid water would be abundant only on Earth-like worlds that were situated at the right distance from their suns—neither too close, where water would boil, nor too far, where it would freeze. In our own solar system, orbital radii greater than that of Venus and less than that of Mars seem right, a region referred to as the Habitable Zone (HZ). For stars dimmer than the Sun, the HZ would be closer in and smaller; for brighter stars, it would be larger and farther out.

This straightforward idea has lately been modified. For one thing, an atmosphere can make a big difference in keeping a planet's surface warm. Mars is cold and dry today, but in the past, when it had a thicker atmosphere of carbon dioxide—an efficient greenhouse gas—there was liquid water gurgling across its landscape. So the extent of the HZ depends on a planet's atmosphere.

In addition, life has been discovered on Earth thriving in decidedly unfriendly environments. Tube worms and bacteria coexist in the inky darkness of ocean deeps. No type of photosynthesis will work in this environment, so the inhabitants of this strange ecosystem take advantage of the chemical nutrients that come churning out of hot water vents (some above 100°C [212°F]) in the ocean crust. Bacteria have also been found in another unexpected environment: kilometers under the ground, where they can live off of chemical nutrients naturally present in rock. The conditions in this environment are brutal: temperatures are high (again, often above 100°C), and the elbowroom, consisting of pores in the rock, is low. A little bit of liquid water in these pressure cooker environments allows these **extremophiles**, as they are called, to survive.

If life can exist in such difficult conditions on Earth, why not in space? These discoveries have challenged scientists with new thoughts about exactly what kinds of worlds are "habitable." The conventional concept of an HZ has been stretched to include icy moons and underground retreats, and this has encouraged scientists to look for life in what were once considered all the wrong places.

extremophiles microorganisms surviving in extreme environments such as high salinity or near boiling water

What Have We Found?

What is the scorecard on the search for life? Broadly speaking, the quest for extraterrestrial biology has been a two-pronged affair: a search for nearby, simple biology (e.g., microbes on Mars) and a hunt for distant, intelligent beings (SETI).

Mars. Of the possible nearby sites for life, Mars has traditionally been everyone's favorite. In the late nineteenth century, some astronomers astounded the world (and their colleagues) by claiming that thin, straight lines could be seen crisscrossing the surface of Mars. These "canals" bespoke the existence of an advanced society on the Red Planet. Unfortunately, the canals turned out to be optical illusions. Nevertheless, of all the worlds in our solar system, Mars is most like Earth. It beckons us with the prospect of nearby, alien life.

In 1976 NASA placed two robot spacecraft on the rusty surface of Mars: the Viking Landers. They were essentially mobile biological laboratories and spent days analyzing the Martian soil for the presence of microbes. They did not look for alien bacteria directly (for example, with a microscope), but searched for organic molecules in the soil, or soaked it with nutrient solu-

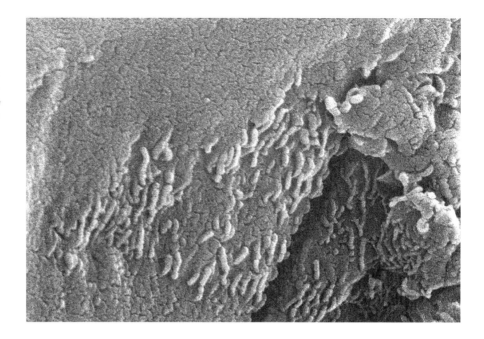

An electron microscope image of unknown tube-like structures on Mars. Some scientists believe that these structures are microscopic fossils of organisms that lived on the planet more than 3.6 billion years ago.

altimeter an instrument designed to measure altitude above sea level

tions and watched for exhaust gases that would betray microbial metabolism. The conclusion of the Viking science team was that the Martian surface was sterile, although it is worth noting that two team members disagreed. This indicates how difficult it may be to design unambiguous experiments to look for extraterrestrial life.

Despite the failure of this sophisticated effort to find Martians directly, there is growing evidence that Mars may once have been a more hospitable environment for life. High-resolution photos from the orbiting Mars Global Surveyor reveal what look like sedimentary rock layers, strongly suggesting that more than 3 billion years ago Mars had lakes—environments that might have spawned life. This same spacecraft had an onboard **altimeter** and discovered an enormous flat region in Mars's northern hemisphere. This may once have been an ocean.

In 1996, NASA scientists examined one of the known Martian meteorites (ALH 84001) and claimed to find several lines of evidence for fossilized microbes within. This evidence included the presence of various chemicals associated with biology, as well as small bits of iron (magnetite) that is commonly found in earthly bacteria. The scientists also made microscope photos of the meteorite's interior, which showed tiny rod- and wormlike structures that look very much like single-celled creatures. Unfortunately, there is great disagreement in the science community about whether this evidence is really due to long-dead Martians or to some inorganic phenomenon.

NASA is planning to send additional orbiters and rovers to Mars in the early years of the twenty-first century. The major goal of these expeditions is to learn more about the history of liquid water on the planet, as this is the key to an improved search for life. Ancient fossils may yet turn up, and some researchers speculate that the descendents of these ancient microbes (if there were any) might still be eking out a dark existence deep under the Martian surface where it is still relatively warm and wet.

Other Solar System Sites. Mars may not be the only solar system site for life other than Earth. Ever since the late 1970s, when the Voyager spacecraft made the first close-up photos of Jupiter's large moons, astronomers have considered whether life might exist even in these cold, dim environs. Europa is the most promising of the moons for biology. Its surface is bright white ice, cracked and glazed like a billiard ball with a bad paint job. The temperature on Europa is −160°C (−256°F), and one might naively assume that no liquid water could exist. But Europa is in a gravitational tug-of-war with its sister moons, and this keeps it in an egg-shaped orbit. The consequence is that Jupiter's changing gravitational pull squeezes and squishes Europa, heating it up the way pastry dough gets warm when kneaded. There is increasing evidence that beneath Europa's granite-hard, icy skin is a 100-kilometer-thick (62-mile-thick) liquid ocean, one that has been there for billions of years. At the bottom of this ocean, vents may spew hot water and chemicals, much as they do on Earth. Needless to say, if this picture of Europa is correct, some simple forms of life may be swimming in these dark, unseen waters.

In 1995, NASA's Galileo spacecraft began taking photos of Europa and other **Jovian** moons. That mission will be followed by an improved orbiter, probably to be launched in 2009, that will carry **radar** equipment to examine the Europan ice. The plan is to find out if the unseen ocean really exists, and if so, whether there any thin spots in the ice where future landers might be able to drill holes and drop equipment down into Europa's briny deep.

Jovian relating to the planet Jupiter

radar a technique for detecting distant objects by emitting a pulse of radio-wavelength radiation and then recording echoes of the pulse off the distant objects

Even Saturn's large moon Titan (which is bigger than Mercury) might conceivably host a bit of biology. Titan sports a substantial atmosphere, one that is denser than Earth's and that seems to be perpetually shrouded in smog. The air on Titan is mostly nitrogen and neon, but hydrocarbons and complex polymers make up the smog, together with a haze of methane (natural gas) crystals and ethane clouds. Some researchers suspect that lakes of liquid ethane, or even a moon-girdling ocean of ethane, methane, and propane, may exist on Titan.

All this hydrocarbon chemistry is discouragingly cold, −180°C (−292°F). Nevertheless, despite resembling an arctic oil refinery gone wild, it is possible that over the course of billions of years, Titan's hydrocarbons have spawned exotic life-forms. In 2004 a probe from the Cassini spacecraft will be dropped into Titan's chilly clouds for the first close-up glimpse of this oddball moon.

SETI. While NASA and other space organizations search for relatively simple living neighbors, SETI scientists turn their large antennas in the directions of nearby stars, hoping to find broadcasts from intelligent beings. The type of signals they look for are called narrowband, which means they are at one spot on the radio dial. Such transmissions could pack a lot of radio energy into a small frequency range, making detection even light-years away much easier. The most sensitive of these searches is Project Phoenix, which uses the 305-meter (1,000-foot) diameter radio antenna at Arecibo, Puerto Rico, to scrutinize about 1,000 Sunlike stars less than 150 light-years distant. Another SETI experiment is called SERENDIP, a project that is less sensitive but searches large tracts of the sky.

While SETI scientists still have not come up with a confirmed, extraterrestrial signal, they are greatly improving their equipment. In the next decades, they will scrutinize as many as a million star systems or more. In addition, new experiments using conventional optical telescopes have been started up. These look for very short (a billionth of a second), very bright laser pulses that an alien civilization might be sending earthward to catch our attention.

The discovery in recent years that many Sun-like stars have planets has greatly encouraged this type of search. It has also prompted space agencies around the world to consider building mammoth space telescopes that could uncover Earth-like planets around other stars. If this is done, then a spectrographic analysis of the atmospheres of these planets might turn up the traces of life—even simple life.

What Finding Extraterrestrial Life Would Mean

As noted earlier, we still have no convincing proof that there are any lifeforms other than those found on Earth. Life is complex, and we still do not understand how it got started on our own planet. But to find living creatures—even microbes—on other worlds would tell us that biology is not some miraculous, extraordinary phenomenon. If SETI succeeds, and we find other intelligence, we might learn much about the universe and long-term survival. In either case, we would know that Earth and its carpet of living things is not the only game in town, but that we share the universe with a vast array of other life. SEE ALSO EXTRASOLAR PLANETS (VOLUME 2); FIRST CONTACT (VOLUME 4); SETI (VOLUME 2); ROBOTIC EXPLORATION OF SPACE (VOLUME 2).

Seth Shostak

Bibliography

Andreas, Athena. *To Seek Out New Life: The Biology of Star Trek.* New York: Crown, 1998.

Darling, David. *The Extraterrestrial Encyclopedia.* New York: Three Rivers Press, 2000.

Goldsmith, Donald, and Tobias Owen. *The Search for Life in the Universe.* Reading, MA: Addison-Wesley, 1992.

Pasachoff, Jay. *Astronomy: From the Earth to the Universe.* Fort Worth, TX: Saunders College Publishing, 1998.

Shostak, Seth. *Sharing the Universe: Perspectives on Extraterrestrial Life.* Berkeley, CA: Berkeley Hills Books, 1998.

Long Duration Exposure Facility (LDEF)

The Long Duration Exposure Facility (LDEF) project, originally called the Meteoroid and Exposure Module, was begun in 1970. Conceived and managed by the National Aeronautics and Space Administration's (NASA) Langley Research Center in Hampton, Virginia, the LDEF was designed as a large structure on which various tests of systems and materials could be carried out. One of its most important functions was to gather data on **meteoroids**, radiation, and other space hazards.

meteoroid a piece of interplanetary material smaller than an asteroid or comet

NASA's Long Duration Exposure Facility was designed to provide long-term data on the space environment. The LDEF was launched in 1984, and spent the next 5.7 years orbiting Earth.

The LDEF was a 9-meter-long (30-foot-long) hexagonal-shaped structure designed to fit snugly into the space shuttle orbiter's cargo bay. The research programs involved corporations, universities, and U.S. and foreign governments. The LDEF was a platform both for engineering and systems development studies and for pure scientific research.

One goal of the LDEF program was to see how a wide variety of materials, such as plastic and glass coverings for solar power (**photovoltaic**) cells, would react to spending a long time in **low Earth orbit** (LEO). The stability of certain plastics was tested. Some of the polymers were found completely unsuitable for use in space.

Another goal was to measure the number and composition of meteoroids, debris, and radiation in LEO. The LDEF was a way for NASA to find out what sort of materials would be needed in any future space stations or satellites that would spend years in LEO. The experiments were mounted on eighty-six separate trays, normally one experiment per tray. Some experiments were carried out using multiple trays, such as the Space Environment Effects on Spacecraft Materials Experiment, which used four

photovoltaic pertaining to the direct generation of electricity from electromagnetic radiation (light)

low Earth orbit an orbit between 300 and 800 kilometers above Earth's surface

different trays, and the High Resolution Study of Ultraheavy Cosmic Rays, which used fifteen.

The LDEF was launched inside space shuttle Challenger on mission STS 41-C in April 1984. Commanded by Robert Crippen, this was the twelfth shuttle mission. The LDEF was placed in orbit at an altitude of 442 kilometers (275 miles) above Earth. It was intended that the LDEF would stay in orbit for just one year, but because of the Challenger disaster in January 1986, the facility was not recovered until January 1990. Thus, it ended up spending five years and seven months in space.

When it was picked up by the space shuttle Columbia, the LDEF was only a few weeks away from falling into Earth's atmosphere and burning up. Over the years, its orbit had decreased to about 280 kilometers (175 miles). As it moved closer to Earth, it also became closer to the upper layers of the planet's atmosphere. Thus, the particles of the atmosphere began to strike it and reduce both its speed and its altitude. The closer it got to Earth the faster it began to fall. The STS-32 mission, commanded by Dan Brandenstein, got there just in time.

Back on Earth, NASA found that the silicon-based adhesives they used on the LDEF spacecraft (as well as on the shuttle) had let off a form of gas that was transformed, by exposure to atomic oxygen, into silicates (SiO_2) and had contaminated some of the surfaces of the LDEF. This showed that there is a danger of silicate contamination of surfaces that have critical optical needs, such as windows, solar cells, and mirrors.

Following completion of the LDEF project, NASA's Langley Research Center built the Modular International Space Station Experiment. This is a test facility, about the size of a suitcase, which will continue the work started by the LDEF. A new wave of Passive Experiment Containers will be attached to the International Space Station and will provide data for the next generation of spacecraft. SEE ALSO CHALLENGER (VOLUME 3); EXPLORATION PROGRAMS (VOLUME 2); SPACE SHUTTLE (VOLUME 3).

Taylor Dinerman

Internet Resources

Long Duration Exposure Facility. NASA Langely Research Center. <http://setas-www.larc.nasa.gov/LDEF/index.html>.

Mars

Mars has fascinated humans throughout history. It appears as a blood-red star in the sky, which led the Romans to name it after their war god. Its motions across the sky helped German astronomer Johannes Kepler (1571–1630) derive his laws of planetary motion, which dictate how celestial bodies move. Two small moons, Phobos and Deimos, were discovered orbiting Mars in 1877. But it is primarily the question of life that has driven scientists to study Mars.

Basic Physical and Orbital Properties

Mars displays a number of Earth-like properties, including a similar rotation period, seasons, polar caps, and an atmosphere. In the 1800s

View of Mars from the Hubble Space Telescope. This image is centered on the dark feature known as Syrtis Major, which was first seen by astronomers in the seventeenth century. To the south of Syrtis is a large circular feature called Hellas, a deep impact crater.

astronomers also noted seasonal changes in surface brightness, which they attributed to vegetation. In 1877 Italian astronomer Giovanni Schiaparelli reported the detection of thin lines crossing the planet, which he called *canali*, Italian for "channels." But the term was mistranslated into English as "canals," which implies waterways constructed by intelligent beings. American astronomer Percival Lowell (1855–1916) popularized the idea of canals as evidence of a Martian civilization, although most of his colleagues believed these features were optical illusions. This controversy continued until the 1960s when spacecraft exploration of the planet showed no evidence of the canals.

Telescopic observations revealed the basic physical and orbital properties of Mars, as well as the presence of clouds and dust storms, which indicated the presence of an atmosphere. Dust storms can be regional or global in extent and can last for months. Global dust storms typically begin in the southern hemisphere around summer solstice because this is also when Mars is closest to the Sun and heating is the greatest. Temperature differences cause strong winds, which pick up the dust and move it around. Astronomers now know that the seasonal variations in surface brightness are caused by a similar movement of dust and not by vegetation.

Spectroscopic analysis suggested that the Martian atmosphere is composed primarily of carbon dioxide (CO_2), and this was confirmed by measurements made by the Mariner 4 spacecraft in 1965. The atmosphere is 96

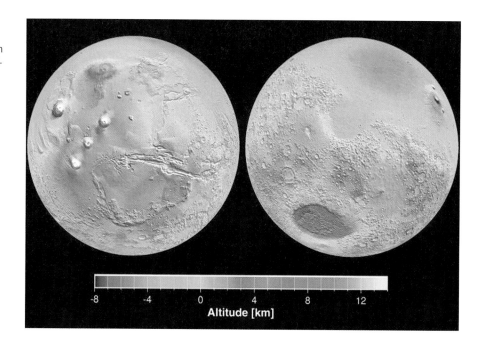

Global false-color topographic views of Mars at different orientations from the Mars Orbiter Laser Altimeter. The right image features the Hellas impact basin (in purple), while the left shows the Tharsis topographic rise (in red and white).

Altitude [km]

percent carbon dioxide, 3 percent nitrogen, and about 1 percent argon, with minor amounts of water vapor, oxygen, ozone, and other substances. The atmosphere is very thin—the pressure exerted by the atmosphere on the surface is only 0.006 bar (the atmospheric pressure at sea level on Earth is 1 bar). This thin atmosphere is unable to retain much heat; hence the Martian surface temperature is always very cold (averaging −63°C [−81°F]). This thin atmosphere also is unable to sustain liquid water on the surface of Mars—any liquid water immediately evaporates into the atmosphere or freezes into ice. Geologic evidence suggests, however, that surface conditions have been warmer and wetter in the past.

A Geologically Diverse Planet

The geologic diversity of Mars was first realized from pictures taken by the Mariner 9 spacecraft in 1971–1972. Three earlier spacecraft (Mariner 4 in 1965 and Mariner 6 and Mariner 7 in 1969) had returned only a few images of the planet as they flew past. These images primarily revealed a heavily cratered surface, similar to the lunar highlands. Mariner 9, however, orbited Mars and provided pictures of the entire planet. Mariner 9 revealed that while 60 percent of the planet consists of ancient, heavily cratered terrain, the other 40 percent (mostly found in the northern hemisphere) is younger. Mariner 9 revealed the existence of the largest volcano in the solar system (Olympus Mons, which is about three times higher than Mt. Everest), a huge canyon system (Valles Marineris) that stretches the distance of the continental United States and is seven times deeper than the Grand Canyon, and a variety of channels formed by flowing water. These channels are not the same thing as the canals—no evidence of engineered waterways has been found on Mars, indicating that the canals are optical illusions. The discovery of channels formed by flowing water, however, reignited the question of whether life may have existed on Mars.

Findings of the Viking Missions

The Viking missions were designed to determine if life currently exists on Mars. Viking 1 and Viking 2 were each composed of an orbiter and a lander. Viking 1's lander set down in the Chryse Planitia region of Mars on July 20, 1976. Viking 2's lander followed on September 4, 1976, in the Utopia Planitia region to the northeast of where the first lander set down. Both landers were equipped with experiments to look for microbial life in the Martian soil as well as cameras to search for any movement of larger life-forms. All the experiments produced negative results, which together with the lack of organic material in the soil led scientists to conclude that no life currently exists on Mars.

The Viking orbiters, meanwhile, were providing the best information of the Martian surface and atmosphere to date. Scientists discovered that seasonal changes in the polar cap sizes are major drivers of the atmospheric circulation. They also discovered that the polar caps are primarily composed of carbon dioxide ice, but that the residual cap that remained at the North Pole even at the height of summer is probably composed of water ice. The frequency, locations, and extents of dust storms were studied in better detail than what Earth-based telescopes could do, providing new information on the characteristics of these events.

Is There Water on Mars?

The surface also continued to reveal new surprises. Fresh **impact craters** are surrounded by fluidized **ejecta** patterns, likely produced by impact into subsurface water and ice. Detailed views of the volcanoes, channels, and canyons provided improved understanding of how these features formed and how long they were active. But most intriguing was the accumulating evidence that liquid water has played a major role in sculpting the Martian surface. Curvilinear features interpreted as shorelines were found along the boundary between the lower northern plains and the higher southern highlands, leading to suggestions that the northern plains were filled with an ocean at least once in Martian history.

Smooth-floored craters whose rims are cut by channels suggest that lakes collected in these natural depressions. The appearance of degraded craters in old regions of the planet suggests erosion by rainfall. Spectroscopic data from Earth-based telescopes as well as the Russian Phobos mission in 1989 indicate that water has affected the mineralogy of the surface materials over much of the planet.

Clearly Mars has been warmer and wetter in the past. Where did all that water go? Some water can be found as vapor in the thin Martian atmosphere and some is locked up as ice in the polar regions. But these two reservoirs contain a small percentage of the total amount of water that scientists believe existed on the planet. Some of the water likely has escaped to space because of Mars' small size and low gravity. But scientists now believe that a large amount of the water is stored in underground ice and water reservoirs. Liquid water, derived from these underground reservoirs, may exist again on the Martian surface in the future because of episodic changes in atmospheric thickness. Scientists now know that the amount of tilt of Mars's rotation axis changes on about a million-year cycle because of grav-

impact craters bowl-shaped depressions on the surfaces of planets or satellites that are the result of the impact of space debris moving at high speeds

ejecta the material thrown out of an impact crater during its formation

View of Mars from Viking I. Although the surface of Mars is dry, rocky, and covered with a thick powdery red soil, it is believed that the planet once had a more extensive atmosphere, allowing for the possible existence of ice and water.

itational influences from other planets. When the Martian poles are tipped more towards the Sun, the poles are exposed to more sunlight and the ices contained in these regions can vaporize to create a thicker atmosphere, which can cause higher surface temperatures by greenhouse warming.

Martian Meteorites

The Viking exploration of Mars ended in 1982, and few spacecraft provided information for the next fifteen years. The United States and Russia launched many spacecraft, but these missions were either failures or only partial successes. Nevertheless, new details were obtained during this time from a different source—meteorites. As early as the 1960s some scientists proposed that some unusual meteorites might be from Mars. These meteorites were volcanic rocks with younger formation ages (about 1 billion years) than typical meteorites (about 4 billion years). There are three major groups of these unusual meteorites: the shergottites, nakhlites, and chassignites (collectively called the SNC meteorites). In 1982 scientists discovered gas trapped in one of these SNC meteorites. When the gas was analyzed it was found to have **isotopic ratios** identical to those found in the Martian atmosphere. This discovery clinched the Martian origin for these meteorites. Scientists believe the meteorites are blasted off the surface of Mars during energetic impact events. The SNCs provide the only samples of the Martian surface that scientists can analyze in their laboratories because none of the Mars missions have yet returned surface material to Earth.

The only Martian meteorite with an ancient formation age (4.5 billion years) was discovered in Antarctica in 1984. Analyses of **carbonate minerals** in the meteorite in 1996 revealed chemical residues that some scientists interpret as evidence of ancient bacteria on Mars. This discovery is still very controversial among scientists but it has raised the question of whether conditions on early Mars were conducive to the development of primitive life. This is a question that many future missions hope to address.

isotopic ratios the naturally occurring proportions between different isotopes of an element

carbonate a class of minerals, such as chalk and limestone, formed by carbon dioxide reacting in water

minerals crystalline arrangements of atoms and molecules of specified proportions that make up rocks

Recent and Future Missions to Mars

Since 1997, spacecraft missions have made several new discoveries about Mars that have continued to support the hypothesis that the planet was warmer, wetter, and more active at times in the past. In 1997 the Mars Pathfinder mission landed on the surface of Mars in the mouth of one of the channels. The mission included a small rover called Sojourner, which was able to analyze a variety of rocks near the landing site. Sojourner revealed that the rocks display a variety of compositions, some of which suggest much more complicated geologic processes than scientists previously believed occurred on Mars. Images from the Mars Pathfinder cameras also suggest that more water flowed through this area than previously believed, increasing the estimates for the amount of water that has existed on the surface of the planet.

In late 1997 the Mars Global Surveyor (MGS) spacecraft began orbiting Mars. This mission is providing new information about atmospheric circulation, dust storm occurrence, and surface properties. MGS has provided scientists with the first detailed topography map of the planet. One of the major results of the topography map is that the northern plains are extremely smooth, a condition encountered on Earth only on sediment-covered ocean floors. This smooth surface, together with better definition of the previously proposed shorelines, lends further support to the idea that an ocean existed in the northern plains. A spectrometer on MGS revealed a large deposit of hematite in the heavily cratered highlands. Hematite is a mineral that is commonly formed by chemical reactions in hot, water-rich areas. Other instruments on MGS have determined that although Mars does not have an active magnetic field today, there was one in the past, as indicated by the remnant magnetization of some ancient rocks. This ancient magnetic field could have protected the early atmosphere from erosion by **solar wind** particles. Finally, the MGS cameras are revealing evidence of sedimentary materials in the centers of old craters and have found gullies formed by recent seepage of groundwater along the slopes of canyons and craters. Crater evidence suggests that some of the volcanoes have been active to more recent times than previously thought, suggesting that heat may be interacting with subsurface water even today. Such **hydrothermal** regions are known to be areas where life tends to congregate on Earth—could Martian biota have migrated underground and formed colonies around similar hydrothermal areas? Scientists do not know but there is much speculation about such a scenario.

The Mars Odyssey spacecraft successfully arrived at Mars in October 2001 and by January 2002 the spacecraft had settled into its final orbit. Its instruments are reporting strong spectroscopic evidence of near-surface ice across most of the planet.

Our view of Mars has changed dramatically from that of a cold, dry, geologically dead world to a warm, wet, oasis where life may have arisen and may yet thrive in certain locations. Several missions are planned in the next few years by the United States, the European Space Agency, Russia, and Japan to further explore Mars. These missions include a variety of orbiters, landers, rovers, and sample-return missions, which will allow scientists to answer additional questions about the history and future of Mars. Eventually humans will likely become directly involved in the exploration of Mars,

solar wind a continuous, but varying, stream of charged particles (mostly electrons and protons) generated by the Sun; it establishes and affects the interplanetary magnetic field; it also deforms the magnetic field about Earth and sends particles streaming toward Earth at its poles

hydrothermal relating to water at high temperature

and colonies may be established so that Mars can become our stepping-stone to further exploration of the universe. SEE ALSO EXPLORATION PROGRAMS (VOLUME 2); GOVERNMENT SPACE PROGRAMS (VOLUME 2); KEPLER, JOHANNES (VOLUME 2); LIFE IN THE UNIVERSE, SEARCH FOR (VOLUME 2); NASA (VOLUME 3); PLANETARY PROTECTION (VOLUME 4); PLANETARY EXPLORATION, FUTURE OF (VOLUME 2); ROBOTIC EXPLORATION OF SPACE (VOLUME 2); SAGAN, CARL (VOLUME 2).

Nadine G. Barlow

Bibliography

Kieffer, Hugh H., Bruce M. Jakosky, Conway W. Snyder, and Mildred S. Matthews. *Mars.* Tucson: University of Arizona Press, 1992.

Raeburn, Paul *Mars: Uncovering the Secrets of the Red Planet.* Washington, DC: National Geographic Society, 1998.

Mercury

Mercury is the innermost and second smallest planet (4,878 kilometers [3,024 miles] in diameter) in the solar system (Pluto is the smallest). It has no known moons. As of the beginning of the twenty-first century, Mariner 10 had been the only spacecraft to explore the planet. It flew past Mercury on March 29 and September 21, 1974, and on March 16, 1975. Mariner 10 imaged only about 45 percent of the surface and only in moderate detail. As a consequence, there are still many questions concerning the history and evolution of Mercury. Two new missions to Mercury will be launched this decade. An American mission called MESSENGER will be launched in March 2004. It will make two **flybys** of Venus and two of Mercury before going into Mercury orbit in April 2009. A European mission called Bepi Colombo, after a famous Italian celestial dynamicist, is scheduled for launch in 2009.

Motion and Temperature

Mercury has the most **elliptical** and inclined (7 degrees) orbit of any planet except Pluto. Its average distance from the Sun is only 0.38 **astronomical unit** (AU). Because of its elliptical orbit, however, the distance varies from 0.3 AU when it is closest to the Sun to 0.46 AU when it is farthest away. Mercury's **orbital velocity** is the greatest in the solar system and averages 47.6 kilometers per second (29.5 miles per second). When it is closest to the Sun, however, it travels 56.6 kilometers per second (35.1 miles per second), and when it is farthest away it travels 38.7 kilometers per second (24 miles per second).

Mercury's rotational period is 58.6 Earth days and its orbital period is 87.9 Earth days. It has a unique relationship between its rotational and orbital periods: It rotates exactly three times on its axis for every two orbits around the Sun. Because of this relationship, a solar day (sunrise to sunrise) lasts two Mercurian years, or 176 Earth days.

Because Mercury is so close to the Sun, has no insulating atmosphere, and has such a long solar day, it experiences the greatest daily range in surface temperatures (633°C [1,171°F]) of any planet or moon in the solar sys-

flyby flight path that takes the spacecraft close enough to a planet to obtain good observations; the spacecraft then continues on a path away from the planet but may make multiple passes

elliptical having an oval shape

astronomical unit the average distance between Earth and the Sun (152 million kilometers [93 million miles])

orbital velocity velocity at which an object needs to travel so that its flight path matches the curve of the planet it is circling; approximately 8 kilometers (5 miles) per second for low-altitude orbit around Earth

Mercury as viewed by
Mariner 10 on its first approach in March 1974.

tem. Mercury's maximum surface temperature is about 450°C (842°F) at the
equator when it is closest to the Sun, but drops to about −183°C (−297°F)
at night.

Interior and Magnetic Field

Mercury's internal structure is unique in the solar system. Mercury's small
size and relatively large mass (3.3×10^{23} kilograms [7.3×10^{23} pounds])

Mercury's opposite hemisphere viewed by Mariner 10 as it left the planet on the first encounter.

uncompressed density the lower density a planet would have if it did not have the force of gravity compressing it

means that it has a very large density of 5.44 grams per cubic centimeters (340 pounds per cubic foot), which is only slightly less than Earth's (5.52 grams per cubic centimeter [345 pounds per cubic foot]) and larger than Venus's (5.25 grams per cubic centimeter [328 pounds per cubic foot]). Because of Earth's large internal pressures, however, its **uncompressed density** is only 4.4 grams per cubic centimeter (275 pounds per cubic foot), compared to Mercury's uncompressed density of 5.3 grams per cubic centimeter (331 pounds per cubic foot). This means that Mercury contains a much larger fraction of iron than any other planet or moon in the solar system. The iron core must be about 75 percent of the planet diameter, or some 42 percent of its volume. Thus, its rocky outer region is only about 600 kilometers (370 miles) thick.

Mercury is the only **terrestrial planet**, aside from Earth, with a significant magnetic field. The maintenance of terrestrial planet magnetic fields is thought to require an electrically conducting fluid outer core surrounding a solid inner core. Therefore, Mercury's magnetic field suggests that Mercury currently has a fluid outer core of unknown thickness.

Exosphere

Mercury has an extremely tenuous atmosphere with a surface pressure a trillion times less than Earth's. This type of tenuous atmosphere is called an exosphere because atoms in it rarely collide. Mariner 10 identified the presence of hydrogen, helium, and oxygen in the atmosphere and set upper limits on the abundance of argon. These elements are probably derived from the **solar wind**. Later Earth-based telescopic observations detected sodium and potassium in quantities greater than the elements previously known. Sodium and potassium could be released from surface rocks by their interaction with **solar radiation** or by impact vaporization of **micrometeoroid** material. Both sodium and potassium show day-to-day changes in their global distribution.

Polar Deposits

High-resolution **radar** observations show highly reflective material concentrated in permanently shadowed portions of craters at the polar regions. These deposits have the same radar characteristics as water ice. Mercury's rotation axis is almost perpendicular to its orbit, and therefore Mercury does not experience seasons. Thus, temperatures in permanently shaded polar areas should be less than $-161°C$ ($-258°F$). At this temperature, water ice is stable, that is, it is not subject to evaporation for billions of years. If the deposits are water ice, they could originate from comet or water-rich asteroid impacts that released the water, which was then cold-trapped in the permanently shadowed craters. Sulfur has also been suggested as a possible material for these deposits.

Geology and Composition

In general, the surface of Mercury can be divided into four major terrains: heavily cratered regions, **intercrater plains**, **smooth plains**, and hilly and lineated terrain. The heavily cratered uplands record the **period of heavy meteoroid bombardment** that ended about 3.8 billion years ago.

The largest relatively fresh impact feature seen by Mariner 10 is the Caloris basin, which has a diameter of 1,300 kilometers (806 miles). The floor structure consists of closely spaced ridges and troughs.

Directly opposite the Caloris basin (the **antipodal** point) is the unusual hilly and lineated terrain that disrupts preexisting landforms, particularly crater rims (see top image on following page). The hilly and lineated terrain is thought to be the result of seismic waves generated by the Caloris impact and focused at the antipodal region.

Mercury's two plains units have been interpreted to be old lava flows. The older intercrater plains are the most extensive terrain on Mercury (see

terrestrial planet a small rocky planet with high density orbiting close to the Sun; Mercury, Venus, Earth, and Mars

solar wind a continuous, but varying, stream of charged particles (mostly electrons and protons) generated by the Sun

solar radiation total energy of any wavelength and all charged particles emitted by the Sun

micrometeoroid any meteoroid ranging in size from a speck of dust to a pebble

radar a technique for detecting distant objects by emitting a pulse of radio-wavelength radiation and then recording echoes of the pulse off the distant objects

intercrater plains the oldest plains on Mercury that occur in the highlands and that formed during the period of heavy meteoroid bombardment

smooth plains the youngest plains on Mercury; they have a relatively low impact crater abundance

period of heavy meteoroid bombardment the earliest period in solar system history (more than 3.8 billion years ago) when the rate of meteoroid impact was very high compared to the present

antipodal at the opposite pole; two points on a planet that are diametrically opposite

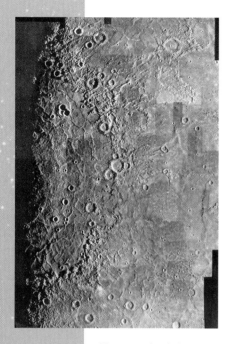

Photomosaic of the 1,300-kilometer (806-mile) diameter Caloris impact basin showing the highly ridged and fractured nature of its floor.

lunar maria one of the large, dark, lava-filled impact basins on the Moon thought by early astronomers to resemble seas

anomalies phenomena that are different from what is expected

bottom image on this page). The intercrater plains were created during the period of late heavy meteoroid bombardment. They are thought to be volcanic plains erupted through a fractured crust. They are probably about 4 to 4.2 billion years old.

The younger smooth plains are primarily associated with large impact basins. The largest occurrence of smooth plains fill and surround the Caloris basin, and occupy a large circular area in the north polar region that is probably an old impact basin about 1,500 kilometers (930 miles) in diameter. They are similar to the **lunar maria** and therefore are believed to be lava flows that erupted relatively late in Mercurian history. They may have an average age of about 3.8 billion years. If so, they are, in general, older than the lunar maria.

Three large radar-bright **anomalies** have been identified on the unimaged side of Mercury. High-resolution radar observations indicate that two of these are similar to the radar signature of a fresh impact crater, and another has a radar signature unlike any other in the solar system. One or both of these craters could account for the polar deposits if they were the result of comets or water-rich asteroid impacts.

Mercury displays a system of compressive **faults** (or **thrust faults**) called **lobate scarps**. They are more-or-less uniformly distributed over the part of Mercury viewed by Mariner 10. Presumably they occur on a global scale. This suggests that Mercury has shrunk. Stratigraphic evidence indicates that the faults formed after the intercrater plains relatively late in Mercurian history. The faults were probably caused by a decrease in Mercury's size due to cooling of the planet. The amount of radius decrease is estimated to have been about 2 kilometers (1.2 miles).

Very little is known about the surface composition of Mercury. A new color study of Mariner 10 images has been used to derive some compositional information of the surface over some of the regions viewed by Mariner 10. The smooth plains have an iron content of less than 6 percent by weight, which is similar to the rest of the regions imaged by Mariner 10. The surface of Mercury, therefore, may have a more homogeneous distribution of elements that affect color than does the Moon. At the least, the smooth plains may be low-iron **basalts**. The MESSENGER mission is designed to accurately determine the composition of the surface.

Geologic History

Knowledge about Mercury's earliest history is very uncertain. The earliest known events are the formation of the intercrater plains (more than 4 billion years ago) during the period of heavy meteoroid bombardment. These plains may have been erupted through **fractures** caused by large impacts in a thin crust. Near the end of heavy bombardment the Caloris basin was formed by a large impact that caused the hilly and lineated terrain from seismic waves focused at the antipodal region. Eruption of lava within and surrounding the large basins formed the smooth plains about 3.8 billion years ago. The system of lobate scarps formed after the intercrater plains, and resulted in a planetary radius decrease of about 2 kilometers (1.2 miles). Scientists will have to await the results of the MESSENGER and Colombo missions to fully evaluate the geologic history of Mercury.

A portion of the hilly and lineated terrain antipodal to the Caloris impact basin. The image is 543 kilometers (337 miles) across.

Origins

How Mercury acquired such a large fraction of iron compared to the other terrestrial planets is not well determined. Three hypotheses have been put forward to explain the enormous iron core. One involves an enrichment of iron due to dynamical processes in the innermost part of the solar system. Another proposes that intense bombardment by solar radiation in the earliest phases of the Sun's evolution vaporized and drove off much of the rocky fraction of Mercury, leaving the core intact. A third proposes that a planet-sized object impacted Mercury and blasted away much of the planet's rocky mantle, again leaving the iron core largely intact. Discriminating among these hypotheses may be possible from the chemical makeup of the surface because each one predicts a different composition. MESSENGER is designed to measure the composition of Mercury's surface, so it may be possible to answer this vital question in the near future. SEE ALSO EXPLORATION PROGRAMS (VOLUME 2); PLANETARY EXPLORATION, FUTURE OF (VOLUME 2); ROBOTIC EXPLORATION OF SPACE (VOLUME 2).

Robert G. Strom

Bibliography

"The Planet Mercury: *Mariner 10* Mission." (various papers and authors) *Journal of Geophysical Research* 80, no. 17 (1975): 2342–2514.

Strom, Robert G. *Mercury: The Elusive Planet.* Washington, DC: Smithsonian Institution Press, 1987.

———. "Mercury: An Overview." *Advances in Space Research* 19, no. 10 (1997): 1,471–1,485.

———. "Mercury." In *Encyclopedia of the Solar System*, eds. Weissman, P. R., L. McFadden, and T. V. Johnson. San Diego: Academic Press, 1999.

Villas, Faith, Clark R. Chapman, and Mildred S. Matthews, eds. *Mercury.* Tucson: University of Arizona Press, 1988.

Meteorites

Most people have looked up into the night sky and seen the fleeting flashes of light that are known as meteors. These flashes are caused by small sand-sized particles that are debris from comets, which melt in the atmosphere and never reach the surface of Earth. Sometimes these flashes come in showers, such as the famous Perseid meteor shower, which occurs from July 23 to August 22 when Earth crosses the debris-strewn orbit of comet Swift-Tuttle.

Meteorites, on the other hand, are extraterrestrial material that have made it to Earth's surface and can weigh many tons. This material is not related to comets but rather to other astronomical bodies. Deceleration of meteorites begins high in the atmosphere where the surface of the incoming body heats up to **incandescence** causing melting and **ablation** and forming a (usually) black fusion crust on the exterior. Whether a meteoroid makes it to Earth's surface (and becomes a meteorite) or not depends on many factors including the mass, initial velocity, angle of entry, composition, and shape of the body. Like the Moon, Earth has been subjected in the past to periods of intense meteorite bombardment, but fortunately many incoming meteoroids disintegrate well up in the atmosphere.

fault a fracture in rock in the upper crust of a planet along which there has been movement

thrust fault a fault where the block on one side of the fault plane has been thrust up and over the opposite block by horizontal compressive forces

lobate scarp a long sinuous cliff

basalts dark, volcanic rock with abundant iron and magnesium and relatively low silica common on all of the terrestrial planets

fractures any break in rock, ranging from small "joints" that divide rocks into planar blocks (such as that seen in road cuts) to vast breaks in the crusts of unspecified movement

incandescence glowing due to high temperature

ablation removal of the outer layers of an object by erosion, melting, or vaporization

The largest meteorite on Earth, weighing some 60 tons, is called Hoba and lies where it fell in Namibia. There are various other meteorite giants, including Chaco (Argentina), weighing 37 tons; Ahnighito (Greenland), 31 tons; and Bacubirito (Mexico), 22 tons.

The orbits of five recovered meteorites are shown in the figure below. Their orbits suggest that their origin lies in the asteroid belt between Mars and Jupiter. These orbits were calculated from photographs taken by networks of cameras in Europe, Canada, and the United States.

In 1982 a 31-gram (1.1-ounce) meteorite discovered in the Allan Hills region of Antarctica was determined to be from the Moon. Since then, more than twenty fragments of the Moon and more than twenty fragments of the planet Mars have been found on Earth.

radioisotope a naturally or artificially produced radioactive isotope of an element

One interesting thing that can be done with meteorites, and one of the reasons why they are so important to science, is to date them by **radioisotope** methods. It turns out that most meteorites have a formation age around 4.56 billion years, when the solar nebula (the hot swirling cloud of dust and gas from which the Sun formed) began to cool enough for solid material, and hence planets, to form. Thus, meteorites represent a fossil record of the early conditions of the solar nebula.

The orbits of five recovered meteorites, showing their connection with the asteroid belt. *Figure adapted from Science Graphics.*

Meteorite Craters

A consequence of large meteorites striking Earth is the formation of craters. This occurs when a large body weighing in excess of 100 tons strikes Earth's

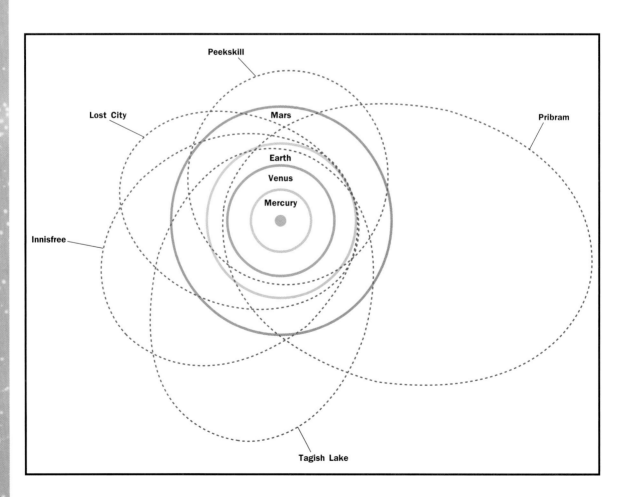

An aerial view of Meteor Crater, which is 1.2 kilometers wide and over 200 meters deep. Tons of nickel-iron meteorite debris have been found in the surrounding area from the original 50-meter-wide impactor.

surface at sufficiently high velocity. The **kinetic energy** of the meteorite is converted to heat, which vaporizes the surrounding rock as well as much of the meteorite, producing an explosion equivalent to a large nuclear device. One of the most famous craters is Meteor Crater (also called Barringer Crater) near Flagstaff, Arizona, which is about 50,000 years old and 1.2 kilometers (0.7 miles) in diameter. The impacting iron mass was approximately 50 meters (164 feet) across, and the consequence of it striking Earth at 60,000 kilometers per hour (37,200 miles per hour) can be seen in the accompanying image.

kinetic energy the energy an object has due to its motion

Meteor Crater is not particularly large among the 150 or so **impact craters** that exist on Earth, and even larger ones can produce global climatic and environmental changes. The asteroid that is thought to have wiped out the dinosaurs was about 10 kilometers (6.2 miles) in diameter and struck off the coast of Yucatan, Mexico, 65 million years ago, producing a crater 300 kilometers (186 miles) in diameter. This event is theorized to have created enormous amounts of dust, which blocked out the Sun, possibly for years, and led to the extinction of 75 percent of all living species.

impact craters bowl-shaped depressions on the surfaces of planets or satellites that result from the impact of space debris moving at high speeds

Can such an event happen in modern times? Since Earth is actually orbiting the Sun through a swarm of solar system debris, the answer has to be yes. In fact, in 1908 there was an enormous atmospheric explosion above Tunguska, Siberia. The resulting blast leveled 2,000 square kilometers (770 square miles) of forest, and the shock wave circled the globe. Such an event is predicted to happen once every few hundred years or so. As recently as 1947, the Sikhote-Alin meteorite crashing north of Vladivostok, Russia, made an array of craters, some of which were one-fourth the size of a football field.

Asteroids Turned Meteorites

Tens of thousands of small bodies called asteroids are found in the asteroid belt, with the largest one, Ceres, discovered in 1801, being about 930 kilometers (575 miles) in diameter. Most of these are in stable orbits around the Sun and are really just small planets, or planetoids. From time to time, these asteroids crash into one another, sending their fragments in all directions. But there are some empty zones in the asteroid belt, known as Kirkwood gaps (after American astronomer Daniel Kirkwood), which are caused by a special gravitational relationship with Jupiter. If some asteroid fragments from a collision are thrown into one of these gaps, Jupiter's enormous gravity has the effect of sending them into a more elliptically shaped orbit (as seen in the figure on page 104) that can intersect Earth's orbit. That is how fragments of the asteroid belt can end up crashing into Earth as meteorites.

One way to study asteroids is to measure the intensity of sunlight at different **wavelengths** reflecting off their surfaces. This is then compared to the light reflected off pulverized meteorites in the laboratory. Reflectance **spectra** from various asteroids can be matched with different types of meteorites, further strengthening the connection between asteroids and meteorites.

There are three basic types of meteorites: stones, stony-irons, and irons. Stones are divided into two main subcategories: chondrites and achondrites. Chondrites are the main type of stony meteorite, constituting 84 percent of all witnessed meteorite falls. Most chondrites are characterized by small spherical globules of silicate, known as chondrules. Interestingly, carbonaceous chondrites also contain organic compounds such as amino acids, which may have contributed to the origin of life on Earth. Chondrites are the most primitive of the meteorites, suffering little change since their origin. Achondrites, on the other hand, come from chondritic parent bodies that have been heated to the melting point, destroying their chondrules and separating heavy and light **minerals** into a core and mantle. These are known as differentiated meteorites. Early volcanism occurred on the surface of their parent bodies forming a thin crust. A subcategory of achondrites called SNC achondrites are believed to have come from Mars.

Stony-irons are a metal-silicate mixture. Meteorites from one subcategory, the pallasites, contain large crystals of the mineral olivine imbedded in a matrix of metal. These are thought to form at the boundary of the molten metal core of an asteroid and its olivine-bearing silicate mantle.

Irons are actually alloys of mostly iron with a small percentage of nickel. As the liquid metal core of an asteroid slowly cools over a period of millions of years, the different alloys of nickel-iron (kamacite and taenite) form an intertwining growth pattern known as a Widmanstätten pattern, which is indicative of extraterrestrial iron meteorites.

Meteorites are true extraterrestrials, valuable not only to science but also to the discoverer. If you happen to find a piece of the Moon lying on the ground (as some people have), you can plan your retirement from that day onward. Today, a thriving market exists as an increasing number of new meteorites are being discovered yearly, many finding their way to the marketplace. A growing number of aficionados eagerly await these new discoveries. SEE ALSO Asteroids (volume 2); Comets (volume 2); Close Encounters (volume 2); Impacts (volume 4); Mars (volume 2); Moon (volume 2).

Joel L. Schiff

Bibliography

Bagnall, Philip M. *The Meteorite and Tektite Collector's Handbook.* Richmond, VA: Willmann-Bell, Inc., 1991.

Hutchison, Robert, and Andrew Graham. *Meteorites.* New York: Sterling Publishing Co., Inc., 1993.

McSween, Harry Y., Jr. *Meteorites and Their Parent Planets*, 2nd ed. Cambridge, UK: Cambridge University Press, 1999.

Norton, O. Richard. *Rocks from Space*, 2nd ed. Missoula, MT: Mountain Press, 1998.

wavelength the distance from crest to crest on a wave at an instant in time

spectra representations of the brightness of objects as a function of the wavelength of the emitted radiation

minerals crystalline arrangements of atoms and molecules of specified proportions that make up rocks

A cut slab of a typical ordinary chondrite showing a field of chondrules.

Microgravity

Gravity is an omnipresent force in our lives. Without it, water from a drinking fountain would simply shoot up from the spout without arcing into the fountain again. Chocolate syrup on a sundae would stay put without dripping down a scoop of ice cream. In fact, gravity, the force of attraction that draws one object to another, is so powerful on Earth that scientists sometimes have to get away from its influence—if only for a short while—to better understand other forces at work in the universe. To do this, they must be in a microgravity environment.

While it may appear that these astronauts are floating, they are actually in a state known as "freefall."

Microgravity, where the effects of gravity are minimized (approximating one millionth that of Earth's normal gravity), is achieved during freefall. At first glance, astronauts working on the International Space Station may appear to be floating. In fact, they are in freefall inside the spacecraft, which is also in freefall. To understand this phenomenon, it may help to think through a mental experiment by English physicist and mathematician Isaac Newton (1642–1727). He understood that the force that causes apples and other objects to fall to the ground is the same force that holds celestial bodies such as the Moon in orbit. If a cannon is fired from atop a high hill, the cannonball will fall to Earth, landing some distance away. If more force is used, the cannonball travels farther before hitting the ground. If the cannonball is propelled with enough force, it will fall all the way around Earth, orbiting the planet, just as the Space Station or any space shuttle does.

Scientists who have conducted experiments in microgravity have discovered countless phenomenon that they would not see in normal gravity. For example, during space shuttle flight STS-95, which carried Senator John Glenn back into orbit in 1998, scientists saw ordered crystals of two different sizes of particles form together in one solution. On Earth, where metals (such as copper and zinc) are melted together to form alloys (such as brass), materials scientists contend with buoyant convection, which is fluid flow that causes denser particles to sink and less dense particles to rise. **Convection** makes it more difficult to blend uniform alloys and other materials.

convection the movement of heated fluid caused by a variation in density; hot fluid rises while cool fluid sinks

Convection also affects how a flame burns. On Earth, gravity pulls cooler, denser air closer to the planet, causing soot and hot, less dense flame gases to rise. This can lead to an unsteady, flickering flame. In microgravity, a candle flame produces minimal soot for a brief time then appears spherical and blue. American combustion researchers found on the Russian space station Mir in 1998 that while a flame in microgravity does need airflow to burn, as it does on Earth, that flow is only a fraction of a centimeter per second, so small one would not feel it. The findings confirmed that materials considered to be flame-resistant on Earth might burn in low-gravity conditions in space.

atrophy to wither, shrink, or waste away

porous allowing the passage of a fluid or gas through holes or passages in the substance

As astronauts learn how physical phenomena are affected in microgravity, they are also finding out how the microgravity environment affects their own bodies. For example, during long-duration flights, such as on the International Space Station, human muscles begin to **atrophy** and bones can become more **porous** as they do in someone with **osteoporosis**. Scientists are researching methods of exercise and bone-replacement therapy that will help astronauts stay in top condition as they continue their discoveries of how forces behave with—or without—gravity. SEE ALSO GRAVITY (VOLUME 2);

osteoporosis the loss of bone density; can occur after extended stays in space

Newton, Isaac (volume 2); Living in Space (volume 3); Living on Other Worlds (volume 4); Long-Duration Spaceflight (volume 3); Zero Gravity (volume 3).

Julie A. Moberly

Bibliography

National Aeronautics and Space Administration, Microgravity Research Division. *Combustion Science.* Huntsville, AL: Author, 1995.

———. *Fluid Physics.* Huntsville, AL: Author, 1995.

Internet Resources

Ducheyne, Paul. "Surface Transformation of Reactive Glass in a Microgravity Environment." *Microgravity Research Division's Online Task Book.* 1999. <http://peer1 .idi.usra.edu/cfpro/peer_review/mtb1_99.cfn?id=269>.

Microgravity Research Program Page. National Aeronautics and Space Administration. <http://microgravity.nasa.gov/>.

An Atlas II rocket takes off from the Cape Canaveral Air Force Station carrying a communications satellite for the United States Navy. The satellite is part of the Navy's global communications network.

ballistic the path of an object in unpowered flight; the path of a spacecraft after the engines have shut down

Military Exploration

Among the different reasons for sending space probes and satellites into orbit is the use of the space environment for defensive purposes. Military equipment such as missiles, rockets, and communications systems were among the first hardware used in the early space programs. Gradually, civil and commercial space projects developed their own purpose-built spacecraft. But the military continues to have a dominant place in the space programs of the United States, China, and Russia. The principal launching sites for rockets in all three countries are military bases, and military ships and planes are used for tracking and communications during rocket launches.

The French commercial launching site in Kourou, French Guiana, had its origins as a French military base in the 1950s. The military forces of Israel, Brazil, India, and North Korea have also been major influences in the origins and evolution of these nations' scientific and commercial space programs.

In the United States, the rockets used in the civil, commercial, and military space programs had their origins as **ballistic** missiles and later were first used for space purposes by the military. The first U.S. space rockets, derived from German V-2 rockets captured by the military following the end of World War II (1939–1945), were tested and flown by the U.S. Army from a military base at White Sands, New Mexico. The rocket that carried the first attempted launch of a U.S. satellite, the Vanguard, was developed by the U.S. Navy. The U.S. Air Force developed subsequent intercontinental ballistic missiles. The R-7 ballistic missile developed by the Soviet military has been adapted as a launching rocket and is still flying today.

Military forces have developed several different types of satellites in various types of orbits in space. These include communications satellites such as the Defense Space Communication System and Milstar, navigation satellites such as the Global Positioning System (GPS), early warning satellites such as the Defense Support Program satellites, and weather satellites such as the Defense Meteorological Satellite Program. During the Gulf

War in 1991, space satellites, including secret **reconnaissance** and surveillance craft, were used by coalition-deployed forces for communicating among force locations and for tracking Scud missiles fired by the Iraqi government.

In 1983 President Ronald Reagan proposed a major expansion of the military use of space in his Strategic Defense Initiative. The project called for the development of a space-based warning, tracking, and intercept system to destroy missiles attacking the United States. The program lasted from 1983 to 1993 and was discontinued following the collapse of the Soviet Union.

The administration of President George H. W. Bush proposed a limited space defense system in 1989 called Global Protection against Limited Strike. This system was to feature a fleet of orbiting attack craft called Brilliant Pebbles. The "Pebbles" carried no explosive equipment but would destroy incoming missiles by colliding with them as they entered space. This project was also canceled when President Bill Clinton entered office in 1993.

A more limited space-based tracking and laser attack system is being researched by the administration of George W. Bush to defend the continental United States from a limited ballistic missile attack from Third World nations. The first test flight of a prototype antimissile space laser is set for 2012. SEE ALSO GOVERNMENT SPACE PROGRAMS (VOLUME 2); MILITARY CUSTOMERS (VOLUME 1); MILITARY USES OF SPACE (VOLUME 4); RECONNAISSANCE (VOLUME 1); SATELLITES, TYPES OF (VOLUME 1).

Frank Sietzen, Jr.

Internet Resources

DefenseLINK. U.S. Department of Defense. <http://www.defenselink.mil/>.

Encylopedia Astronautica. <http://www.astronautix.com/>.

Federation of American Scientists. <http://www.fas.org/>.

Moon

Our solitary and prominent Moon orbits Earth at a mean distance of only 382,000 kilometers (236,840 miles). The nearest planet, Venus, is never closer than 40 million kilometers (25 million miles). The Moon's mass is just under one-eightieth that of Earth, its volume just over one-fiftieth; the difference mainly stems from the Moon lacking a large metallic iron core and therefore having a much lower overall density than Earth. Its low mass is responsible for the low surface gravity (one-sixth that at Earth's surface), popularly recognized in the jumping, bouncing gait of Apollo astronauts. The mass is much too low for the Moon to hold any significant atmosphere—it is essentially in a **vacuum**—or for its surface to have liquid water.

The surface area of the Moon is only about four times that of the land area of the United States. The Moon is not as large as any planet other than distant little Pluto but is of the same scale as the Galilean satellites of Jupiter. These moons are much smaller in comparison with the planet they orbit.

reconnaissance a survey or preliminary exploration of a region of interest

vacuum an environment where air and all other molecules and atoms of matter have been removed

Basins formed on the Moon 4 billion years ago, and were subsequently filled with basaltic lava flows. The Orientale basin, 600 miles across, is near the center, while the area at the upper right is the large, dark Oceanus Procellarum.

Earth's Moon is very different in chemical composition and structure—and probably origin—from any other body in the solar system.

Orbit and Rotation

The 29.53-day orbit provides us with the lunar phases, as well as the occasional eclipses of the Sun and the more frequent eclipses of the Moon. The orbit is tilted only slightly (5.1°) from the plane of the **ecliptic**, but because Earth itself has a tilted axis of rotation (23.5°), the Moon's orbit is tilted substantially with respect to Earth's equator. The Moon's own axial rotation period is exactly the same as its orbital period, and so it shows almost the same face to Earth continuously. It is not exactly the same face because of the tilt of the Moon's rotational axis (1.5°) to its orbital plane around Earth, and the slight ellipticity of that orbit (the position of the observer on Earth also has a slight effect). Altogether, only 41 percent of the Moon's surface is permanently invisible to observers on Earth.

ecliptic the plane of Earth's orbit

The gravitational pull of the Moon provides the twice-daily tides on Earth as Earth spins under the Moon. The Moon is gradually receding because of the tidal effects. As the Moon recedes, its **angular momentum** increases, compensated by a decrease in the spin rate of Earth. Thus, Earth's day is increasing in length; 600 million years ago it was only about eighteen hours long. The Moon stabilizes the tilt of Earth's own axis of rotation over long periods of time, and this has been important for stabilizing climate and thus life habitats.

angular momentum the angular equivalent of linear momentum; the product of angular velocity and moment of inertia (moment of inertia = mass × radius²)

This is a false-color composite image of the Moon, photographed through three color filters by the Galileo spacecraft. The colors aid in the interpretation of the satellite's surface soil composition: red areas typically correspond to lunar highlands; orange to blue shades suggest ancient volcanic lava flow or lunar sea; and purple sections indicate pyroclastic deposits.

The Exploration of the Moon

Even to the naked eye the Moon's face has darker and lighter patches. Italian mathematician and astronomer Galileo Galilei used a telescope in 1610 to discover its rugged, varied, and essentially unchanging features. He distinguished the brighter areas as higher and more rugged, the darker as lower, flatter, and smoother. He called the former "terra" (meaning "land"; pl. "terrae") and the latter "mare" (meaning "sea"; pl. "maria"), although that is not what they are.

For three centuries the Moon remained an object of astronomical study, with the collection of data about its shape, size, movements, and surface physical properties, as well as mapping. Not until the middle of the twentieth century were observations and a combination of natural and terrestrial analogs advanced enough that the volcanic origin of its dark plains and the impact origin of its craters and basins could be considered as settled. In the 1960s, a program of geological mapping, using techniques such as crater counting and overlapping relationships, confirmed and elucidated the nature of geological units and the order in which they were produced.

The study of the Moon reached peak activity in the space age, when spacecraft sent back detailed information from orbiters, **hard-landers**, and **soft-landers** (mainly from 1959 to 1970), and Apollo astronauts conducted experiments and made observations from **equatorial orbit** and at the surface (from 1968 to 1972). Six Apollo missions and three robotic sample-

hard-lander a spacecraft that collides with the planet or satellite, making no attempt to slow its descent; also called crash-landers

soft-lander spacecraft that uses braking by engines or other techniques (e.g., parachutes, airbags) such that its landing is gentle enough that the spacecraft and its instruments are not damaged, and observations at the surface can be made

equatorial orbit an orbit parallel to a body's geographic equator

111

BASIC DATA ABOUT THE MOON

Greatest distance from Earth	406,697 km
Shortest distance from Earth	356,410 km
Eccentricity of orbit	0.0549
Rotation period (synodic month)	29.53 Earth days
Rotation period (sidereal month)	27.32 Earth days
Mean orbital inclination to ecliptic	5° 08' 43"
Inclination of rotation axis to orbit plane	1° 32'
Mean orbital velocity	1.68 km/s
Period of revolution of perigee	3,232 Earth days
Regression of the nodes	18.60 years
Mass	7.35×10^{22} kg
Mean Density	3.34 g/cc
Surface gravity	1.62 m/s²
Escape velocity	2.38 km/s
Mean diameter	3,476 km
Mean circumference	10,930 km
Surface area	37,900,000 km²
Albedo (fraction light reflected) terrae	0.11–0.18
Albedo (fraction light reflected) mare	0.07–0.10
Mean surface temperature day	107°C
Mean surface temperature night	−153°C
Mean surface temperature at poles: light	−40°C
Mean surface temperature at poles: dark	−230°C

radiogenic isotope techniques use of the ratio between various isotopes produced by radioactive decay to determine age or place of origin of an object in geology, archaeology, and other areas

flyby flight path that takes the spacecraft close enough to a planet to obtain good observations; the spacecraft then continues on a path away from the planet but may make multiple passes

✴ The Galileo mission successfully used robots to explore the outer solar system. This mission used gravity assists from Venus and Earth to reach Jupiter, where it dropped a probe into the atmosphere and studied the planet for nearly seven years.

return vehicles collected samples of the Moon (from 1970 to 1976). Samples are particularly useful for understanding the processes that created the rocks and for the dating of events using **radiogenic isotope techniques**. Two **flybys** by the Galileo mission ✴ to Jupiter (in 1990 and 1992), the Clementine lunar polar orbiter (in 1994) and the Lunar Prospector polar orbiter (in 1998) have provided substantially more global imaging, topographic, chemical, and mineralogical data.

Global and Interior Characteristics

The Moon is nearly homogeneous, as shown by its motions in space, and by the fact that rocks near the surface are not much different in density from the Moon as a whole. Nonetheless, samples show that the Moon was thoroughly heated at its birth about 4.5 billion years ago, possibly to the point of total melting, and then quickly solidified to produce a comparatively thin (60 to 100 kilometers [37 to 62 miles]) crust of slightly lighter material. This structure was confirmed by seismic experiments performed on the early Apollo missions. There may be an iron core, but if so it is very tiny, and there is no significant magnetic field.

Samples show that the Moon is very depleted in volatile elements (those that form gases and low-temperature boiling-point liquids), to the extent that it lacks any water of its own at all, even bonded into rocks. Water delivered to the Moon by cometary impact might exist, frozen in crater floors near the poles. The Moon is very reduced chemically, such that iron metal exists, but rust (oxidized, ferric iron) does not. The Moon is very depleted in the siderophile elements ("iron-loving") that go with metallic iron into a core, except for the surface rubble to which such elements have been delivered by eons of meteorite impact.

The Uppermost Surface of the Moon

The Moon has been bombarded by meteorites ranging in size from numerous tiny dust particles to rare objects hundreds of kilometers in diame-

Astronaut and geologist Harrison H. Schmitt working at the Taurus-Littrow landing site, where he first spotted orange soil. The lunar surface is covered everywhere with a thin fragmental layer ("regolith") that consists mainly of ground-up and remelted rocks.

ter. The surface is covered everywhere with a thin fragmental layer (known as soil, or "regolith") that consists mainly of ground-up and remelted lunar rocks, with an average grain size of less than 0.1 millimeters (0.004 inch). This soil contains pebbles, cobbles, and even boulders of lunar rocks. A small percentage of the regolith consists of the meteoritic material that did the bombarding. The regolith is about 5 meters (16.5 feet) thick on **basalts** that were poured out about 3 billion years ago, while older surfaces have even

basalts dark, volcanic rock with abundant iron and magnesium and relatively low silica common on all of the terrestrial planets

cosmic radiation high energy particles that enter Earth's atmosphere from outer space causing cascades of mesons and other particles

solar wind a continuous, but varying, stream of charged particles (mostly electrons and protons) generated by the Sun; it establishes and affects the interplanetary magnetic field; it also deforms the magnetic field about Earth and sends particles streaming toward Earth at its poles

minerals crystalline arrangements of atoms and molecules of specified proportions that make up rocks

thicker regoliths. This regolith layer, exposed to **cosmic radiation** and the **solar wind**, contains materials, such as hydrogen, that do not reach the surface of Earth because of its protection by both a magnetic field and an atmosphere.

The Older Crust of the Moon

Much of the crust consists of material that formed within a few tens of millions of years of the Moon's origin, partly by the floating of light (in both density and color) feldspar **minerals**, which crystallized from a vast ocean of silicate magma. The magma formed because of the Moon's rapid formation, and because of the generation of radioactive heat, which was greater then than now. Continued melting and remelting added to the crust, and the final dregs of the crystallizing magma ocean, richer in those elements that do not easily fit into common crystallizing minerals (feldspar, pyroxene, and olivine), also ended up in the crust. The rocks from the dregs are commonly called "KREEP"-rich because they are richer in potassium (K), rare Earth elements (REE) such as lanthanum, and phosphorus (P) than are typical rocks. Most, though not all, of this crust was in place by 4.3 billion years ago.

At its birth and at about 3.9 billion years ago (what happened in the time between remains somewhat unknown) the Moon was subjected to enormous bombardments that created deep basins as well as numerous small craters, partly disrupting the crust. This crust is somewhat thinner on the front side (about 60 kilometers [37 miles]) than on the farside (about 100 kilometers [62 miles]).

The Younger Crust of the Moon

mare dark-colored plains of solidified lava that mainly fill the large impact basins and other low-lying regions on the Moon

spherules tiny glass spheres found in and among lunar rocks

Impacts decreased substantially after 3.8 billion years ago, to a level close to that of today by about 3.2 billion years ago. The Moon's deep basins, partly filled with overlapping thin flows of **mare** basalt, formed from the melting of small amounts of the lunar interior. These basins (150 kilometers [93 miles] to perhaps 500 kilometers [310 miles] deep) are prominent as the dark plains—the maria—of the Moon and show many signs of volcanic flow. Some of the volcanic lava erupted as fiery fountains, forming heaps of glass **spherules**. These lavas comprise only about 1 percent of the crust, but as the latest, topmost rocks, least affected by impacts, they remain clearly visible. They are much less abundant on the lunar farside, and everywhere their formation had ceased by 2 billion years ago. The Moon is now magmatically dead, and its uppermost crust is being continually gardened and converted into regolith.

The Origin of the Moon

isotope ratios the naturally occurring proportions between isotopes of an element

Earth and the Moon show an identical relationship of oxygen **isotope ratios** (oxygen being the most common element in both planets), a relationship that is different from all other measured solar system objects (including Mars) except yEH chondrites. This indicates that Earth and the Moon formed in the same part of the solar system and gives credence to ideas that the Moon formed from Earth materials.

The pre-Apollo ideas of either capture, fission from Earth (by rapid spinning), or formation together as a double planet are not consistent with what scientists now know from geological or sample studies, nor with the orbital and angular momentum constraints. Thus a new concept was developed in the 1980s: Earth collided during its growth with an approximately Mars-sized object, producing an Earth-orbiting disk of material that accumulated to form the Moon. This idea can account for many features, including the chemistry of the Moon, its magma ocean, and even the tilt of Earth's axis. It is compatible with concepts of how planets develop by accumulation of solid objects. One of the implications of this theory is that the Moon actually must have accumulated very rapidly, on the order of days to years, rather than older ideas of tens of millions of years, and this explains the early melting of the Moon. SEE ALSO APOLLO (VOLUME 3); APOLLO LUNAR LANDING SITES (VOLUME 3); EXPLORATION PROGRAMS (VOLUME 2); GALILEI, GALILEO (VOLUME 2); LUNAR BASES (VOLUME 4); LUNAR OUTPOSTS (VOLUME 3); LUNAR ROVERS (VOLUME 3); NASA (VOLUME 3); PLANETARY EXPLORATION, FUTURE OF (VOLUME 2); ROBOTIC EXPLORATION OF SPACE (VOLUME 2); SHOEMAKER, EUGENE (VOLUME 2).

Graham Ryder

Bibliography

Heiken, Grant H., David Vaniman, Bevan M. French, and Jack Schmidt, eds. *The Lunar Sourcebook: A User's Guide to the Moon.* Cambridge, UK: Cambridge University Press, 1991.

Ryder, Graham. "Apollo's Gift: The Moon." *Astronomy* 22, no. 7 (1994):40–45.

Spudis, Paul D. "An Argument for Human Exploration of the Moon and Mars." *American Scientist* 80, no. 3 (1992):269–277.

———. *The Once and Future Moon.* Washington, DC, and London: Smithsonian Institution Press, 1996.

———. "The Moon." In *The New Solar System*, eds. J. Kelly Beatty, Carolyn C. Peterson, and Andrew Chaikin. Cambridge, UK: Cambridge University Press, 1999.

Taylor, G. Jeffrey. "The Scientific Legacy of Apollo." *Scientific American* 271, no. 1 (1994):26–33.

Wilhelms, Donald E. *To a Rocky Moon: A Geologist's History of Lunar Exploration.* Tucson: University of Arizona Press, 1993.

National Aeronautics and Space Administration
See NASA (Volume 3).

Neptune

Neptune is the most distant giant planet, circling the Sun at an average distance of almost 6 billion kilometers (3.7 billion miles; thirty-nine times the distance from Earth to the Sun). Neptune is a near twin to Uranus in size (with a radius of 24,764 kilometers [15,354 miles] at the equator), in composition (about 80 percent hydrogen, 15 percent helium, and 3 percent methane, with other trace elements), and in internal structure (a rocky core surrounded by a methane- and ammonia-rich watery mantle topped by a thick atmosphere).

The icy particles in the upper cloud decks of Neptune differ slightly from those of Uranus. Their color, combined with the atmospheric methane that absorbs red light, gives Neptune a rich sky-blue tint compared with the more greenish Uranus. Neptune has the strongest internal heat source of all the giant planets, radiating almost three times more heat than one would expect. Like Jupiter and Saturn, which radiate about twice as much energy than expected, Neptune is thought to have excess heat from the time of the

Neptune as seen through green and orange camera filters from a Voyager 2 flyby. Visible planetary features include: the Great Dark Spot (planet center); the bright, fast moving feature Scooter (to the west); and the little dark spot (below).

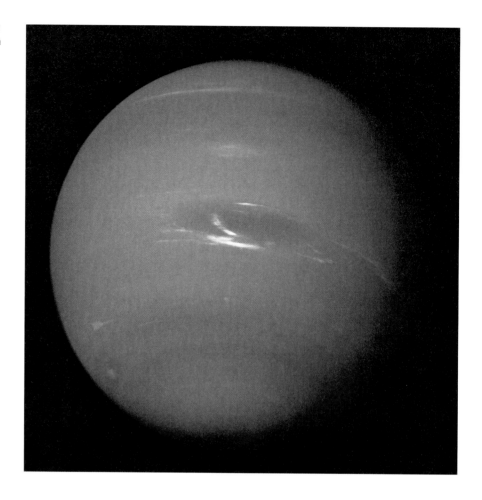

gravitational contraction the collapse of a cloud of gas and dust due to the mutual gravitational attraction of the parts of the cloud; a possible source of excess heat radiated by some Jovian planets

planet's formation and from continued **gravitational contraction**. Neptune's rotational axis is inclined only 29 degrees, compared with Uranus's more than 90 degrees.

A Saga of Discovery

The discovery of Neptune was a mathematical triumph and a political nightmare. After Uranus was discovered in 1781, astronomers inferred the presence of another planet from the shape of the Uranian orbit. In England, astronomer John Adams made meticulous but unpublished calculations of the planet's likely position in 1845. Shortly thereafter, French astronomer Urbain Leverrier independently determined the suspected planet's position, which nearly matched Adams's prediction. After Leverrier's work was published in 1846, English astronomers realized that Adams's work warranted a more serious look. But by then, the French astronomer had sent his prediction to observers in Berlin. Almost immediately, German astronomer Johann Galle discovered Neptune near the predicted location. For years, debates raged across national boundaries over who deserved credit for the discovery of Neptune. We now credit both Leverrier and Adams for the prediction and recognize Galle for the actual observation.

Unusual Cloud Features

In 1989, Voyager 2 flew by Neptune and detected numerous cloud features. The biggest was the Great Dark Spot, a hurricane-like storm that was about half the size of Earth. The next to be discovered was a small white spot, which appeared to race rapidly around the planet when compared with the lumbering Great Dark Spot. It was named the "Scooter." Many more spots were found, many of which were rotating even faster than Scooter. A small dark spot in the south developed a bright core, and a bright clump near the south pole was observed to be composed of many fast-moving bright patches.

Rotation and Magnetic Field

Voyager 2 measured Neptune's 16.11-hour internal rotation period by monitoring the planet's magnetic field. The atmosphere rotates with periods ranging from over 18 hours near the equator to faster than 13 hours near the poles. In fact, the winds of Neptune are among the fastest in the solar system; only Saturn's high-speed equatorial jet is faster. Like the Uranian magnetic field, Neptune's magnetic field is also offset from the planet's center and significantly tilted with respect to the planet's rotation axis. Neptune's field is about 60 percent weaker than that of Uranus.

The Moons of Neptune

Neptune's largest moon, Triton, has a **retrograde** and highly inclined orbit. This suggests the moon may have been captured rather than formed around Neptune. Triton has a thin atmosphere of primarily nitrogen gas, thought to be in equilibrium with the nitrogen ice covering Triton's surface. Because of Triton's unusual orbit, however, the surface ice is thought

retrograde having the opposite general sense of motion or rotation as the rest of the solar system; that is, clockwise as seen from above Earth's north pole

Triton partially obscures Neptune during this Voyager 2 flyby from August 1989. Triton, at 2,706 kilometers (1,683 miles) in diameter, is Neptune's largest moon.

impact craters bowl-shaped depressions on the surfaces of planets or satellites that result from the impact of space debris moving at high speeds

elliptical having an oval shape

prograde having the same general sense of motion or rotation as the rest of the solar system, that is counter-clockwise as seen from above Earth's north pole

shepherding small satellites exerting their gravitational influence to cause or maintain structure in the rings of the outer planets

to change with time, leading to the possibility that Triton's atmosphere also varies. Recent occultations (observations of stars glimmering through Triton's tenuous atmosphere) suggest that Triton's atmosphere may have expanded by nearly a factor of two since the Voyager 2 encounter. Triton's northern hemisphere looks much like the surface of a cantaloupe. The southern hemisphere is dominated by a polar ice cap, probably composed of nitrogen. In the highest resolution images, active geysers (ice volcanoes) were seen spewing columns of dark material many kilometers into the thin atmosphere. Triton's surface has relatively few **impact craters**, suggesting that it is young.

Nereid (the only other Neptune moon known prior to the Voyager 2 mission) also has an unusual orbit that is highly **elliptical** and tilted nearly 30 degrees, again suggesting a capture origin. Little is known about it other than its irregular shape. Voyager 2 discovered six additional moons around Neptune. These are all in circular **prograde** orbits near Neptune's equatorial plane, and they probably formed in place. One of these, Proteus, is larger than Nereid; it had not been discovered prior to the Voyager 2 encounter because it is so close to Neptune. Proteus is irregular in shape. A particularly large impact crater suggests that it came close to destruction in an earlier collision.

The Rings of Neptune

Astronomers used occultations to search for rings around Neptune, because that technique had been successful for discovering the rings of Uranus. The results were odd: some events seemed to clearly show rings, but others clearly did not. The Voyager 2 encounter solved the puzzle. There were three complete rings, but the rings were variable in their thicknesses (the three distinct rings were named Adams, Leverrier, and Galle, after the astronomers who were involved in the discovery saga). The thickest parts—dubbed rings arcs—were seen during occultations; the other parts of the rings were too thin to be detected. Scientists are not sure what causes Neptune's rings arcs. Some of the smallest moons appear to "shepherd" the inner edges of two of the rings, but no moons were found at locations that would explain the clumps through a **shepherding** mechanism. Despite their clumpiness, Neptune's rings are very circular, unlike the rings of Uranus.

Recent Hubble Space Telescope images have continued to show remarkable changes in Neptune's atmosphere: the Great Dark Spot discovered by Voyager 2 in 1989 had disappeared, and a new Great Dark Spot developed in the northern hemisphere. From the dynamics of Neptune's clouds, to the expanding Triton atmosphere, to the forces creating the clumpy rings, many interesting puzzles remain to be solved in the Neptune system. SEE ALSO EXPLORATION PROGRAMS (VOLUME 2); NASA (VOLUME 3); ROBOTIC EXPLORATION OF SPACE (VOLUME 2); URANUS (VOLUME 2).

Heidi B. Hammel

Bibliography

Cruikshank, Dale P., ed. *Neptune and Triton.* Tucson: University of Arizona Press, 1995.

Moore, Patrick. *The Planet Neptune.* Chichester, UK: Ellis Horwood, 1988.

Standage, Tom. *The Neptune File: A Story of Astronomical Rivalry and the Pioneers of Planet Hunting.* New York: Walker and Company, 2000.

Newton, Isaac

British Physicist and Mathematician
1642–1727

Considered one of the greatest scientists of all time, Isaac Newton was a British physicist and mathematician. Born in 1642, the year Italian mathematician and astronomer Galileo Galilei died, Newton's astounding list of contributions include discovering the law of gravity, designing a novel type of reflecting telescope, and writing the landmark work, *The Mathematical Principles of Natural Philosophy* (1687). This seminal volume spelled out the law of gravity, the laws of motion, and the universality of the gravitational force. It was Newton who first realized that white light is made up of the colors of the rainbow, made visible through the prism.

Newton's brilliance is very much in evidence today. For instance, the newton is a unit for force named after him. The Newtonian telescope is a type of reflecting telescope still in popular use. Newton's law of gravity and his laws of motion are at work, evidenced by the trajectory of a spacecraft circling Earth and by the behavior of all astronomical objects, such as the planets within the solar system. Newton's first law of motion is called the law of inertia; a second law concerns acceleration; while a third law states that for every action there is an equal and opposite reaction. Newton died in 1727, receiving recognition at the time for his brilliance. SEE ALSO EINSTEIN, ALBERT (VOLUME 2); GALILEI, GALILEO (VOLUME 2); GRAVITY (VOLUME 2).

Leonard David

Sir Isaac Newton formulated a single law of gravitation based on Kepler's three laws of planetary motion.

Bibliography

Newton, Isaac. *The Principia: Mathematical Principles of Natural Philosophy*, trans. I. Bernard Cohen and Anne Whitman. Berkeley: University of California Press, 1999.

Westfall, Richard. *The Life of Isaac Newton.* New York: Cambridge University Press, 1993.

Observatories, Ground

Astronomers study the universe by measuring electromagnetic radiation—**gamma rays**, **X rays**, optical and **infrared radiation**, and radio waves—emitted by planets, stars, galaxies, and other distant objects. Because Earth's atmosphere is transparent to optical and infrared radiation and to radio waves, these types of radiation can be studied from ground-based observatories. Astronomers must launch telescopes into space in order to study X rays, gamma rays, and other radiation that is blocked by absorption in Earth's atmosphere.

Astronomers make use of ground-based observatories whenever they can. It is about 1,000 times cheaper to build a telescope of a given size on the ground than to launch it into space, so it is much more economical to operate on the surface of Earth.

gamma rays a form of radiation with a shorter wavelength and more energy than X rays

infrared radiation radiation whose wavelength is slightly longer than the wavelength of light

X rays a form of high-energy radiation just beyond the ultraviolet portion of the spectrum

This telescope is housed inside the interior dome of the Mauna Kea Observatory, Hawaii. Situated on a volcano, Mauna Kea is considered one of the world's best sites for optical and infrared astronomy.

A telescope can be thought of as a bucket that collects light or radio waves and brings them to a focus. More light can be gathered with a larger bucket. Since most astronomical sources of light are very faint, it is desirable to build telescopes as large as possible. Given current technology, we can build much larger telescopes on the ground than we can in space, which is another reason that ground-based observatories remain very important.

The Locations of Ground-Based Observatories

The best ground-based sites for optical and infrared astronomy are Mauna Kea, a volcano on the Big Island of Hawaii that is 4,205 meters (13,796 feet) high, and mountain peaks in the desert in northern Chile. Other good sites are in the Canary Islands and the southwestern United States. The following are the characteristics that astronomers look for when they select a site for an optical/infrared telescope:

1. *Clear skies.* The best sites in the world are clear about 75 percent of the time. Most types of astronomical observations cannot be carried out when clouds are present.

2. *Dark skies.* The atmosphere scatters city lights, making it impossible to see faint objects. The best sites are therefore located far away from large cities. (Even with the naked eye, one can see quite clearly the difference between what can be seen in the night sky in a city and in the country.)

MAJOR RADIO OBSERVATORIES OF THE WORLD

Observatory	Location	Description	Web Site
Individual Radio Dishes			
Arecibo Telescope (National Astron. & Ionospheric Center)	Arecibo, Puerto Rico	305-m fixed dish	www.naic.edu
Greenbank Telescope (National Radio Astron. Observ.)	Green Bank, West Virginia	100- x 110-m steerable dish	www.gb.nrao.edu/GBT/GBT.html
Effelsberg Telescope (Max Planck Institute für Radioastronomie)	Bonn, Germany	100-m steerable dish	www.mpifr-bonn.mpg.de/ effberg.html
Lovell Telescope (Jodrell Bank Radio Observat.)	Manchester, England	76-m steerable dish	www.jb.man.ac.uk/
Goldstone Tracking Station (NASA/JPL)	Barstow, California	70-m steerable dish	gts.gdscc.nasa.gov/
Australia Tracking Station (NASA/JPL)	Tidbinbilla, Australia	70-m steerable dish	tid.cdscc.nasa.gov/
Parkes Radio Observatory	Parkes, Australia	64-m steerable dish	www.parkes.atnf.csiro.au/
Arrays of Radio Dishes			
Australia Telescope	Several sites in Australia	8-element array (seven 22-m dishes plus Parkes 64-m)	www.atnf.csiro.au/
MERLIN	Cambridge, England and other British sites	Network of 7 dishes (the largest of which is 32 m)	www.jb.man.ac.uk/merlin/
Westerbork Radio Observatory	Westerbork, the Netherlands	12-element array of 25-m dishes (1.6-km baseline)	www.nfra.nl/wsrt
Very Large Array (NRAO)	Socorro, New Mexico	27-element array of 25-m dishes (36-km baseline)	www.nrao.edu/doc/vla/html/ VLAhome.shtml
Very Long Baseline Array (NRAO)	Ten U.S. sites, Hawaii to Virgin Islands	10-element array of 25-m dishes (9000)-km baseline	www.nrao.edu/doc/vlba/html/ VLBA.html
Very–Long–Baseline–Interferom. Space Observ. Program (VSOP)	Connect a satellite to network on Earth	Japanese HALCA 8-m dish in orbit and ≈ 40 dishes on Earth	sgra.jpl.nasa.gov/
Millimeter–Wave Telescopes			
IRAM	Granada, Spain	30-m steerable mm-wave dish	iram.fr/
James Clerk Maxwell Telescope	Mauna Kea, Hawaii	15-m steerable mm-wave dish	www.jach.hawaii.edu/JCMT/ pages/intro.html
Nobeyama Cosmic Radio Observatory	Minamimaki-Mura, Japan	6-element array of 10-m mm-wave dishes	www.nro.nao.ac.jp/~nma/ index-e.html
Hat Creek Radio Observatory (University of California)	Cassel, California	6-element array of 5-m mm-wave dishes	bima.astro.umd.edu/bima

3. *High and dry.* Water vapor in Earth's atmosphere absorbs infrared radiation. Fortunately, water vapor is concentrated at low altitudes, and so infrared observatories are best located at high altitudes.

4. *Stable air.* Light rays are distorted when they pass through turbulent air, with the result that the image seen through a telescope is distorted and blurred. The most stable air occurs over large bodies of water such as oceans, which have a very uniform temperature. Therefore, the best sites are located in coastal mountain ranges (e.g., in northern Chile or California) or on isolated volcanic peaks in the middle of oceans (e.g., Mauna Kea).

The Hubble Space Telescope is above Earth's atmosphere, so its images are much clearer and sharper than the distorted images that are ob-

Stephen Hinman, Mirror Lab manager at the Steward Observatory Mirror Laboratory at the University of Arizona, examines a spun cast telescopic mirror.

adaptive optics the use of computers to adjust the shape of a telescope's optical system to compensate for gravity or temperature variations

ultraviolet radiation electromagnetic radiation with a shorter wavelength and higher energy than visible light

served from the ground. Astronomers are, however, devising techniques called **adaptive optics** that can correct atmospheric distortions by changing the shapes of small mirrors hundreds of times each second to compensate precisely for the effects of Earth's atmosphere. Even when this technique is perfected, space observatories will still be needed to observe gamma rays, X rays, **ultraviolet radiation**, and other wavelengths that are absorbed by Earth's atmosphere before they reach the ground.

The requirements for radio observatories are not nearly so stringent as for optical/infrared telescopes, and many types of radio observations can be made through clouds. Therefore, countries that do not have good optical/

infrared sites, such as Great Britain, Japan, the Netherlands, and Germany, have concentrated on radio astronomy.

While they are not bothered much by clouds or city lights, radio telescopes are affected by electrical interference generated by cell phones, radio transmitters, and other artifacts of civilization. Therefore, radio telescopes are often located far away from large population centers in special radio-quiet zones. Also, certain radio wavelengths are reserved for the use of radio astronomy and cannot be used to transmit human signals.

Optical and Infrared Telescopes

There are two main types of telescopes: refracting telescopes, which use lenses to gather the light and form an image; and reflecting telescopes, which use mirrors to accomplish the same purpose. Telescopes are described by the size of the largest lens or mirror that they contain. The largest refracting telescope ever built is the Yerkes 40-inch (1-meter) telescope, which is located in southeastern Wisconsin. Refractors are limited to fairly small sizes for two reasons. First, since the light must pass through a lens to be focused, the lens must be supported around its outside edge, not from behind. Large lenses tend to sag and distort in shape because of the effects of gravity, and the focused image is not as sharp as it should be. Second, because the light passes through the lens, the glass must be entirely free of bubbles or other defects that would distort the image. It is difficult and costly to make large pieces of perfect glass.

Reflecting telescopes make use of mirrors. Since the light is reflected from the front surface, mirrors can be supported from behind and can therefore be made as large as several meters in diameter. The front surface is coated with highly reflective (shiny) aluminum or silver. Since the light in a reflector never passes through the mirror, the glass can contain a few bubbles or other flaws. For these reasons the largest telescopes in the world are reflectors.

Reflecting telescopes are used for both infrared and optical astronomy. Because glass does not transmit infrared radiation very efficiently, refracting telescopes are unsuitable for most kinds of infrared astronomy.

New Technology Telescopes

For about forty years after its completion in 1948, the Palomar 5-meter (16.7 feet) reflector in southern California was the largest telescope in the world. The 5-meter Palomar mirror is very thick and is therefore rigid enough not to change shape when the telescope tracks stars as they rise in the east and set in the west. The Palomar mirror weighs about 20 tons, and a very large steel structure (weighing about 530 tons) is required to hold it. The Palomar telescope is near the limit in size of what can be built for a reasonable cost with a massive, rigid mirror.

In the 1990s, many countries took advantage of developments in technology to build telescopes with diameters of 6.5 to 10 meters (21 to 33 feet). It is now possible to use thin telescope mirrors, which do change shape when they are pointed in different directions. High-speed computers calculate the forces that must be applied to the flexible mirrors to produce the correct shape. These restoring forces can be adjusted many times each second if

LARGE OPTICAL TELESCOPES BEING BUILT OR IN OPERATION

Aperture (m)	Telescope Name	Location	Status	Web Address
16.4	Very Large Telescope (four 8.2–m telescopes)	Cerro Paranal, Chile*	First telescope completed 1998	www.eso.org/vlt/
11.8	Large Binocular Telescope (two 8.4–m telescopes)	Mount Graham, Arizona	First light 2002–2003	medusa.as.arizona.edu/btwww/tech/lbtbook.html
10.0	Keck I	Mauna Kea, Hawaii	Completed 1993	astro.caltech.edu/mirror/keck/index.html
10.0	Keck II	Mauna Kea, Hawaii	Completed 1996	astro.caltech.edu/mirror/keck/index.html
9.9	Hobby–Eberly (HET)	Mount Locke, Texas	Completed 1997	www.astro.psu.edu/het/overview.html
8.3	Subaru (Pleiades)	Mauna Kea, Hawaii	First light 1998	www.naoj.org/
8.0	Gemini (North)	Mauna Kea, Hawaii†	First light 1999	www.gemini.edu
8.0	Gemini (South)	Cerro Pachon, Chile†	First light 2000	www.gemini.edu
6.5	Multi-Mirror (MMT)	Mount Hopkins, Arizona	First light 1998	sculptor.as.arizona.edu.edu/foltz/www/
6.5	Magellan	Las Campanas, Chile	First light 1997	www.ociw.edu/~johns/magellan.html
6.0	Large Alt-Azimuth	Mount Pastukhov, Russia	Completed 1976	—
5.0	Hale	Palomar Mountain, California	Completed 1948	astro.caltech.edu/observatories/palomar/public/index.html
4.2	William Herschel	Canary Islands, Spain	Completed 1987	www.ast.cam.ac.uk/ING/PR/pr.html
4.2	SOAR	Cerro Pachon, Chile	First light 2002	www.noao.edu/
4.0	Blanco Telescope (NOAO)	Cerro Tololo, Chile†	Completed 1974	www.ctio.noao.edu/ctio/html
3.9	Anglo-Australian (AAT)	Siding Spring, Australia	Completed 1975	www.aao.gov.au/index.html
3.8	NOAO Mayall	Kitt Peak, Arizona†	Completed 1973	www.noao.edu/noao.html
3.8	United Kingdom Infrared (UKIRT)	Mauna Kea, Hawaii	Completed 1979	www.jach.hawaii.edu/UKIRT/home.html
3.6	Canada-France-Hawaii (CFHT)	Mauna Kea, Hawaii	Completed 1979	www.cfht.hawaii.edu/
3.6	ESO	Cerro La Silla, Chile*	Completed 1976	www.ls.eso.org/
3.6	ESO New Technology	Cerro La Silla, Chile*	Completed 1989	www.ls.eso.org/
3.5	Max Planck Institut	Calar Alto, Spain	Completed 1983	www.mpia-hd.mpg.de/CAHA/
3.5	WIYN	Kitt Peak, Arizona†	Completed 1993	www.noao.edu/wiyn/wiyn.html
3.5	Astrophysical Research Corp.	Apache Point, New Mexico	Completed 1993	www.apo.nmsu.edu/
3.0	Shane (Lick Observatory)	Mount Hamilton, California	Completed 1959	www.ucolick.org/
3.0	NASA Infrared (IRTF)	Mauna Kea, Hawaii	Completed 1979	irtf.ifa.hawaii.edu

*Part of the European Southern Observatory (ESO).
†Part of the U.S. National Optical Astronomy Observatories (NOAO).

necessary. A lightweight thin mirror can be supported by a lightweight steel structure, and telescopes double the size of the Palomar telescope are affordable with this new technology.

At the dawn of the twenty-first century, the largest single mirror that has been manufactured to date is 8.4 meters (27.5 feet) in diameter, and it is scheduled to be installed in a telescope in southern Arizona in 2003. This is probably about the largest single mirror that is feasible. Given the width of highways and tunnels, it would be impossible to transport a much larger mirror from where it was manufactured to a distant mountaintop.

Currently the largest telescopes in the world are the twin 10-meter (33-foot) Keck telescopes on Mauna Kea. These telescopes do not contain a single mirror that is 10 meters in diameter. Rather, each consists of thirty-six separate hexagonal-shaped mirrors that are 1.8 meters (6 feet) in diameter. These mirrors are positioned so precisely relative to one another that they can collect and focus the light as efficiently as a continuous single mirror.

Radio Telescopes

Radio astronomy is a young field relative to optical astronomy. Italian mathematician and astronomer Galileo Galilei used the first optical telescope, a refractor, in 1610. By contrast, American electrical engineer Karl Jansky first detected astronomical radio waves in 1931. Astronomical radio waves cannot be heard. Like light, radio waves are a form of electromagnetic radiation. Unlike light, however, we cannot sense radio waves directly but must use electronic equipment. Radio waves are reflected by surfaces that conduct electricity, just as light is reflected by a shiny aluminum or silver surface. Accordingly, a radio telescope consists of a concave metal reflector that focuses the radio waves on a receiver.

Interferometry

Resolution refers to the fineness of detail that can be seen in an image. The larger the telescope, the finer the detail that can be observed. One way to see finer detail is to build a larger single telescope. Unfortunately, there are practical limits to the size of a single telescope—currently about 10 meters (33 feet) for optical/infrared telescopes and about 100 meters (330 feet) for radio telescopes. If, however, astronomers combine the signals from two or more widely separated telescopes, they can see the fineness of detail that would be observed if they had a single telescope of that same diameter. Telescopes working in combination in this way are called **interferometers**. For example, infrared radiation falling on the two 10-meter Keck telescopes, which are about 85 meters (279 feet) apart, has been combined, allowing astronomers to obtain the kind of detailed image that they would observe if they had a single telescope 85 meters in diameter.

interferometers devices that use two or more telescopes to observe the same object at the same time in the same wavelength to increase angular resolution

Radio interferometry is easier than optical and infrared interferometry because radio waves have much longer wavelengths than optical or infrared radiation. The equipment used to measure radio waves need not be built to the same precision as optical telescopes, and radio waves are not distorted very much by turbulence in Earth's atmosphere. For these reasons, radio astronomers have been able to build whole arrays of telescopes separated by thousands of kilometers to conduct interferometry. For example, U.S. astronomers operate the Very Long Baseline Array, which consists of ten telescopes located across the United States and in the Virgin Islands and Hawaii. When combined with a telescope in Japan, this array of radio telescopes has the same resolution as a telescope with the diameter of Earth.

The Future of Ground-Based Observatories

By 2003, fourteen mirrors with diameters larger than 6.5 meters (21.3 feet) will have been installed in optical/infrared telescopes. During the early twenty-first century, these telescopes are likely to produce many impressive discoveries. But astronomers are already planning for the next generation of large telescopes. These will truly be "world" telescopes. The costs, which are estimated to be several hundred million dollars each, are beyond the reach of any single country. Therefore, the new, very large telescopes will be built through international consortia involving many countries.

Astronomers in Europe are exploring the feasibility of building an optical/infrared telescope that is 100 meters (330 feet) in diameter—about the length of a football field. This telescope is called the OWL telescope, which stands for Overwhelmingly Large Telescope. The mirror would be built in the same way as the Keck mirrors, that is, by combining literally thousands of smaller mirrors to form a single continuous surface. This telescope would be powerful enough to study objects present when the universe was only a few million years old. The current age of the universe is about 14 billion years, and so with a telescope such as OWL astronomers could observe directly the evolution of the universe throughout nearly all of its history.

In radio astronomy, the next major project is likely to be the Atacama Large Millimeter Array (ALMA). The project will be an interferometer that detects radio radiation with wavelengths between 0.350 and 10 millimeters (0.014 and 0.4 inches). The facility will consist of sixty-four radio antennas, each 12 meters (39 feet) in diameter, with the separations between antennas varying from 150 meters (490 feet) to 10 kilometers (6.2 miles). ALMA will be located at one of the driest spots on Earth—a large plateau at an altitude of 5,000 meters (16,400 feet) in the Atacama Desert in northern Chile. Water vapor in Earth's atmosphere absorbs much of the millimeter wavelength radiation that astronomers would like to detect, and so it is important to select an extremely dry site. The facility will be particularly useful for studying how stars and planets form and what galaxies were like when the universe was very young. SEE ALSO ASTRONOMER (VOLUME 2); ASTRONOMY, HISTORY OF (VOLUME 2); ASTRONOMY, KINDS OF (VOLUME 2); CAREERS IN ASTRONOMY (VOLUME 2); HUBBLE SPACE TELESCOPE (VOLUME 2); OBSERVATORIES, SPACE-BASED (VOLUME 2).

Sidney C. Wolff

Bibliography

Florence, Ronald. *The Perfect Machine: Building the Palomar Telescope.* New York: Harper Perennial, 1994.

Fugate, Robert, and Walter Wild. "Untwinkling the Stars." *Sky and Telescope* 87, no. 6 (1994):24.

Pilachowski, Caty, and Mark Trueblood. "Telescopes of the 21st Century." *Mercury* 27, no. 5 (1998):10–17.

Preston, Richard. *First Light.* New York: Atlantic Monthly Press, 1987.

Tarenghi, Massimo. "Eyewitness View: First Sight for a Glass Giant." *Sky and Telescope* 96, no. 5 (1998):46–55.

Wakefield, Julie. "Keck Trekking." *Astronomy* 26, no. 9 (1998):52–57.

Observatories, Space-Based

Space-based observatories are telescopes located beyond Earth, either in orbit around the planet or in deep space. Such observatories allow astronomers to observe the universe in ways not possible from the surface of Earth, usually because of interference from our planet's atmosphere. Space-based observatories, however, are typically more complicated and expensive than Earth-based telescopes. The National Aeronautics and Space Administration (NASA) and other space agencies have been flying space observatories of one type or another since the late 1960s. While the Hubble Space Telescope is the most famous of the space observatories, it is just one of many

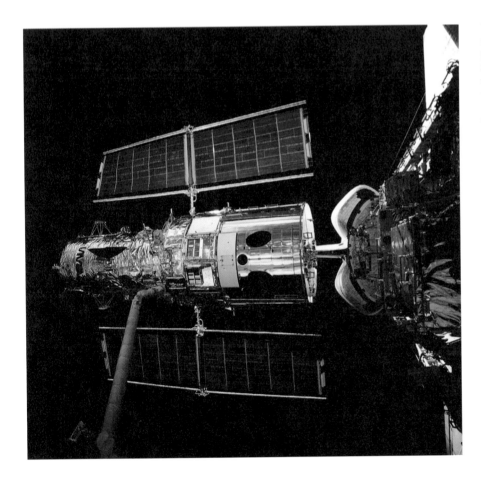

The Hubble Space Telescope (HST) is lifted into position by the Remote Manipulator System (RMS) from its berth in the cargo bay of the Earth-orbiting space shuttle Discovery.

that have provided astronomers with new insights about the solar system, the Milky Way galaxy, and the universe.

The Advantages and Disadvantages of Space-Based Telescopes

Observatories in space have a number of key advantages. Telescopes in space are able to operate twenty-four hours a day, free of both Earth's day-night cycle as well as clouds and other weather conditions that can hamper observing. Telescopes above the atmosphere can also observe portions of the **electromagnetic spectrum** of light, such as **ultraviolet radiation**, **X rays**, and **gamma rays**, which are blocked by Earth's atmosphere and never reach the surface. Telescopes in space are also free of the distortions in the atmosphere that blur images. These factors increase the probability that space telescopes will be more productive and useful than their ground-based counterparts.

Space-based observatories also have some disadvantages. Unlike most ground-based telescopes, space observatories operate completely automatically, without any humans on-site to fix faulty equipment or deal with other problems. There are also limitations on the size and mass of objects that can be launched, as well as the need to use special materials and designs that can withstand the harsh environment of space, creating limitations on the types of observatories that can be flown in space. These factors, as well as current high launch costs, make space observatories very expensive: the

electromagnetic spectrum the entire range of wavelengths of electromagnetic radiation

ultraviolet radiation electromagnetic radiation with a shorter wavelength and higher energy than visible light

X rays a form of high-energy radiation just beyond the ultraviolet portion of the electromagnetic spectrum

gamma rays a form of radiation with a shorter wavelength and more energy than X rays

With the Earth as a backdrop, astronauts John M. Grunsfeld and Steven L. Smith replace gyroscopes inside the Hubble Space Telescope during extravehicular activity.

largest observatories, such as the Hubble Space Telescope, cost over $1 billion, whereas world-class ground-based telescopes cost less than $100 million. In many cases, though, there is no option other than to fly a space observatory, because ground-based telescopes cannot accomplish the required work.

The History of Space-Based Telescopes

The first serious study of observatories in space was conducted in 1946 by astronomer Lyman Spitzer, who proposed orbiting a small telescope. In the late 1960s and early 1970s NASA launched four small observatories under the name Orbiting Astronomical Observatories (OAO). Two of the OAO missions were successful and conducted observations, primarily in ultraviolet light, for several years. NASA followed this up with a number of other small observatories, including the International Ultraviolet Explorer in 1978 and the Infrared Astronomy Satellite in 1983.

While NASA was developing and launching these early missions, it was working on something much larger. In the 1960s it started studying a proposal to launch a much larger observatory to study the universe at visible, ultraviolet, and infrared **wavelengths**. This observatory was originally known simply as the Large Space Telescope, but over time evolved into what became known as the Hubble Space Telescope. Hubble was finally launched by the space shuttle Discovery in April 1990. After astronauts corrected a problem with the telescope's optics in 1993, Hubble emerged as

wavelength the distance from crest to crest on a wave at an instant in time one of the best telescopes in the world. Hubble is scheduled to operate at least through 2010.

The Great Observatories

While the Hubble Space Telescope may be the most famous space observatory, it is far from the only major one. NASA planned for Hubble to be the first of four "Great Observatories" studying the universe from space, each focusing on a different portion of the spectrum. The second of the four Great Observatories, the Compton Gamma Ray Observatory (CGRO), was launched by the space shuttle Atlantis on mission STS-37 in April 1991. The telescope was named after Arthur Holly Compton, a physicist who won the Nobel Prize in 1927 for his experimental efforts confirming that light had characteristics of both waves and particles.

The purpose of CGRO, also known as Compton, was to study the universe at the wavelengths of gamma rays, the most energetic form of light. CGRO carried four instruments that carried out these observations. Data from these instruments led to a number of scientific breakthroughs. Astronomers discovered through CGRO data that the center of our galaxy glows in gamma rays created by the annihilation of matter and **antimatter**. Observations by CGRO of hundreds of mysterious gamma-ray bursts showed that the bursts are spread out evenly over the entire sky and thus likely originate from far outside our own galaxy. Astronomers used CGRO to discover a new class of objects, known as blazars: **quasars** that generate gamma rays and jets of particles oriented in our direction.

CGRO was intended to operate for five years but continued to work for several years beyond that period. In early 2000 one of Compton's three gyroscopes, used to orient the spacecraft, failed. Because the spacecraft was so heavy—at 17 tons it weighed more than even Hubble—NASA was concerned that if the other gyroscopes failed the spacecraft could reenter Earth's atmosphere uncontrolled and crash, causing damage and injury. To prevent this, NASA deliberately reentered Compton over the South Pacific on June 4, 2000, scattering debris over an empty region of ocean and ending the spacecraft's nine-year mission.

The third spacecraft in NASA's Great Observatories program is the Chandra X-Ray Observatory. The spacecraft, originally called the Advanced X-Ray Astrophysics Facility but today known simply as Chandra, was launched by the space shuttle Columbia on mission STS-93 in July 1999. Chandra was the largest spacecraft ever launched by the space shuttle. The spacecraft is named after Subrahmanyan Chandrasekhar, an Indian-American astrophysicist who won the Nobel Prize for physics in 1983 for his studies of the structure and evolution of stars.

Chandra carries four instruments to study the universe at X-ray wavelengths, which are slightly less energetic than gamma rays, at up to twenty-five times better detail than previous spacecraft missions. To carry out these observations Chandra is in an unusual orbit: Rather than a circular orbit close to Earth, as used by Hubble and Compton, it is in an **eccentric** orbit that goes between 10,000 and 140,000 kilometers (6,200 and 86,800 miles) from Earth. This **elliptical** orbit allows Chandra to spend as much time as possible above the charged particles in the **Van Allen radiation belts** that would interfere with the observations.

antimatter matter composed of antiparticles, such as positrons and antiprotons

quasars luminous objects that appear star-like but are highly redshifted and radiate more energy than an entire ordinary galaxy; likely powered by black holes in the centers of distant galaxies

eccentric the term that describes how oval the orbit of a planet is

elliptical having an oval shape

Van Allen radiation belts two belts of high energy charged particles captured from the solar wind by Earth's magnetic field

The Chandra X-Ray Observatory, just prior to deployment from space shuttle Columbia's payload bay. The 50,162 pound observatory is named after astrophysicist Subrahmanyan Chandrasekhar, who won the Nobel Prize for physics in 1983 for his studies of the structure and evolution of stars.

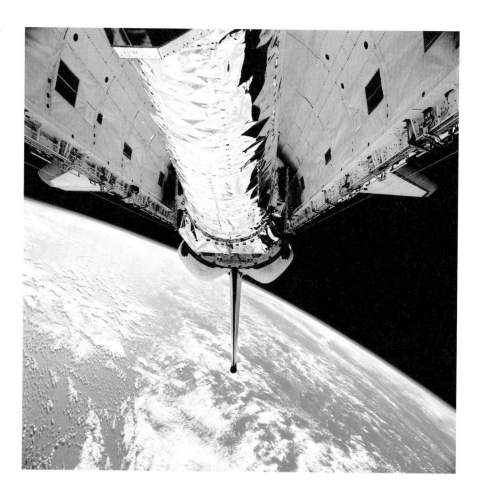

dark matter matter that interacts with ordinary matter by gravity but does not emit electromagnetic radiation; its composition is unknown

black holes objects so massive for their size that their gravitational pull prevents everything, even light, from escaping

supernova an explosion ending the life of a massive star; caused by core collapse or the sudden onset of nuclear fusion

brown dwarfs star-like objects less massive than 0.08 times the mass of the Sun, which cannot undergo thermonuclear process to generate their own luminosity

Although Chandra has been in orbit only a relatively short time, it has provided astronomers with a wealth of data. Astronomers have used Chandra to learn more about the **dark matter** that may make up most of the mass of the universe, study **black holes** in great detail, witness the results of **supernova** explosions, and observe the birth of new stars. Chandra's mission is officially scheduled to last for five years but will likely continue so long as the spacecraft continues to operate well.

The final spacecraft of the Great Observatories program is the Space Infrared Telescope Facility (SIRTF). SIRTF will probe the universe at infrared wavelengths of light, which are longer and less energetic than visible light. SIRTF is scheduled for launch in January 2003 on an unpiloted Delta rocket. Rather than go into Earth orbit, SIRTF will be placed in an orbit around the Sun that gradually trails away from Earth; this will make it easier for the spacecraft to perform observations without interference from Earth's own infrared light. Astronomers plan to use SIRTF to study planets, comets, and asteroids in our own solar system and look for evidence of giant planets and **brown dwarfs** around other stars. SIRTF will also be used to study star formation and various types of galaxies during its five-year mission.

Other Space Observatories

Besides NASA's Great Observatories, there have been many smaller, space-based observatories that have focused on particular objects or sections of the electromagnetic spectrum. A number of these missions have made major contributions. NASA's Cosmic Background Explorer (COBE) spacecraft was launched in 1989 on a mission to observe **cosmic microwave background** radiation, light left over from shortly after the **Big Bang**. COBE's instruments were able to measure small variations in the background, providing key proof for the Big Bang model of the universe. NASA launched a new mission, the Microwave Anisotropy Probe (MAP), in June 2000 to measure the variations in the microwave background in even greater detail.

NASA is not the only space agency to launch space observatories. The European Space Agency (ESA) has launched a number of its own observatories to study the universe. The Infrared Space Observatory provided astronomers with unprecedented views of the universe at infrared wavelengths in the mid- and late 1990s. In 1999 ESA launched XMM-Newton, an orbiting X-ray observatory similar to NASA's Chandra spacecraft. XMM-Newton and Chandra serve complementary purposes: Whereas Chandra is designed to take detailed X-ray images of objects, XMM-Newton focuses on measuring the **spectra** of those objects at X-ray wavelengths.

Japan has also contributed a number of small space observatories. The Advanced Satellite for Cosmology and Astrophysics spacecraft was launched in 1993 and continues to operate in the early twenty-first century, studying the universe at X-ray wavelengths. The Yohkoh spacecraft was launched in 1991 to study the Sun in X rays. The Halca spacecraft, launched in 1997, conducts joint observations with radio telescopes on Earth. The Soviet Union also flew several space observatories, including the Gamma gamma-ray observatory and the Granat X-ray observatory. Since the collapse of the Soviet Union, however, Russia has been unable to afford the development of any new orbiting telescopes.

Future Space Observatories

The success of past and present space observatories has led NASA, ESA, and other space agencies to plan a new series of larger, more complex spacecraft that will be able to see deeper into the universe and in more detail than their predecessors. Leading these future observatories is the Next Generation Space Telescope (NGST), the successor to the Hubble Space Telescope. Scheduled for launch in 2009, NGST will use a telescope up to 6.5 meters (21.3 feet) in diameter (Hubble's is 2.4 meters [7.9 feet] across), which will allow it to observe dimmer and more distant objects. The telescope will be located at the Earth-Sun L-2 point, 1.5 million kilometers (930,000 miles) away, to shield it from Earth's infrared radiation. NASA is also supporting the development of other new space observatories, including GLAST, a gamma-ray observatory scheduled for launch in 2006.

ESA is developing several space observatories that will observe the universe at different wavelengths. Integral is a gamma-ray observatory scheduled for launch in 2002. Planck, scheduled for launch in 2007, will build upon the observations of the cosmic microwave background made by COBE and MAP. Herschel, also scheduled for launch in 2007, will observe the universe at far-infrared wavelengths. ESA is also collaborating with NASA on development of the NGST.

In the future, space observatories may consist of several spacecraft working together. Such orbiting arrays of telescopes could allow astronomers to

cosmic microwave background ubiquitous, diffuse, uniform, thermal radiation created during the earliest hot phases of the universe

Big Bang name given by astronomers to the event marking the beginning of the universe, when all matter and energy came into being

spectra representations of the brightness of objects as a function of the wavelength of the emitted radiation

get better images without the need to build extremely large and expensive single telescopes. One such mission, called Terrestrial Planet Finder (TPF), would combine images from several telescopes, each somewhat larger than Hubble, to create a single image. A system of this type would make it possible for astronomers to directly observe planets the size of Earth orbiting other stars. TPF is tentatively scheduled for launch no sooner than 2011. NASA is also studying a similar proposal, called Constellation-X, which would use several X-ray telescopes to create a virtual telescope 100 times more powerful than existing ones.

In the more distant future, astronomers have proposed developing large telescopes, and arrays of telescopes, on the surface of the Moon. The far-side of the Moon is an ideal location for a radio telescope, because it would be shielded from the growing artificial radio noise from Earth. However, there are as of yet no detailed plans for lunar observatories. SEE ALSO ASTRONOMER (VOLUME 2); ASTRONOMY, HISTORY OF (VOLUME 2); ASTRONOMY, KINDS OF (VOLUME 2); HUBBLE SPACE TELESCOPE (VOLUME 2); OBSERVATORIES, GROUND (VOLUME 2).

Jeff Foust

Bibliography

Chaisson, Eric. *The Hubble Wars.* New York: HarperCollins, 1994.

Smith, Robert W. *The Space Telescope.* Cambridge, UK: Cambridge University Press, 1993.

Tucker, Wallace H., and Karen Tucker. *Revealing the Universe: The Making of the Chandra X-Ray Observatory.* Cambridge, MA: Harvard University Press, 2001.

Internet Resources

The Chandra X-Ray Observatory Center. Harvard-Smithsonian Center for Astrophysics. <http://chandra.harvard.edu/>.

The Hubble Space Telescope. Space Telescope Science Institute. <http://hst.stsci.edu/>.

Next Generation Space Telescope. NASA Goddard Space Flight Center. <http://ngst.gsfc.nasa.gov/>.

Space Infrared Telescope Facility Science Center. California Institute of Technology. <http://sirtf.caltech.edu/AboutSirtf/index.html>.

Terrestrial Planet Finder. NASA Jet Propulsion Laboratory. <http://tpf.jpl.nasa.gov/>.

Oort Cloud

The Oort cloud is a vast swarm of some 2 trillion comets orbiting our star in the most distant reaches of our solar system, extending from beyond the orbits of Neptune and Pluto out to 100,000 times the Earth-Sun distance—nearly one-third the distance to the nearest star. While the planets are confined to a flattened disk in the solar system, the Oort cloud forms a spherical shell centered on the Sun, which gradually flattens down to an extended disk in the inner region, called the Kuiper belt. Bright comets observed through telescopes or with the unaided eye get perturbed out of the Oort cloud or Kuiper belt, and become visible when they get close to enough so that the Sun's energy can transform the surface ices into gases. These gases drag off the embedded dust, and we see the light reflected from the dust as a tail.

Comets are the leftover icy building blocks from the time of planet formation, which formed in the region of the outer planets. Essentially these

The Oort Cloud, which may contain as many as two trillion comets, orbits the outer reaches of the solar system.

comets are dirty snowballs, composed primarily of water ice, with some carbon monoxide and other ices, in addition to **interstellar** dust. When their orbits passed close enough to the giant planets to be affected, some were thrown toward the Sun and some were tossed outward toward the distant reaches of the solar system, the spherical swarm we now call the Oort cloud. Some of the comets sent inward hit the inner rocky planets, and probably contributed a significant amount of ocean water and organic material, the building blocks of life, to Earth. Comets that live in the Oort cloud are especially important scientifically because they have been kept in a perpetual deep freeze since the formation of our solar system 4.6 billion years ago. This means that they preserve, nearly intact, a record of the chemical conditions during the first few million years of the solar system's history, and can be used to unravel our solar system's origins much like an archaeologist uses artifacts to decipher an ancient civilization.

interstellar between the stars

The Oort cloud was "discovered" by Dutch astronomer Jan Hendrik Oort in 1950, not through telescopic observations, but through a theoretical study of the orbits of long-period comets (comets with periods greater than 200 years). Long-period comets can have orbits ranging from **eccentric** ellipses to parabolas to even modest hyperbolas. While trying to explain the distribution of these orbits (which were mostly nearly parabolic or hyperbolic), Oort concluded that the only explanation was that the source of these comets had to be a massive cloud of comets surrounding the solar

eccentric the term that describes how oval the orbit of a planet is

system. These comets would be fed into the region of the planets as the motion of the solar system through the galaxy caused the solar system to pass relatively close to stars. The slight change in the gravitational acceleration from these stars was enough to send some distant comets into orbits that brought them into the inner solar system.

Oort's remarkable discovery was made with only a few handfuls of comet observations. Since then, precise observations of comet orbits and new modern computer models have shown not only that his ideas were correct but also that the Oort cloud can be divided into different regions: the outer Oort cloud, acted upon by passing stars; the inner Oort cloud, which is close enough to the Sun (perhaps 2,000 to 15,000 Earth-Sun distances) that the comets are not affected by gravitational interactions, and finally the flattened innermost region—the Kuiper belt. Kuiper belt comet orbits can be perturbed through interactions with the outer planets, and these comets then become observable as short-period comets. Because the Kuiper belt is much closer to the Sun, the world's largest telescopes began directly observing these comets in the late-twentieth century. The first Kuiper belt object was discovered in 1993, and by 2002 more than 500 such objects were discovered. SEE ALSO COMETS (VOLUME 2); KUIPER BELT (VOLUME 2); ORBITS (VOLUME 2); PLANETESIMALS (VOLUME 2); SMALL BODIES (VOLUME 2).

Karen J. Meech

Bibliography

Jewitt, D. C., and J. X. Luu. "Physical Nature of the Kuiper Belt." In *Protostars and Planets IV*, ed. V. Mannings, A. P. Boss, and S. S. Russell. Tucson: University of Arizona Press, 2000.

Oort, Jan H. "The Structure of the Cloud of Comets Surrounding the Solar System and a Hypothesis Concerning Its Origin." *Bulletin of the Astronomical Institute of the Netherlands* 11 (1950):91–110.

Weissman, P. R. "The Oort Cloud and the Galaxy: Dynamical Interactions." In *The Galaxy and the Solar System*, ed. R. Smoluchowski, J. N. Bahcall, and M. S. Matthews. Tucson: University of Arizona Press, 1986.

Orbits

Orbits are the pathways taken by objects under the influence of the gravity of another object. These **trajectories** are governed by the fundamental laws of gravity and the motion of the object. The ability to calculate the orbit of an object, be it a planet, moon, asteroid, or spacecraft, makes it possible to predict where it will be in the future. Both solar system objects and spacecraft can be found in a wide range of orbits, including specific types of Earth orbits that are particularly useful for some types of spacecraft.

All objects in space are attracted to all other objects by the force of gravity. In the case of an object orbiting a planet or other celestial body, where the mass of the object is much less than the mass of the planet, the object will fall towards the planet. However, if the object has some initial **velocity**, it will not fall straight towards the planet, as its trajectory will be altered by gravity. If the object is going fast enough, it will not hit the planet, because the planet's surface is curving away underneath it. Instead, it will keep "falling" around the planet in a trajectory known as an orbit. Orbits require a specific range of speeds. If the object slows down below a mini-

trajectories paths followed through space by missiles and spacecraft moving under the influence of gravity

velocity speed and direction of a moving object; a vector quantity

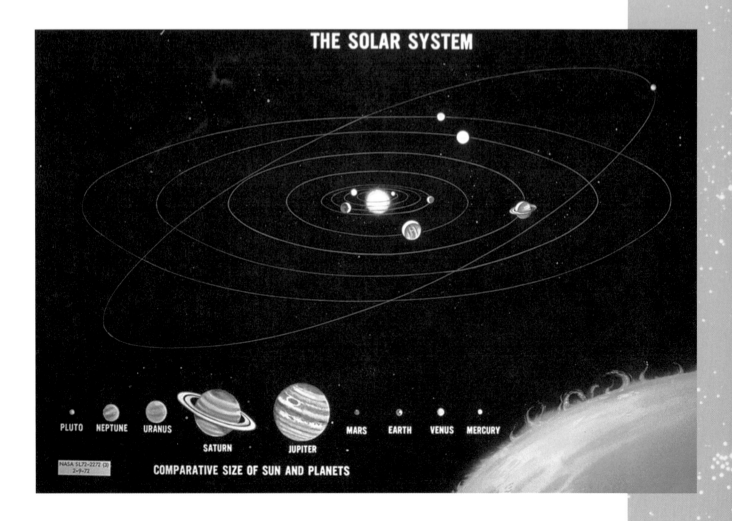

THE SOLAR SYSTEM

PLUTO   NEPTUNE   URANUS   SATURN   JUPITER   MARS   EARTH   VENUS   MERCURY

NASA SL72-2272 (3)
2-9-72

COMPARATIVE SIZE OF SUN AND PLANETS

A graphical diagram of the solar system showing the nine planets and their comparative sizes, as well as each planet's orbital track around the Sun.

mum orbital velocity, it will hit the planet; if it speeds up beyond a maximum escape velocity, it will move away from the planet permanently.

Orbital Elements

The shape of an orbit around a planet (or another body) can be defined by several key factors, known as orbital elements. One is the orbit's mean radius, also known as the semimajor axis. The second is the eccentricity of the orbit, or the degree by which the orbit differs from a circle. The third factor is the inclination of the orbit, or the angle between the plane of the orbit and the plane of Earth's orbit. An inclination of 0 degrees would mean the orbit is perfectly aligned with Earth's orbital plane. Three other factors, known as the right ascension of the ascending node, argument of periapsis, and true anomaly, further refine the orientation of the orbit as well as the position of the object in orbit at a given time.

Calculating these orbital elements requires a minimum of three measurements of the position of the object at different times. Additional observations help refine the calculation of the orbit and reduce errors. Once the orbital elements are known, other key parameters of the orbit can be computed, such as its period, and the closest and farthest the object is in its orbit (known respectively as periapsis and apoapsis). For objects orbiting Earth, periapsis and apoapsis are known as perigee and apogee; for

The Moon's orbit around Earth is slightly tilted and moves around Earth counterclockwise as seen from above the North Pole.

perihelion the point in an object's orbit that is closest to the Sun

aphelion the point in an object's orbit that is farthest from the Sun.

objects orbiting the Sun, these locations are known as **perihelion** and **aphelion**.

Types of Orbits

Solar system objects travel in a wide variety of orbits. Most planets go around the Sun in nearly circular, low-inclination orbits. The exception is Pluto, which has an inclined orbit that is so eccentric that it is closer to the Sun than Neptune is for twenty years out of each 248-year orbit. Asteroid orbits can be more eccentric, particularly for those objects whose orbits have been altered by the gravity of Jupiter or another planet. Comet orbits, however, can be extremely eccentric, especially for long-period comets that pass through the inner solar system only once every hundreds, or thousands, of years.

Spacecraft orbiting Earth can be found in several different types of orbits based on their altitude and orientation. Many spacecraft, including the space shuttle and International Space Station, are in low Earth orbit, flying a few hundred kilometers above Earth and completing one orbit in about ninety minutes. These orbits are the easiest to get into and are particularly useful for spacecraft that observe Earth. Higher orbits, extending out to altitudes of tens of thousands of kilometers, are used by specific types of communications, navigation, and other spacecraft. At these higher orbits it can take many hours to complete a single orbit.

Special Classes of Orbits

There are several special classes of orbits of particular interest. The best-known special orbit is geostationary orbit, a circular orbit 36,000 kilometers (22,320 miles) above Earth. At this altitude it takes a satellite twenty-four hours to complete one orbit. To an observer on the ground a satellite in this orbit would appear motionless in the sky, hence the name geostationary. Geostationary orbit is also known as a Clarke orbit, after science fiction writer Arthur C. Clarke, who first proposed the concept in 1945. This orbit is used today by hundreds of communications and weather satellites.

Satellites in geostationary orbit do not work well for people in high latitudes, because the satellites appear near the horizon. To get around this limitation, the Soviet Union placed communications satellites in highly inclined, **elliptical** orbits, so that they appeared to hang nearly motionlessly high in the sky for hours at a time. Such orbits are known as Molniya orbits, after the class of spacecraft that launched them.

elliptical having an oval shape

Another special type of orbit is Sun-synchronous orbit. This nearly polar orbit is designed such that the spacecraft's orbital path moves at the same apparent rate as the Sun. This allows the spacecraft to pass over different regions of Earth at the same local time. Sun-synchronous orbits are used primarily by **remote sensing** satellites that study Earth because these orbits make comparisons between different regions of Earth and different times of the year easier. The Mars Global Surveyor and Mars Odyssey spacecraft orbiting Mars also use versions of Sun-synchronous orbit. SEE ALSO ASTEROIDS (VOLUME 2); COMETS (VOLUME 2); GRAVITY (VOLUME 2); SATELLITES, TYPES OF (VOLUME 1); TRAJECTORIES (VOLUME 2).

remote sensing the act of observing from orbit what may be seen or sensed below Earth

Jeff Foust

Bibliography

Szebehely, Victor G., and Hans Mark. *Adventures in Celestial Mechanics.* New York: John Wiley & Sons, 1997.

Internet Resources

Braeunig, Robert A. "Orbital Mechanics." <http://users.commkey.net/Braeunig/space/orbmech.htm>.

Graham, John F. "Orbital Mechanics." <http://www.space.edu/projects/book/chapter5.html>.

Important Satellite Orbits. University Corporation for Atmospheric Research. <http://www.windows.ucar.edu/spaceweather/types_orbits.html>.

Satellite Orbits. <http://www.factmonster.com/ce6/sci/A0860928.html>.

Planet X

Ever since the discovery of the ninth planet, Pluto, astronomers have speculated about whether a still more distant, tenth planet may exist. This thinking was initially based on two lines of reasoning. First, astronomers had been

successful at discovering Uranus, and then Neptune, and then Pluto. This led them to believe that additional planets might await discovery farther out, if they only searched hard enough for them. Second, the search for Pluto, like the search for Neptune before it, had been based on the apparent tug of a giant, unseen world affecting the orbit of its predecessor. Yet after Pluto's discovery, it became clear that the newly found planet was very small, smaller than the United States. Such a small world could not have tugged significantly on Neptune.

Not long after the discovery of Pluto, however, it was discovered that the discrepancies in the orbit of Neptune that led astronomers to search for Pluto had been fictitious—they were simply measurement errors made by old telescopes. Pluto's discovery had been a lucky accident. With this finding, most astronomers concluded in the early 1980s that the train of logic leading to suspicions of a "Planet X" was faulty, and that it was unlikely a large Planet X existed beyond Pluto.

More recently, however, the tide has begun to swing back to a general consensus that there may indeed be planets orbiting the Sun beyond Pluto. Indeed, there may be not just one (i.e., Planet X), but many. Why this change? For one thing, astronomers discovered the Kuiper belt, a teeming ensemble of miniature worlds within which Pluto orbits. Objects half as large as Pluto have already been discovered among the hundreds of Kuiper belt objects found since 1992, and most astronomers expect that still larger objects, probably including some larger than Pluto itself, will eventually be identified.

Moreover, it has become clear from computer-generated solar system formation models that during the final stage of the formation of giant planets, a significant number of larger, "runner-up" objects to the giant planets (some perhaps even larger than Earth) may have been ejected to orbits in the Oort cloud of comets lying far beyond Pluto.

Do such objects actually exist? We will not know until observational searches either find them or rule them out. Such a search is difficult. Because such objects will be farther out than Pluto, and therefore dimmer, locating them will be rather akin to finding a needle in a haystack. Searches now underway and planned for the first decade of the twenty-first century may well settle the question. Until then, however, the subject of Planet X (and planets Y, Z, and so forth) will remain a subject of ongoing scientific debate. SEE ALSO Extrasolar Planets (volume 2); Kuiper Belt (volume 2); Oort Cloud (volume 2).

S. Alan Stern

Bibliography

Stern, S. Alan, and Jacqueline Mitton. *Pluto and Charon: Ice Worlds on the Ragged Edge of the Solar System.* New York: John Wiley & Sons, 1998.

Planetariums and Science Centers

Planetariums, museums, and science centers can be found in most major cities around the world. Science centers are an outgrowth of the original planetarium theaters, which, at their inception in the 1930s, were the only

places the general public could learn about science in general and astronomy in particular.

A planetarium consists of a hemispherical-domed theater in which a specialized star projector can display the night sky at any time of the year. Central to the experience is the planetarium projector. There are two principle types of star projectors: opto-mechanical and digital.

The sphere of the Hayden Planetarium is part of the Rose Center for Earth and Science at New York City's American Museum of National History. The planetarium contains the largest and most powerful virtual reality simulator in the world.

Traditional Planetarium Projectors

Opto-mechanical planetarium projectors typically have two spheres incorporating carefully drilled metal plates that, when illuminated, reproduce the constellation patterns on the reflective dome surface. In addition, the positions of planets hundreds of years into the past or future can be displayed. There are four principle manufacturers of this kind of projector: Zeiss (Germany), Minolta (Japan), Gotoh (Japan), and Spitz, Inc. (United States). Walther Bauersfeld of the Zeiss company built the first opto-mechanical star projector in 1923.

Digital Planetarium Projectors

A digital projector called "Digistar®," short for digital stars, was introduced in 1980 by Evans and Sutherland, Inc. Because of its low profile, the projector

The entrance to the Omnimax theater of the Henry Crown Space Center, at the Museum of Science and Industry in Chicago.

is effectively hidden in the center of the theater. Digistar® is built around a cathode-ray tube with a large fish-eye lens. A computer database of 9,000 stars can be projected onto the dome. Since the system is computer-based, it can recreate the night sky as seen from other nearby stars in addition to the view from Earth. Constellations can be "flown around," and the solar system becomes a dynamic projection that can be rotated as if the audience were in space. The system can also be used to draw any three-dimensional object as a "wire-frame," such as an architectural model, a mathematical shape, or a spacecraft.

The Spread of Planetariums

Since the invention of the planetarium projector, many major cities have built planetarium theaters. The first theaters were built in Vienna, Rome, and Moscow, all prior to 1930. The first planetarium in the United States was the Adler Planetarium in Chicago, which opened in 1930.

During the late 1960s and early 1970s many new planetariums were built. Installations in schools and colleges were encouraged by the introduction of cheaper star projectors by Spitz, based in Pennsylvania. This expansion reflected a growth in the teaching of science during a time of great exploration of space, which culminated in humans landing on the Moon in 1969.

Evolution of Science Centers

As an outgrowth of the tremendous interest in science generated by planetariums, and responding to a greater public desire to learn more, the concept of a science center developed. Science centers provide hands-on activities that engage people in science. Some science centers combine locally built exhibits that match local interests with traveling exhibits that focus on particular issues or subjects, such as "Missions to Mars" or "Global Warming." Science centers develop their own educational programs. Today, most cities have a science center or a planetarium. People working in various departments, such as education, marketing, technical support, administration and presentations, staff these large informal science learning centers. The larger science centers often incorporate a large-format film theater in addition to the popular planetarium theaters.

Planetariums of the Twenty-First Century

Planetarium theaters come in many shapes and sizes. Some serve as unique classrooms and belong to schools or colleges. Classes held in planetariums teach the basics of astronomy, the night sky, the seasons, and other topics related to the science curriculum. Larger theaters, such as the Adler Planetarium in Chicago, are stand-alone facilities. Others are part of larger science centers. For example, the Buhl Planetarium is part of the Carnegie Science Center in Pittsburgh, Pennsylvania. These larger facilities play to a general public audience, and their goals are very different from school-based theaters. They encourage the spark that may ignite a child's interest in science, and strive to develop public's understanding of space and astronomy.

At the dawn of the twenty-first century, traditional planetarium projectors are being replaced with modern digital projectors. Planetariums are changing from places to view the night sky as seen from Earth, into amazing domed theaters where audiences are immersed in three-dimensional digital images and can be transported to a new universe full of realism. Elaborate productions using synchronized sound and narration elevate the planetarium theater to new heights. The changes have been reflected in the role of the audiences. Previously, passive audiences watched a show from beneath the dome as if in a glorified lecture theater. Now, audiences are transported to places they previously could imagine only in their dreams. Faster and more powerful computers and real-time image generation have added new capabilities to the modern planetarium theater, allowing audience interaction and making the visitor an integral part of the show.

A modern planetarium theater combines full-color video-graphics with stars. Recent advances in video technology allow full-dome, full-motion video scenes to be created. The most modern advance consists of a real-time image generator that creates images that can fill the whole dome, and five-button keypads at each seat that allow the audience to control part of the show, meaning that audience members are no longer passive participants in the immersive experience. The first two facilities in the world to house this system were Adler Planetarium in Chicago and the Boeing CyberDome at Exploration Place in Wichita, Kansas.

Career Options

The options for a career in the planetarium and science center industry are wide and varied. Staffing requirements include educators, scientists, computer graphic artists, teachers, exhibit designers, writers, marketing staff, and administrators.

International Planetarium Society

The International Planetarium Society, founded in 1970, is the organization for planetarium professionals. Representatives from regional planetarium associations from around the world form its council, and biennial conferences provide opportunities for exchanging ideas and experiences.

Association of Science and Technology Centers

Founded in 1973, the Association of Science and Technology Centers now numbers more than 550 members in forty countries. Members include not only science and technology centers and science museums but also nature centers, aquariums, planetariums, zoos, botanical gardens, space theaters, and natural history and children's museums. SEE ALSO ASTRONOMY, KINDS OF (VOLUME 2); CAREERS IN ASTRONOMY (VOLUME 2); CAREERS IN SPACE SCIENCE (VOLUME 2).

Martin Ratcliffe

Bibliography

Wilson, Kenneth, ed. *So You Want to Build a Planetarium.* Rochester, NY: International Planetarium Society, 1994. Also available at <http://www.ibiblio.org/ips/sywtbap3.html>.

Internet Resources

Association of Science and Technology Centers Page. <http://www.astc.org/>.

Chartrand, Mark R. "A Fifty Year Anniversary of a Two Thousand Year Dream." *The Planetarian.* September 1973. <http://www.griffithobs.org/IPSDream.html>.

International Planetarium Society Page. <http://www.ips-planetarium.org/>.

Planetary Exploration, Future of

The first artificial satellites launched into Earth orbit were part of an international scientific program called the International Geophysical Year. They returned data on Earth and its space environment. Before long, the United States and the Soviet Union began sending spacecraft to study the Moon and, later, other planets.

Sending Spacecraft to the Planets

The U.S. Ranger spacecraft (1961–1965) were designed to crash into the Moon, transmitting television images right up to the moment of impact. The Surveyor spacecraft (1966–1968) soft-landed on the Moon, verifying that the lunar surface would support an **Apollo** lander, taking pictures of the surface surroundings, and performing the first crude geochemical analyses of lunar rocks. A series of Lunar Orbiter spacecraft photographed the Moon, developed the film onboard, scanned the developed images, and transmitted the scans to Earth. The Orbiter photographs constituted the primary database for planning the Apollo landings and comprised the only global set of pictures available to scientists for over twenty-five years.

A Soviet Zond spacecraft (1965) returned the first pictures of the farside of the Moon, the side that always faces away from Earth. Although of poor quality, the pictures showed features that the Soviets were allowed to name through international agreements. Thus, there are names such as

Apollo American program to land men on the Moon. Apollo 11, 12, 14, 15, 16, and 17 delivered twelve men to the lunar surface between 1969 and 1972 and returned them safely back to Earth

Lunar Orbiter I launched September 6, 1966. The mission went on to gather data on the Moon's surface features.

Gagarin and Tsiolkovsky for craters on the farside. The Soviet Luna series of spacecraft performed several landings on the Moon, returning pictures and other data. Three of those spacecraft, Luna 16 (1970), Luna 20 (1972), and Luna 24 (1976), acquired lunar surface material by drilling and returned the samples to Earth. No other robotic spacecraft has ever collected extraterrestrial material and returned it to Earth.

The United States explored the inner planets (sometimes called the terrestrial planets) with the Mariner series of spacecraft (1962–1973). The tiny Pioneers 10 (1973) and 11 (1974) were sent hurtling past Jupiter and Saturn, taking crude pictures and measuring magnetic fields and charged particles. These Lilliputian explorers are still flying far beyond the planets, returning data on the farthest reaches of the Sun's influence over **interstellar** space.

interstellar between the stars

An artist's rendition of a laser power station that might one day propel spacecraft through the solar system by tapping the energy stores of the local environment.

Following Pioneer were the larger and more capable Voyagers 1 (September, 1977) and 2 (August, 1977), which completed remarkable journeys, passing all of the large outer planets—Jupiter, Saturn, Uranus, and Neptune. The Voyager mission took advantage of a rare alignment of the outer planets that allowed the spacecraft to receive gravitational boosts at each planet, which were necessary to complete the journey to the next planet.

Until recently, the Soviet Union was the only other nation to attempt planetary exploration. Although all Soviet missions to Mars have failed, the Soviets achieved a unique and amazing success by landing two Venera spacecraft (1981) on the surface of Venus. These landers survived the hellish surface environment long enough to return pictures and send back geochemical data on surrounding rocks.

Exploring a Planet in Stages

reconnaissance a survey or preliminary exploration of a region of interest

The exploration of a planet by spacecraft can be characterized in terms of several stages or levels of completeness: **reconnaissance**, orbital survey, surface investigation, sample return, and human exploration. The first stage, reconnaissance, is accomplished through a **flyby** of the planet. As the spacecraft passes, pictures are taken, measurements of the environment are made, and navigation data is accumulated for future missions.

In the second stage, orbital survey, a spacecraft is placed in orbit around the planet. A planetary photographic database is accumulated and **remote sensing** observations are conducted. Depending on the sensors aboard the spacecraft, data may be collected on planetary surface composition; atmospheric composition, structure, and dynamics; the nature of the gravity field, which can yield information about the internal structure of the planet; and the nature of the magnetic field. A series of orbiters may be flown over time, observing different phenomena or improving the level of detail and resolution of the data.

Eventually, a landing is made on the planet for surface investigation. Two Viking landers settled onto the surface of Mars in 1976 to test for signs of biological activity in the Martian soil. Actual soil samples were placed in special chambers on the lander. Similarly, certain geochemical or geological measurements cannot be made remotely. A lander can also observe the planet at small scales that cannot be imaged from orbit. In some cases, a **rover** can leave the lander and explore the surroundings. Such was the case of the small rover named Sojourner on the Pathfinder mission to Mars (1996).

As scientists accumulate knowledge about a planet, they seek answers to questions of increasing complexity. At some point, the measurements required are too complex and demanding to be carried out on a robotic spacecraft of limited capability. A sample return mission can bring pieces of the planet to laboratories on Earth, where the most qualified experts using equipment of the highest sophistication can examine them. Preservation of the scientific integrity of the sample has the highest priority—during collection on the planet, during transit to Earth, and after delivery to a special facility for curation. If the samples are cared for appropriately, they become treasures for future scientists with ever more advanced analytic techniques.

The final stage of study is human exploration. Astronaut explorers, aided by robotic assistants, can observe, experiment, innovate, and adapt to changing conditions in ways that cannot be duplicated by machines. Besides, why should robots have all the fun?

Future Exploration of the Solar System

By the beginning of the twenty-first century, reconnaissance had been completed for the inner solar system and for the giant planets of the outer solar system. Orbital surveys have been accomplished at the Moon (Lunar Orbiters, Clementine, Lunar Prospector), Venus (Magellan), Mars (Mariner 9, Viking, Mars Global Surveyor), and Jupiter (Galileo). The Cassini spacecraft was en route to Saturn, with exploration of that planet expected to begin in 2004. Some asteroids have been photographed by passing spacecraft, and the NEAR Shoemaker spacecraft engaged in a rendezvous with the asteroid Eros. Samples have been returned only from the Moon. **Meteorites** collected on Earth are samples from (unknown) asteroids, from the Moon, and from Mars. Human exploration has occurred briefly and only on the Moon.

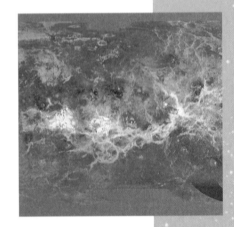

Lava flows on the surface of Venus, as viewed from the spacecraft Magellan (1990–1994). The Magellan mapped 98 percent of Venus' surface, thus revolutionizing our understanding of the planet, particularly its geology.

flyby flight path that takes a spacecraft close enough to a planet to obtain good observations; the spacecraft then continues on a path away from the planet but may make multiple passes

remote sensing the act of observing from orbit what may be seen or sensed below Earth

rover name of a vehicle used to move about on a surface

meteorite any part of a meteoroid that survives passage through Earth's atmosphere

The National Aeronautics and Space Administration (NASA) has future plans for more orbital surveys of Mars as well as landings and sample return missions. Human exploration of Mars is being discussed, as is further human exploration of the Moon, without definite commitments. The European Space Agency (ESA), as well as individual European nations, will join NASA in exploring Mars. European and Japanese spacecraft will visit the Moon. India and China are discussing possible Moon missions. Some private companies have plans to land on the Moon through profit-seeking ventures. ESA is planning an orbital survey of Mercury. The Japanese space agency, the Institute of Space and Aeronautical Science, is working on a sample return from an asteroid.

Planetary exploration is becoming an international activity. Equally exciting is the prospect of planetary missions sponsored by institutions other than the traditional government agencies. The future may hold surprises for all of us. SEE ALSO APOLLO (VOLUME 3); APOLLO LUNAR LANDING SITES (VOLUME 3); EARTH—WHY LEAVE? (VOLUME 4); EXPLORATION PROGRAMS (VOLUME 2); MARS MISSIONS (VOLUME 4); RECONNAISSANCE (VOLUME 1); ROBOTIC EXPLORATION OF SPACE (VOLUME 2).

Wendell Mendell

Bibliography

Kluger, Jeffrey. *Journey Beyond Selene: Remarkable Expeditions to the Ends of the Solar System.* New York: Simon & Schuster, 1999.

Moore, Patrick. *Mission to the Planets: The Illustrated Story of Man's Exploration of the Solar System.* New York: Norton, 1990

Morrison, David, and Tobias Owen. *The Planetary System,* 2nd ed. Reading, MA: Addison-Wesley, 1996.

Neal, Valerie, Cathleen S. Lewis, and Frank H. Winger. *Spaceflight: A Smithsonian Guide.* New York: Macmillan USA, 1995.

Planetary Protection *See Planetary Protection (Volume 4).*

Planetesimals

Planetesimals are the fundamental building blocks of the planets as well as the ancestors of asteroids and comets. To understand them and their importance, one must first understand how planets form.

The solar system formed 4.6 billion years ago from an interstellar cloud of gas and dust. When this gaseous cloud became unstable, it collapsed under the force of its own gravity and became a flattened, spinning disk of hot material. The region with the greatest concentration of mass became the Sun. The rest of the mass, perhaps only a little more than the mass of the Sun, eventually cooled enough to allow solid grains to condense, with rocky ones close to the Sun and icy ones farther away. The grains settled near the midplane of the disk, where mutual collisions allowed them to slowly grow into pebble-sized objects. At this point, the story is less clear. Some astronomers claim particle velocities in the disk remained low enough to allow the pebbles to stick to one another. Others argue that pebbles are generally not very sticky, and gravitational forces alone could cause concentrated swarms of pebbles to coalesce. In either case, the process eventually produced planetesimals, which measured a few kilometers across.

Models of planetesimal disks suggest that low relative velocities between the bodies produce accretion (objects hit and stay together) rather than fragmentation (objects break up and disperse). As planetesimals grow still larger, their gravitational attraction increases, allowing them to become even more effective at accreting nearby planetesimals. This process, called runaway growth, allows some planetesimals to reach the size of the Moon or even Mars. These so-called protoplanets were the precursors of the current planets of the solar system. For reference, 2 billion planetesimals, each one being 10 kilometers (6 miles) in diameter, are needed to make an Earth-sized planet.

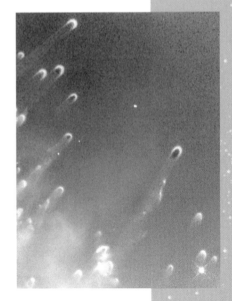

Planetesimal clouds formed in the Helix Nebula.

At this point in solar system evolution, the disk mass is dominated by protoplanets, planetesimals, and gas. Mutual gravitational interactions force most protoplanets to collide and merge, eventually producing small planets such as Earth. If a protoplanet grows large enough, however, it can also gravitationally capture enormous amounts of the remaining gas. This explains why Jupiter and Saturn are so much larger than Earth.

The same interactions that cause protoplanets to collide also stir up the remaining planetesimals. Most of these objects end up impacting existing protoplanets or are thrown out of the solar system. The leftovers that managed to stay in the stable regions of the solar system until planet formation ended are now called asteroids and comets. The asteroid belt is a population of rocky planetesimals located between the orbits of Mars and Jupiter. The Kuiper belt and Oort cloud are populations of icy planetesimals located beyond the orbit of Neptune. Even now, mutual collisions between and among asteroids and comets as well as planetary interactions cause pieces of the survivors to escape their small body reservoirs. A few of these multikilometer or smaller objects strike Earth. Small impactors deliver meteorites, while large ones infrequently wreak global devastation. SEE ALSO ASTEROIDS (VOLUME 2); COMETS (VOLUME 2); GRAVITY (VOLUME 2); IMPACTS (VOLUME 4); KUIPER BELT (VOLUME 2); METEORITES (VOLUME 2); OORT CLOUD (VOLUME 2); ORBITS (VOLUME 2); SMALL BODIES (VOLUME 2).

William Bottke

Bibliography

Safronov, Viktor S. *Evolution of the Protoplanetary Cloud and Formation of the Earth and the Planets.* Moscow: Nauka Press, 1969. Trans. NASA TTF 677, 1972.

Ward, Wilham R. "Planetary Accretion." In *Completing the Inventory of the Solar System,* eds. Terrence W. Rettig and Joseph M. Hahn. San Francisco: Astronomical Society of the Pacific, 1996.

Pluto

Pluto is the only planet in the solar system still unvisited by a spacecraft. Its status as the only planet in our Sun's family still studied purely by telescope is unique—and frustrating—to planetary scientists trying to uncover its secrets.

Pluto's Strange Orbit

American astronomer Clyde Tombaugh discovered Pluto in 1930. Despite astronomers' best efforts, Pluto's faintness and star-like appearance allowed the planet to keep most of its secrets. For twenty-five years, we could only refine our knowledge of its strange orbit, finding it on old photographs and taking new ones. Pluto's orbit is more **eccentric** and more tilted (inclined) than any other planet, taking 248.8 years to make one trip around the Sun. At perihelion (closest approach, which last occurred in 1996), it is only 60 percent as far from the Sun as at aphelion (farthest approach). So at perihelion, Pluto is closer to the Sun than Neptune ever gets. Yet, Pluto and Neptune cannot collide for two reasons. First, the relative inclination of the two orbits means their paths do not intersect. Second, Pluto is in a 2:3 orbit-orbit resonance with Neptune. This means that for every two trips Pluto makes around the Sun, Neptune makes exactly three. When Pluto is at perihelion, Neptune is on the other side of the Sun.

The Significance of Brightness Measurements

In 1955, photometry (brightness measurements) of Pluto showed a repetition of 6.38 days—the length of Pluto's day. Two trends in the evolution of the brightness have since been found. First, its **amplitude** has increased from about 10 percent to a current value of 30 percent. This tells us that the **subsolar point** has been moving equatorward, and that the planet's spin axis must be severely tilted. Second, the average brightness has faded over the years, evidence that Pluto's poles are likely brighter than its equator. Decades of photometry have been interpreted to derive maps of Pluto's surface reflectance, or albedo. These are comparable in detail with what the Hubble Space Telescope has been able to reveal.

The Size and Composition of Pluto and Its Moon

Little regarding Pluto's size or composition was known until recently. In 1976 the absorption of methane was discovered in Pluto's spectrum. This implied a bright, icy planet, and therefore a small radius. In 1978 James Christy, then an astronomer at the United States Naval Observatory, discovered Pluto's satellite, which was named Charon. Orbiting Pluto with the same 6.38726-day period as Pluto's spin, Charon was the key to unlocking Pluto's secrets. By timing the orbital period and measuring the estimated separation between the two, astronomers could compute the total mass of the system—about 0.002 Earth masses. Charon orbits **retrograde**, and Pluto spins backwards (just like Venus and Uranus).

Charon's orbital plane above Pluto's equator was seen edge-on in 1988. This produced a series of **occultations** and eclipses of and by the satellite, each half-orbit, from 1985 to 1992. Timing these "mutual events" allows calculation of the radii for both bodies—approximately 1,153 kilometers (715 miles) for Pluto and 640 kilometers (397 miles) for Charon. The sum is about the radius of the Moon. When Charon hid behind the planet, Pluto's spectrum could be observed uncontaminated by its moon. This spectrum, when subtracted from a combined spectrum of the pair taken a few hours before or after, yields the spectrum of Charon. Pluto's spectrum showed methane frost: the gas we use for cooking is frozen solid on its surface! Charon's spectrum revealed nothing but dirty water ice. (Independent mea-

eccentric the term that describes how oval the orbit of a planet is

amplitude the height of a wave or other oscillation; the range or extent of a process or phenomenon

subsolar point the point on a planet that receives direct rays from the Sun

retrograde having the opposite general sense of motion or rotation as the rest of the solar system, clockwise as seen from above Earth's north pole

occultations phenomena that occurs when one astronomical object passes in front of another

surements show the amount of methane on Pluto varies with longitude. Bright regions have more methane than dark regions.) When Charon passed between Pluto and Earth, it (and its shadow) selectively hid different portions of its **primary**. Interpretation of these measurements is complicated but has allowed refined albedo (or reflectivity) maps of one hemisphere of Pluto to be extracted.

An artist's rendering of Pluto, partially obscured by its satellite, Charon. Pluto is the only planet in the solar system still unvisited by a spacecraft.

Surface and Atmospheric Readings

The surface temperature of Pluto is currently under debate. Two results have been published: about 40°K (−233°C; −388°F) and about 55°K (−218°C; −361°F). The first value is similar to the temperature on Triton, Neptune's largest moon; the latter is more consistent with Pluto's lower albedo. In either case, it is very cold. Water ice on Pluto is harder than steel is at room temperature! Misconceptions exist about how dark it would seem for an astronaut on Pluto. Despite the planet's remote distance, the Sun would appear to have the brightness of about 70 full Moons on Earth. Combine this with the bright, icy surface and one would have no problems navigating the surface.

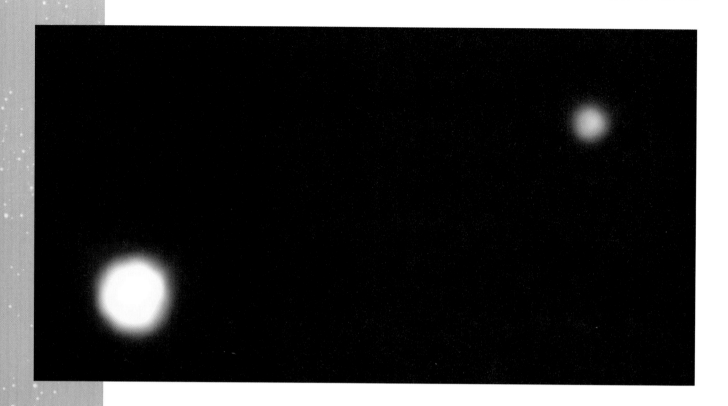

Pluto (lower left) and its moon, Charon (top right), as seen through the Hubble Space Telescope, February 21, 1994. At the time this image was taken, Pluto was 4.4 billion kilometers (2.6 billion miles) from Earth.

primary the body (planet) about which a satellite orbits

adaptive optics the use of computers to adjust the shape of a telescope's optical system to compensate for gravity or temperature variations

On June 9, 1985, Pluto passed in front of a star. Rather than blinking out, the starlight gradually dimmed due to refraction by an atmosphere. Too dense to be methane alone, the atmosphere was suspected to contain nitrogen and carbon monoxide. Both have since been identified on Pluto's surface, with nitrogen comprising about 97 percent of the ground material. From details of precisely how the starlight faded, scientists believe there is a temperature increase close to the surface, much like on Earth. Pluto's atmospheric pressure is only a few millionths that of Earth, and the atmosphere actually may "frost out" with increasing distance from the Sun.

The Hubble Space Telescope has been used to measure the size of Charon's orbital radius, about 19,500 kilometers (12,090 miles, or approximately 1.5 Earth diameters). Densities have also been calculated: 1.8 to 2.0 grams per cubic centimeter (112 to 125 pounds per cubit foot) for Pluto and 1.6 to 1.8 grams per cubic centimeter (100 to 112 pounds per cubit foot) for Charon. From the density, scientists can infer the internal composition, a roughly 50-50 mix of rock and ice.

Future Spacecraft Visit?

Efforts to learn more continue. New large Earth-based telescopes equipped with **adaptive optics** and fast computers will allow the blurring effects of our atmosphere to be nullified, surpassing the resolution of Hubble's rather small 2.4-meter (4.9-foot) mirror. In contrast, the "faster, better, cheaper" policy of the National Aeronautics and Space Administration (NASA) has led to a halt of the Pluto–Kuiper Express spacecraft. A new mission profile, called the New Horizons Pluto–Kuiper Belt Mission, was approved by Congress in 2001. However, funding for this mission is not in the President's proposed budget for 2002. Launch must happen by 2006 or Jupiter will no longer be in position to slingshot the craft towards Pluto with a **gravity assist**, and the trip to Pluto will take years longer. We will have to wait the better part of a Jupiter orbit (11.8 years) until the geometry repeats itself. By then, Pluto's atmosphere may have frozen out. Until the task is taken seriously, Pluto will remain the only planet unvisited by a spacecraft. SEE ALSO HUBBLE SPACE TELESCOPE (VOLUME 2); KUIPER BELT (VOLUME 2); NASA (VOLUME 3); ORBITS (VOLUME 2); PLANET X (VOLUME 2); PLANETARY EXPLORATION, FUTURE OF (VOLUME 2); TOMBAUGH, CLYDE (VOLUME 2).

Robert L. Marcialis

Bibliography

Binzel, Richard P. "Pluto." *Scientific American* 262, no. 6 (1990):50–58.

Marcialis, Robert L. "The First Fifty Years of Pluto-Charon Research." In *Pluto and Charon*, ed. S. Alan Stern and David J. Tholen. Tucson: University of Arizona Press, 1997.

Internet Resources

PLUBIB: A Pluto-Charon Bibliography. Ed. Robert L. Marcialis. University of Arizona. <http://www.lpl.arizona.edu/∽umpire/science/plubib_home.html>.

Pulsars

The discovery of pulsars in 1967 was a complete surprise. Antony Hewish and his student Jocelyn Bell (later Bell Burnell) were operating a large radio antenna in Cambridge, England, when they detected a celestial source of radio waves that pulsed every 1.3373 seconds. Never before had a star or **galaxy**, or any other astronomical phenomenon, been observed to tick like a clock.

Hewish and Bell considered a number of exotic explanations for the speed and regularity of the pulsing radio source, including the possibility that it was a beacon from an extraterrestrial civilization. Within a few years, the correct explanation emerged, which is no less exotic. A pulsar is a city-sized spinning ball of ultradense material that emits beams of radiation, which flash Earth-like lighthouse beams, as it spins.

How Pulsars Are Created

Pulsars are produced when certain types of stars stop producing energy and collapse. The attractive force of gravity is always trying to contract the material of a star into an ever-smaller ball, but a star can maintain its size for billions of years because of the heat and pressure produced by nuclear reactions within it. When a star finally exhausts its supply of nuclear fuel, it collapses. An ordinary star (such as the Sun) will quietly contract into an Earth-sized glowing ember called a white dwarf. A more massive star will explode violently in an event called a supernova. It is within the detritus of such explosions that pulsars are born.

The reason for the explosion is that when the star collapses all the way down to a diameter of about 20 kilometers (12 miles), its atoms are packed so closely that their **protons** and **electrons** merge to form **neutrons**, which repel each other by nuclear forces and oppose further shrinkage. The collapsing material suddenly rebounds, producing a huge expanding fireball. In

gravity assist using the gravitational field of a planet during a close encounter to add energy to the motion of a spacecraft

galaxy a system of as many as hundreds of billions of stars that have a common gravitational attraction

protons positively charged subatomic particles

electrons negatively charged subatomic particles

neutrons subatomic particles with no electrical charge

Interior view of the Crab Nebula showing the Crab Pulsar, formed from a supernova explosion over 900 years ago.

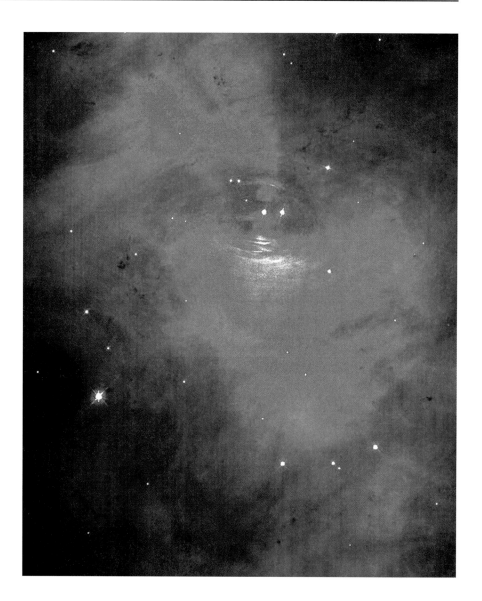

some cases, the neutron matter is obliterated by the blast, or it keeps shrinking all the way down to a single point, forming a black hole. Sometimes, however, the dense nugget of neutrons survives the explosion, in which case it becomes what is called a neutron star.

Pulsars and Neutron Stars

Neutron stars are unfathomably dense. A marble with the same density as a neutron star would weigh as much as a boulder 400 meters (0.25 mile) across. Because the rotation of the star is amplified during its collapse (much as an ice-skater spins faster by pulling in her arms), neutron stars are born spinning quickly, as fast as 100 revolutions per second. They also have the most intense magnetic fields known in the universe. If Earth had a magnetic field as strong, it would erase credit cards as far away as the Moon. This powerful magnetism causes intense beams of radio waves to be launched from both magnetic poles of at least some neutron stars; the poles swing around as the star rotates and may flash Earth if they happen to be oriented in the right direction.

In 1934, just two years after the discovery of the neutron, astronomers Walter Baade and Fritz Zwicky predicted that neutron stars should exist. Five years later, Robert Oppenheimer and George Volkoff published a detailed theory of neutron stars. But none of these scientists knew whether neutron stars could ever be observed with telescopes. They were expected to be so dim as to be invisible; nobody predicted they would emit focused beams of radiation. Even now that more than 1,500 pulsars have been discovered, nobody understands the details of how the radiation beams are produced.

Whatever the mechanism, a pulsar keeps pulsing for millions of years. Although the pulse rate is remarkably steady, it does slow down by a tiny but measurable amount. For example, the pulsar at the center of the Crab **Nebula** (the site of a supernova that occurred in 1054 C.E. in the constellation Taurus) blips once every 33 milliseconds, but this pulse period is slowing by 0.013 milliseconds every year. Eons from now, the pulsar will spin too slowly to produce radiation beams bright enough to observe, and will spend the rest of eternity as a quiet neutron star.

nebula clouds of interstellar gas and/or dust

X-ray Pulsars

A neutron star can be rescued from this oblivion, and resume its identity as a pulsar, if it happens to have a companion star. Stars are often found in pairs (or even triplets or quadruplets) and when one star explodes in a supernova, the other may survive. Eventually the intense gravity of the pulsar may rip material away from the giant star. As the material swirls down to the pulsar's surface, it heats up to millions of degrees and glows brightly in **X rays**. The swirling matter may be funneled by the neutron star's magnetic field onto a hot spot on the neutron star's surface; as this spot rotates with the neutron star, astronomers see pulses of X rays, and the neutron star regains the limelight as an "X-ray pulsar."

X rays a form of high-energy radiation just beyond the ultraviolet portion of the electromagnetic spectrum

Millisecond Pulsars

It is also possible that the swirling matter will cause the neutron star to spin faster and faster, like a top being spun up. The rotation period can become as short as a few thousandths of a second, which is enough to reactivate the radio pulses, and the neutron star is reborn as a "millisecond pulsar." The fastest known millisecond pulsar spins 642 times per second, which is impressive for something bigger than London and more massive than the Sun.

Areas of Future Research

Despite all of this knowledge, the life cycles of pulsars are still a subject of research. Many of the unanswered questions are about young pulsars: How often do supernovas produce them? Can they be created in other ways? Are they always born spinning quickly?

In particular, due to recent advances in X-ray astronomy, a new category of young pulsars has been discovered consisting of objects that spin relatively slowly and emit X rays rather than radio waves, even though they do not have a stellar companion. A consensus is developing that these unusual pulsars should be called "magnetars," because they seem to have magnetic fields hundreds of times larger than the already enormous fields of "ordinary" pulsars. If proven to be accurate, this would be yet another surprising development in the history of pulsar science. SEE ALSO ASTRONOMY, KINDS OF (VOLUME 2); BLACK HOLES (VOLUME 2); EINSTEIN, ALBERT (VOLUME 2); GRAVITY (VOLUME 2); STARS (VOLUME 2); SUPERNOVA (VOLUME 2).

Joshua N. Winn

Bibliography

Greenstein, George. *Frozen Star*. New York: Freundlich Books, 1983.

Kaspi, Victoria M. "Millisecond Pulsars: Timekeepers of the Cosmos." *Sky and Telescope* 89, no. 4 (1995):18–23.

Lyne, Andrew G., and Francis Graham-Smith. *Pulsar Astronomy*, 2nd ed. Cambridge, UK: Cambridge University Press, 1998.

Nadis, Steve. "Neutron Stars with Attitude." *Astronomy* 27, no. 3 (1999):52–56.

Winn, Joshua N. "The Life of a Neutron Star." *Sky and Telescope* 98, no. 1 (1999): 30–38.

Robotic Exploration of Space

flyby flight path that takes the spacecraft close enough to a planet to obtain good observations; the spacecraft then continues on a path away from the planet but may make multiple passes

crash-lander or hard-lander: a spacecraft that collides with the planet, making no (or little) attempt to slow down; after collision, the spacecraft ceases to function because of the (intentional) catastrophic failure

orbiter spacecraft that uses engines and/or aerobraking, and is captured into circling the planet indefinitely

atmospheric probe a separate piece of a spacecraft that is launched from it and separately enters the atmosphere of a planet on a one-way trip, making measurements until it hits a surface, burns up, or otherwise ends its mission

In January 1959, only a little more than a year after the launching of Sputnik 1, the Soviet Union's Luna 1 flew 5,955 kilometers (3,692 miles) above the surface of the Moon, thus quickly heralding the age of planetary exploration. Since then, nearly 100 successful robotic missions to obtain closer looks at the planets, their moons, and asteroids have been launched, mainly by the United States and the Soviet Union. Among the planets, only Pluto, so far away that a signal from it would take five hours to reach Earth, has not had a spacecraft fly at least close by it. During the more than four decades of planetary exploration, continued improvements and miniaturization in electronics and computers, and in rocketry and instrument techniques in general, have been used to gain scientific knowledge.

The term "robotic" is used for any spacecraft without a human pilot. In addition to the obvious method of sending instructions directly from Earth, control of robotic spacecraft can also be programmed precisely in advance or programmed to react to the environment. Missions have included **flybys**, **crash-landers**, **orbiters**, **atmospheric probes**, and **soft-landers**. Some robotic spacecraft have multiple components, for instance an orbiter that released a lander (Viking 1 and Viking 2, at Mars) or an atmospheric probe (Galileo, at Jupiter). Robotic spacecraft have so far been essentially in the business of sending information or data about other planetary bodies back to Earth. Only spacecraft to the Moon have returned samples; in addition to the crewed Apollo missions, three Soviet spacecraft also brought back samples.

The Challenge of Robotic Planetary Exploration

Lunar and planetary exploration is a proposition far different from orbiting Earth. A launch vehicle can throw only about one-fifth as much mass out of Earth's gravitational field as it is capable of putting into Earth orbit. Time becomes more and more a factor with distance: To send a signal to and then from the Moon takes only two seconds. For Mars it takes between eighteen and forty-five minutes, depending on the relative positions of Mars and Earth. Everything has to be planned and programmed, well in advance.

An artist's rendition of the Mars Rover Sojourner (foreground) and the Pathfinder Lander in operation on the Martian surface. These robots work in environments that may be harmful to humans or in situations where sending a human crew would be too costly.

Interplanetary spacecraft must last for years in space. It takes a few minutes to reach Earth orbit but almost a year to get to Mars. Galileo was launched in October 1989, and arrived at Jupiter through a complex route more than six years later. Once there, it began orbiting and sending back data, continuing to do so into the early twenty-first century. Spacecraft have to be remarkably reliable, because there is no means of replacing or repairing parts. Interplanetary space has dangers from solar and **cosmic radiation**, and there is the potential for damage from **solar wind**, dust, and even larger chunks of material. Spacecraft have to be resilient to the range of cold and hot temperatures, and the uneven temperatures, to which they are exposed.

Spacecraft: Getting There

The first lunar probe, Luna 1, went on to become the first artificial object to orbit the Sun. It was equipped to measure solar and cosmic radiation, interplanetary magnetic fields, the **micrometeoroid flux**, and the composition of gases. During the same year, Luna 2 became the first artificial object

soft-lander spacecraft that uses braking by engines or other techniques (e.g., parachutes, airbags) such that its landing is gentle enough that the spacecraft and its instruments are not damaged, and observations at the surface can be made

cosmic radiation high energy particles that enter Earth's atmosphere from outer space causing cascades of mesons and other particles

solar wind a continuous, but varying, stream of charged particles (mostly electrons and protons) generated by the Sun; it establishes and affects the interplanetary magnetic field; it also deforms the magnetic field about Earth and sends particles streaming toward Earth at its poles

micrometeoroid flux the total mass of micrometeoroids falling into an atmosphere or on a surface per unit of time

aerobraking the technique of using a planet's atmosphere to slow down an incoming spacecraft; its use requires the spacecraft to have a heat shield, because the friction that slows the craft is turned into intense heat

Gemini the second series of American-piloted spacecraft, crewed by two astronauts; the Gemini missions were rehearsals of the spaceflight techniques needed to go to the Moon

payload any cargo launched aboard a rocket that is destined for space, including communications satellites or modules, supplies, equipment, and astronauts; does not include the vehicle used to move the cargo or the propellant that powers the vehicle

parking orbit placing a spacecraft temporarily into Earth orbit, with the engines shut down, until it has been checked out or is in the correct location for the main burn that sends it away from Earth

to crash into the Moon, and Luna 3 the first to circle behind the Moon. Luna 3 also sent back the first human glimpses of the farside of the Moon. These earliest missions illustrate most of the basic needs of robotic spacecraft: launch vehicle, power, communications, and scientific instruments. What they lacked that most subsequent missions (apart from the Ranger series) had was a propulsion system or a navigation system; they were just thrown at the Moon.

The earliest missions to any planets have been flybys, with some impacts. A flyby has the advantage of seeing more of the planet; an impact allows a closer look but there is then an immediate loss of the spacecraft. An orbiter allows for a longer, more complete look at a planet but requires an engine and fuel to slow the spacecraft and insert it into orbit. These engines have to be controllable and thus cannot be solid fuel rockets that cannot be turned off. If the planet has an atmosphere, it can be used for **aerobraking** to slow down the spacecraft, and this has been done at Mars. In many cases the engine is used several times to change the orbit. A controllable engine is also used for midcourse corrections, which are necessary to reach exactly the required destination.

An engine is needed for soft-landings, and in rare cases, for taking off again. Soft-landing techniques depend on whether the planet has an atmosphere or not. If it does, then parachutes, air bags, and other devices can be used to get hardware safely to the surface, after first slowing the spacecraft with engines and aerobraking. If there is no atmosphere, descent must be under the control of rocket engines.

The smallest spacecraft have weighed a little over 200 kilograms (440 pounds), the largest nearly 6,000 kilograms (13,230 pounds), as they left Earth orbit. For comparison, the mass of the Mercury spacecraft in orbit was about 1,400 kilograms (3,080 pounds), whereas a **Gemini** spacecraft weighed about 3,800 kilograms (8,370 pounds). A huge amount of mass on the launch pad is needed to get a few kilograms of scientific instruments to their destination. An Atlas-Agena space rocket and its load on the launch pad weighed about 125,000 kilograms (275,575 pounds)—about 90 percent of which was fuel—and at its destination the Ranger spacecraft was less than 400 kilograms (880 pounds). On the launch pad, the Luna 16 launch vehicle and its **payload** had a mass of about 1 million kilograms (2.2 million pounds). The spacecraft on the way to the Moon was about 5,600 kilograms (12,345 pounds); the empty lander, 1,900 kilograms (4,185 pounds); and the little sphere that eventually was parachuted back to Earth containing its precious cargo of only about 100 grams (3.5 ounces) of lunar soil, less than 40 kilograms (88 pounds).

Generally a spacecraft is first placed in an Earth **parking orbit**, and from there is given another boost to give it the appropriate interplanetary trajectory. In some cases, such as the Magellan mission to Venus and the **Galileo mission** to Jupiter, the spacecraft was carried first to **low Earth orbit** on a space shuttle. While early lunar **trajectories** were fairly direct, many later missions had more complex journeys. Mariner 2 to Venus was launched in the direction opposite of Earth's orbit, then gradually fell in towards the Sun, overtaking Earth, and catching up with its target. Voyager 2 used **gravitational assists** consecutively to fly past Jupiter, Saturn, Uranus, and Neptune, with ten years between the first and last encounters. Galileo,

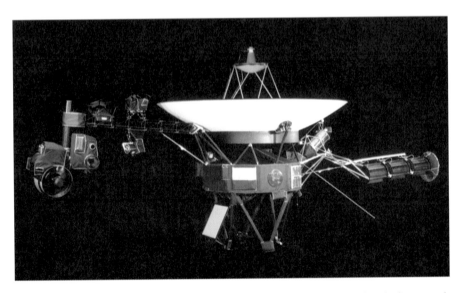

The spacecraft Voyager was developed in the late 1970s, when NASA mission designers realized that the giant outer planets—Jupiter, Saturn, Uranus and Neptune—would soon align in such a way that single spacecraft could theoretically use gravity assists to hop from one planet to the next.

now in orbit around Jupiter, flew by Venus and twice by Earth to obtain gravitational assists, and also flew by two asteroids on its journey. Most of a spacecraft's flight is **ballistic**, that is, it is not powered but is pulled by gravity, with engines needed for course corrections.

All spacecraft must have power, to run instruments and controls, and communications systems, to receive and send information. In the inner solar system, including Mars, at least some of the power can be obtained from solar panels. At the outer planets there is insufficient sunlight for solar panels. For these, spacecraft are designed around large-dish antennae powered with **radioisotope thermoelectric generators**, which generate heat by natural radioactive decay. Spacecraft have been designed to operate on very little power. Cassini's generators produced 815 watts at launch and will still be producing over 600 watts at the end of its mission at Saturn (by comparison, a typical household light bulb is 100 watts).

Missions and Scientific Instruments

Spacecraft instruments have used much of the **electromagnetic spectrum** to observe the planets and their surroundings, including low-energy radar waves, **infrared**, visible light (with which we are most familiar), **ultraviolet**, and high-energy **X rays**. Most instruments are passive but some are active, including **laser-pulsing** to measure distance (and hence topography) and radar sounding. Particles and dust have also been measured directly. A wide variety of instruments have been carried on the nearly 100 missions undertaken through the early twenty-first century, but some have been more commonly used than others.

All spacecraft carry a radio transmitter, used for transmitting both data about the spacecraft itself and about the scientific measurements. As a spacecraft goes behind a planet with an atmosphere, the changes in the radio signal provide information on the atmosphere, such as its density and thickness.

Galileo mission successful robot exploration of the outer solar system; this mission used gravity assists from Venus and Earth to reach Jupiter, where it dropped a probe into the atmosphere and studied the planet for nearly seven years

low Earth orbit an orbit between 300 and 800 kilometers above Earth's surface

trajectories paths followed through space by missiles and spacecraft moving under the influence of gravity

gravitational assist the technique of flying by a planet to use its energy to "catapult" a spacecraft on its way, thus saving fuel, and thus mass and cost of a mission; gravitational assists typically make the total mission duration longer, but they also make things possible that otherwise would not be possible

ballistic the path of an object in unpowered flight; the path of a spacecraft after the engines have shut down

radioisotope thermo-electric generator device using solid state electronics and the heat produced by radioactive decay to generate electricity

electromagnetic spectrum the entire range of wavelengths of electromagnetic radiation

infrared portion of the electromagnetic spectrum with wavelengths slightly longer than visible light

ultraviolet the portion of the electromagnetic spectrum just beyond (having shorter wavelengths than) violet

X rays a form of high-energy radiation just beyond the ultraviolet portion of the electromagnetic spectrum

laser-pulsing firing periodic pulses from a powerful laser at a surface and measuring the length of time for return to determine topography

Lunar Orbiter a series of five unmanned missions in 1966 and 1967 that photographed much of the Moon at medium to high resolution from orbit

magnetometer an instrument used to measure the strength and direction of a magnetic field

wavelength the distance from crest to crest on a wave at an instant in time

minerals crystalline arrangements of atoms and molecules of specified proportions that make up rocks

The signal is used to track the spacecraft, and variations in the spacecraft motion are informative. The overall gravitational tug provides information about the mass of the planet. Tiny measurable changes in the motion of the spacecraft tell about variation in the gravity field, which in turn tell about how mass is distributed in the planet. This standard technique of planetary study was used in the analysis of **Lunar Orbiter** tracking to determine mass concentrations beneath many of the large circular basins on the Moon.

A **magnetometer** has been carried on many spacecraft because they are small, reliable, and useful instruments. Even Luna 1 carried a magnetometer and found that the Moon produces no significant magnetic field, quite unlike Earth. The magnetic fields around planets and asteroids are informative, like gravity, about what the interior is like at the present time. They also provide information about what things may have been like in the past. The magnetometer carried on the Mars Global Surveyor has shown areas that were magnetized at some time in the past. Magnetometers discovered the powerful magnetic fields produced by Jupiter.

Many spacecraft carry a camera providing images that approximate what human eyes see. These come in many different varieties, some providing wide views, some having very high resolutions. The Ranger missions to the Moon carried cameras as the prime instrument, to see what the Moon was like as we saw it from closer and closer. With cameras we have seen the volcanoes, craters, and valleys of Mars, the cratered Moonlike surface of Mercury, the smooth but cratered surfaces of some asteroids, and the cracked icy crusts of some of Jupiter's moons. Images have provided huge amounts of geological information about the planets. Geologists are still using images from the Moon collected in the 1960s. For Venus, which has a thick atmosphere, such images are no good for surface studies, and instead radar techniques have been used to map and understand the surface features.

Chemical and mineralogical data about both surfaces and atmospheres have been obtained from orbit. Natural X-ray and gamma-ray sources provide direct chemical information but are blocked by thick atmospheres. They have been used extensively for the Moon, however, and relevant instruments are part of a planned mission to Mercury. The Martian atmosphere is thin enough that such instruments can be used there, but the spacecraft carrying them, Mars Observer, failed. X-ray fluorescence (carrying their own X-ray sources) or other chemistry instruments were carried by the Viking landers on Mars, the Surveyor landers on the Moon, and the Venera landers on Venus, to measure the composition of rocks and soils, as well as to search for traces of life. Spectral reflectance observations, in the ultraviolet through visible through infrared **wavelengths**, have been used extensively to understand mineralogy. Specific **minerals** absorb particular wavelengths, so measurements are made that show these absorptions, and in this way the mineralogy can be inferred. Such absorptions also show the presence of phases, such as water, in atmospheres.

Some missions are much more complicated, involving different kinds of instrumentation. Galileo is not only observing Jupiter and its moons from orbit, but it also launched a probe into Jupiter's atmosphere, to measure its chemical composition and physical properties (e.g., density). Atmospheric probes have also been used at Venus, and the Cassini mission to Saturn is carrying a probe to be launched into the atmosphere of Titan, that planet's largest moon. While the Soviet Union used primitive rovers on two lunar missions in the 1960s, the only other **rover** yet used to add to our knowledge of planets was flown on Mars Pathfinder. Its rover Sojourner made small forays to investigate rocks and soils near the lander.

rover name of a vehicle used to move about on a surface

Spacecraft: The Future

Spacecraft have flown by every major planet, and most of their important moons, in the solar system. We can hardly say that we know enough about any of them as yet, both from the point of view of types of mission or of instruments flown, or of how much has been seen before a mission ends. Mercury has only been flown by, as have Saturn and the more distant planets. Cassini is on a mission to orbit Saturn, while a Discovery-class mission will orbit Mercury. We have barely looked at comets; the small Discovery mission Stardust is on its way to comet Wild 2. It will meet its coma at 20,000 kilometers per hour (12,400 miles per hour), six times faster than a speeding bullet, collect small particles, and bring them to Earth. Because of the potential relationship with understanding the origin of life, many people think it highly desirable to find out more about the properties of Europa, one of Jupiter's large moons, and its potential sub-ice ocean.

Two types of mission are likely to become more common in the future. One is a sample return, particularly from Mars. These missions are inherently complex. The other category consists of rover missions. A fixed lander obviously has limited capabilities, and the extension of its senses by adding mobility has tremendous advantages. Sojourner demonstrated the usefulness of such machines, but in both a real as well as a metaphorical sense, it only scratched the surface. SEE ALSO EXPLORATION PROGRAMS (VOLUME 2); GOVERNMENT SPACE PROGRAMS (VOLUME 2); NASA (VOLUME 3); PLUTO (VOLUME 2); ROBOTICS TECHNOLOGY (VOLUME 2).

Graham Ryder

Bibliography

Adler, Robert. "To the Planets on a Shoestring." *Nature* 408 (2000):510–512.

Baker, David, ed. *Janes Space Directory, 1999–2000*, 15th ed. Alexandria, VA: Jane's Information Group, 1999.

Binder Alan. "Lunar Prospector: Overview." *Science* 281 (1998):1,475–1,476.

Burrows, William E. *Exploring Space: Voyages in the Solar System and Beyond*. New York: Random House, 1990.

———. *This New Ocean: The Story of the First Space Age*. New York: Random House, 1998.

Kraemer, Robert S. *Beyond the Moon: A Golden Age of Planetary Exploration, 1971–1978*. Washington, DC, and London: Smithsonian Institution Press, 2000.

Milstein Michael. "Hang a Right at Jupiter." *Air and Space* (December/January 2000): 67–71.

Nicks, Oran W. *Far Travellers: The Exploring Machines*. Washington, DC: National Aeronautics and Space Administration, 1985.

Nozette, Stewart, et al. "The Clementine Mission to the Moon: Scientific Overview." *Science* 266 (1994):1,835–1,839.

Pieters, Carlé M., and Peter A. J. Englert, eds. *Remote Geochemical Analysis: Elemental and Mineralogical Composition*. Cambridge, UK: Cambridge University Press, 1993.

Spilker, Linda J., ed. *Passage to a Ringed World: The Cassini-Huygens Mission to Saturn and Titan*. Washington, DC: National Aeronautics and Space Administration, 1997.

Robotics Technology

The word "robot" was coined in 1934 by the Czech playwright Karel Čapek from the Czech word *robota*, meaning "compulsory labor." While this original meaning still applies to most Earth-bound robots, robots in space have broken through the tedium to become great explorers. They work in environments that may be harmful to humans or in situations where sending a human crew would be too costly. They have been sent as advanced guards to measure the temperature, evaluate the atmosphere, and analyze the soil of other worlds to determine what human explorers can expect to find.

What, exactly, is a robot? A broad definition considers any mechanism guided by automatic controls to be a robot; a very narrow definition requires a robot to be a humanoid mechanical device capable of performing complex human tasks automatically. Robots in space have fallen somewhere in between these extremes. They generally involve a mechanical arm—resembling part of a human, at least—attached to a stationary planetary landing module or to a mobile rover that must perform complex tasks, such as recognizing and avoiding dangerous obstacles in its path. But the evolution to humanoid robots is well under way with the Robonaut being developed by the National Aeronautics and Space Administration (NASA).

Early Space Robots

The first robot in space was a motor-driven mechanical arm equipped with a scoop on the Surveyor 3, which landed on the Moon on April 20, 1967. Acting on signals sent from engineers on Earth, the arm extended and the scoop dug four trenches in the lunar soil, up to 18 centimeters (7 inches) deep. It then placed the samples in front of a camera for scientists on Earth to see. Later Surveyor missions carried analytical instrumentation to determine the chemical composition of the soil samples.

Following the successful human Moon landings that began in 1969 with Apollo 11, NASA began to prepare for piloted missions to Mars. They launched two spacecraft called Viking 1 and Viking 2, which landed on Mars in 1976 on July 20 and September 3, respectively. The Viking landers transmitted pictures of the rock-strewn, rusty-red landscape of Mars back to Earth for the first time. Because there had long been speculation about life on Mars, the Viking landers carried three biological experiments onboard. When the robotic arm of Viking 1 put a sample of the Martian soil into one of the experimental chambers, an excessive amount of oxygen was generated—a possible indication of some form of plant life in the soil. But, to the dismay of the scientists, when the same experiment was performed by Viking 2, no signs of life were found. The question of whether there is life on Mars remains unanswered.

A different type of robot called an "aerobot" was used by Soviet and French scientists to analyze the atmosphere of Venus as part of the Vega balloon mission in 1985. Two Teflon-coated balloons (aerobots) carrying scientific instrumentation floated through the thick Venusian atmosphere for forty-eight hours while researchers recorded temperature, pressure, vertical wind **velocity**, and visibility measurements. Separate landing modules carried analytical instrumentation to determine the composition of the atmosphere and of the surface on landing. More advanced aerobot technol-

velocity speed and direction of a moving object; a vector quantity

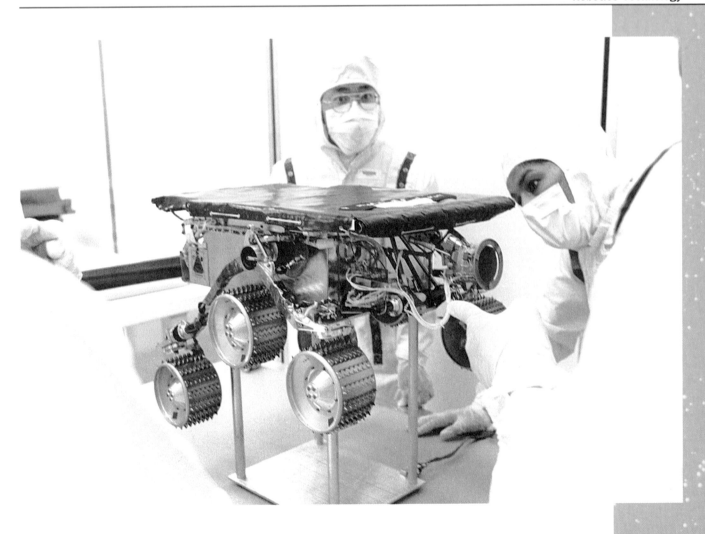

ogy is being developed for NASA's Mars Aerobot Technology Experiment, scheduled for April 2003.

Space Shuttle–Era Robots

The space shuttle was developed as a reusable spacecraft to replace the costly one-time-use-only vehicles that marked the Apollo era. On its second mission in November 1981, astronauts aboard the space shuttle Columbia tested the Remote Manipulator System (RMS), a robotic arm located in the cargo bay. The RMS is 15 meters (50 feet) long 38 centimeters (15 inches) in diameter and weighs 411 kilograms (905 pounds). It has a shoulder (attached to the cargo bay), a lightweight boom that serves as the upper arm, an elbow joint, a lower arm boom, a wrist, and an "end effector" (a gripping tool that serves as a hand) that can grab onto a **payload**. The RMS was designed to lift a satellite weighing up to 29,500 kilograms (65,000 pounds) from the payload bay of the shuttle and release it into space. It can also retrieve defective satellites in orbit for the astronauts to repair. Perhaps the greatest achievement of the RMS has been the retrieval and repair of the Hubble Space Telescope (HST), whose initially flawed primary mirror produced blurry pictures. After it was hauled in by the RMS and repaired using corrective optics in 1993, the HST began delivering the high-quality photographs that astronomers had long awaited.

Technicians examine the Microrover Sojourner, the first robotic roving vehicle sent to Mars. Made with a six-wheel chassis and a rotating joint suspension system instead of springs, the design of Sojourner provides greater obstacle-crossing reliability with full unit stability.

payload any cargo launched aboard a rocket that is destined for space, including communications satellites or modules, supplies, equipment, and astronauts; does not include the vehicle used to move the cargo or the propellant that powers the vehicle

The Remote Manipulator System (RMS) of the space shuttle Atlantis moves the Destiny laboratory from its storage bay for future mission use.

rover a vehicle used to move about on a surface

alpha proton X-ray spectrometer analytical instrument that bombards a sample with alpha particles (consisting of two protons and two neutrons); the X rays are generated through the interaction of the alpha particles and the sample

infrared portion of the electromagnetic spectrum with wavelengths slightly longer than visible light

After two decades of debate about the need to explore Earth's nearest neighbor in the solar system, the Mars Pathfinder landed on the Red Planet on July 4, 1997, and deployed a six-wheeled robotic **rover** called Sojourner to explore the terrain. Standing only 30 centimeters (1 foot) tall and resembling a rolling table with its flat solar panels facing skyward to soak up energy from the Sun, Sojourner roamed short distances to take pictures of interesting rock formations. It used two stereoscopic cameras mounted on its front to see the terrain in three dimensions, just like we do with our slightly separated stereoscopic eyes. A laser beam continuously scanned the area immediately in front of Sojourner to avoid collisions with objects the cameras might have missed. Sojourner analyzed the chemical composition of fifteen rocks using its **alpha proton X-ray spectrometer**. NASA plans to land a pair of advanced rovers on Mars in 2003.

Robonaut and Beyond

Engineers are starting to think of robots on a more human scale again. Since the space shuttle and the International Space Station are designed on a human scale, having robots built to the same scale would be advantageous in working on these spacecraft. NASA is currently developing the Robonaut, a humanoid robotic astronaut about the size of a human astronaut, with a head mounted on a torso, a primitive electronic brain that allows it to make decisions relating to its work, four cameras for eyes, a nose with an **infrared** thermometer to determine an object's temperature, two arms containing 150 sensors each, and two five-fingered hands for dexterous manipulation of objects. It will work alone or alongside human astronauts on space walks to build or repair equipment.

Robotics engineers are also working on a personal satellite assistant, which is a softball-size sphere that would hover near an astronaut in a spacecraft, monitoring the environment for oxygen and carbon monoxide concentrations, bacterial growth, and air temperature and pressure. It will also provide additional audio and video capabilities, giving the astronaut another set of eyes and ears. SEE ALSO EXPLORATION PROGRAMS (VOLUME 2); ROBOTIC EXPLORATION OF SPACE (VOLUME 2).

Tim Palucka

Bibliography

Asimov, Isaac, and Karen A. Frenkel. *Robots: Machines in Man's Image*. New York: Harmony Books, 1985.

Masterson, James W., Robert L. Towers, and Stephen W. Fardo. *Robotics Technology*. Tinley-Park, IL: Goodheart-Willcox, 1996.

Moravec, Hans. *Robot: Mere Machine to Transcendent Mind*. New York: Oxford University Press, 1999.

Thro, Ellen. *Robotics: The Marriage of Computers and Machines*. New York: Facts on File, 1993.

Yenne, Bill. *The Encyclopedia of U.S. Spacecraft*. New York: Exeter Books, 1985.

Internet Resources

Aerobot. National Aeronautics and Space Administration. <http://robotics.jpl.nasa.gov/tasks/aerobot/background/when.html>.

Robonaut. National Aeronautics and Space Administration. <http://vesuvius.jsc.nasa.gov/er_er/html/robonaut/robonaut.html>.

2003 Mars Mission. National Aeronautics and Space Administration. <http://mars.jpl.nasa.gov/missions/future/2003.html>.

Sagan, Carl

American Astronomer, Author, and Educator
1934–1996

Carl Sagan was a Pulitzer-prize winning author, visionary educator, and devoted scientist. He worked to extend humankind's reach into the solar system, and to help people understand the importance and meaning of the scientific method. Born on November 9, 1934, Sagan conducted his undergraduate work at Harvard University, and earned doctorates in astronomy and astrophysics at the University of Chicago. He was named a professor at Cornell University in Ithaca, New York, in 1971. Sagan's academic research concentrated on biology, evolution, astrophysics, planetary science, and anthropology.

Sagan was the author of more than 600 academic papers, twenty books, and a television miniseries called *Cosmos*. His novel about contact with an extraterrestrial civilization, *Contact*, was made into a popular Hollywood film in 1997. Much of Sagan's life was devoted to debunking scientific

Carl Sagan pioneered the study of exobiology—the study of possible alien life forms and biochemistry.

elliptical having an oval shape

stratosphere a middle portion of a planet's atmosphere above the tropopause (the highest place where convection and "weather" occurs)

magnetosphere the magnetic cavity that surrounds Earth or any other planet with a magnetic field. It is formed by the interaction of the solar wind with the planet's magnetic field

misconceptions and advocating clear thinking and better appreciation for the basics of science and its importance in everyday life. He was also a strong advocate for space exploration, especially robotic exploration of the solar system and beyond. Sagan also supported nuclear disarmament and urged the United States and the then–Soviet Union to undertake a joint mission to explore Mars.

Sagan was cofounder of the Planetary Society, a nonprofit organization supporting the exploration of space. He died on December 20, 1996, in Seattle, Washington, at the age of sixty-two. SEE ALSO ASTRONOMY, KINDS OF (VOLUME 2); LITERATURE (VOLUME 1).

Frank Sietzen, Jr.

Bibliography

Poundstone, William. *Carl Sagan: A Life in the Cosmos.* New York: Henry Holt, 2000.

Internet Resources

Carl Sagan. <http://www.carlsagan.com>.

Saturn

Saturn, the sixth planet from the Sun, revolves around the Sun in a slightly **elliptical** orbit at a mean distance of 1.4294 billion kilometers (888,188,000 miles) in 29.42 years. Perhaps best known for its rings, Saturn also has a large collection of moons orbiting around it.

Physical and Orbital Properties

One of four gas giant outer planets (along with Jupiter, Uranus, and Neptune), Saturn is the second most massive planet in the solar system. It has a mass equivalent to 95.159 times Earth's and possesses an atmosphere composed primarily of the gases hydrogen and helium (by mass, comprising approximately 78 percent and 22 percent of the atmosphere, respectively).

It is the trace elements and their compounds that give the planet its golden color and the faint banded structure of the cloud tops in its lowermost **stratosphere**. Methane, ethane, other carbon compounds, and ammonia are observed in the atmosphere. Winds can exceed 450 meters per second (1,000 miles per hour). There is no solid surface beneath the clouds. With depth, the atmosphere slowly thickens from gas to liquid. At very great depths, liquid hydrogen may be compressed enough to become metallic. Saturn has a molten core of heavy elements including nickel, iron, silicon, sulfur, and oxygen, which totals as much as three Earth-masses.

Saturn's magnetic field is much like the field of a simple bar magnet and similar to the planetary magnetic fields of Earth, Jupiter, Uranus, and Neptune. But its near-perfect alignment with the planet's rotation axis makes its origin mysterious. The magnetic field governs Saturn's huge, tadpole-shaped **magnetosphere**, the volume of space controlled by Saturn rather than by the interplanetary magnetic field.

Saturn is the second largest planet in the solar system. Its equatorial diameter is 120,660 kilometers (74,975 miles). Saturn rotates rapidly, having

a day lasting only 10 hours and 39.9 minutes. The **centrifugal** force of this rapid rotation forces the planet to look slightly squashed: its polar diameter is 108,831 kilometers (67,624 miles). Saturn's axis of rotation is inclined to the plane of its orbit by 25.2 degrees, much like Earth's inclination of 23.4 degrees. Like Earth, Saturn has seasons and it constantly changes its presentation to Earth over its long orbit. Weather on Saturn is controlled not by its seasons or the Sun but by the flow of heat from inside the planet. This outward heat flow exceeds the heat received from the Sun by a factor of about three. Its origin is still being investigated.

The combination of Saturn's mass and volume leads to an average density unique in the solar system: at 0.70 grams per cubic centimeter it is less dense than water (1 gram per cubic centimeter). Because of the planet's large size, the force of gravity at Saturn's cloud tops is only 1.06 times Earth's. Nevertheless, to escape from Saturn, a rocket launched from its cloud tops would have to achieve a speed of 35.5 kilometers per second (22 miles per second), more than three times Earth's escape velocity of 11.2 kilometers per second (7 miles per second).

This picture of the Saturnian system was prepared from a collection of images taken by Voyager 1 and 2 spacecraft. The orbit of Saturn's moons and their distinctly different compositions make the Saturnian satellites a small-scale version of the solar system.

centrifugal directed away from the center through spinning

Saturn's rings are composed primarily of water ice particles, and range in size from micrometers to meters.

The Rings of Saturn

Italian mathematician and astronomer Galileo Galilei noted Saturn's odd telescopic appearance in 1610, but Dutch astronomer Christiaan Huygens, who had discovered Saturn's largest moon, Titan, in 1655, was the first to identify it as a ring in 1659. Huygens also demonstrated how the ring plane was tilted, explaining the odd behavior seen over the previous decades.

Italian-born French astronomer Giovanni Domenico Cassini noted a gap within Huygens's single ring in 1675. Now called the Cassini division, this gap separates the outer A ring from the inner B ring. The C ring, inside the others, was discovered in 1850. More than a century later, hints of the D ring were found (and then confirmed by the spacecraft Voyager 1 in 1980), and in 1966 the E ring was observed. The Pioneer 11 spacecraft discovered the F and G rings in 1979. In order outward from the planet, the rings are D, C, B, A, F, G, E. (See table below.)

While Saturn's main rings span a huge distance, they are less than 1 kilometer (0.6 mile) thick and their plane is slightly warped. Ring particles in the main rings range in size from a few tens of meters across down to the size of smoke particles, about 1 micrometer (10^{-6} meter). The E ring is different, being composed of small particles that orbit within a much thicker volume.

The Satellite System of Saturn

Saturn's system of satellites (moons) is notable, ranging from inside the A ring to almost 13 million kilometers (about 8 million miles) from the planet. The classical nine largest moons were discovered between 1655 (Titan) and 1898 (Phoebe). With the rings nearly invisible during the ring plane crossing of 1966, two additional co-orbital (sharing an orbit) moons were discovered, situated between the F and G rings.

Observations in 1980–1981 by the Voyager spacecraft added more moons. Besides an A-ring shepherd moon (which limits the outer edge of the ring) and one in the A ring's Encke gap, small moons trapped in grav-

THE RINGS OF SATURN

Ring Designation	Distance from Saturn		
	km	R_s	
Saturn Radius, R_s	60,330	1.00	
D (inner edge)	66,970	1.11	
C (inner edge)	74,510	1.24	
B (inner edge)	92,000	1.53	
B (outer edge) (Cassini Division)	117,580	1.95	
A (inner edge)	122,170	2.03	
A (ring gap center)	133,400	2.21	
A (outer edge)	136,780	2.27	
F (center)	140,180	2.32	(width 50 km)
G (center)	170,180	2.82	(width variable)
E (inner edge)	~181,000	~3	
E (outer edge)	~483,000	~8	

Green, violet and ultraviolet filtered images were combined to create this image of Saturn, July 12, 1981. This image was taken by Voyager 2 at a distance of 43 million kilometers from Saturn.

itationally stable points (called Lagrangian points, L4 and L5) in the orbits of two of the larger moons were discovered. By 1990 Saturn's satellite count had reached eighteen.

State-of-the-art telescopes and techniques increased Saturn's moon count during the last half of 2000. Twelve additional, tiny outlying satellites were discovered, with additional ones awaiting confirmation. Saturn's total moon count thus reached thirty and was likely to increase further. Some of these small, distant, outer moons orbit Saturn backwards compared to its rotation direction, as Phoebe does, whereas others move in the same direction as the rotation but have orbits highly inclined to Saturn's equator.

Among the classical set of icy satellites, Enceladus and Iapetus are particularly noteworthy. Enceladus, with a diameter of only 498 kilometers (310 miles), is the most reflective solid body in the solar system. Surprisingly for a small, cold moon, the Voyager spacecraft showed that large areas of its surface have recently (over a small fraction of the age of the solar system) melted. Interestingly, the E ring has its maximum density at the same orbital distance as Enceladus.

Iapetus, second largest of the icy moons (and third overall, at 1,436 kilometers [892 miles]), has one hemisphere that reflects as well as snow, whereas its other hemisphere is blacker than asphalt.

In a class by itself is the giant moon Titan. Its diameter of 5,150 kilometers (3,200 miles) exceeds that of the planet Mercury. It has a nitrogen (plus methane) atmosphere, like Earth's (nitrogen plus oxygen), but with a surface pressure about 1.5 times Earth's air pressure at sea level. Titan may be a deep-frozen copy of what Earth was like shortly after its formation.

Beginning in 2004, the Cassini spacecraft and Huygens probe will explore Saturn and Titan. Our understanding of the fascinating and mysterious Saturnian system will increase enormously. SEE ALSO CASSINI, GIOVANNI DOMENICO (VOLUME 2); EXPLORATION PROGRAMS (VOLUME 2); GALILEI, GALILEO (VOLUME 2); HUYGENS, CHRISTIAAN (VOLUME 2); JUPITER (VOLUME 2); NASA (VOLUME 3); ROBOTIC EXPLORATION OF SPACE (VOLUME 2); PLANETARY EXPLORATION, FUTURE OF (VOLUME 2).

Stephen J. Edberg

Bibliography

Bishop, Roy, ed. *Observer's Handbook, 2000.* Toronto: Royal Astronomical Society of Canada, 1999.

Edberg, Stephen J., and Lori L. Paul, eds. *Saturn Educators Guide.* Washington, DC: NASA, 1999. Also available at <http://www.jpl.nasa.gov/cassini/english/teachers/guides/educatorguide>.

Spilker, Linda J., ed. *Passage to a Ringed World.* Washington, DC: National Aeronautics and Space Administration, 1997.

Internet Resources

The Cassini Mission to Saturn (fact sheet). Pasadena, CA: Jet Propulsion Laboratory 400-842, rev. 1, 1999. <http://saturn.jpl.nasa.gov/cassini/english/teachers/factsheets/casini_msn.pdf>.

Saturnian Satellite Fact Sheet. National Space Science Data Center. <http://nssdc.gsfc.nasa.gov/planetary/factsheet/saturniansatfact.html>.

Sensors

Satellites and space probes are launched on their missions to a wide variety of destinations. Some satellites are sent into Earth orbit to look down at Earth's surface or atmosphere; others aim outward, to study the Moon, other planets, or the universe itself. Probes are sent beyond Earth orbit to pass near, land on, or orbit other planets or the Moon. No matter what their eventual destination, the primary objective of sending these craft into space is to gather information and relay it back to Earth in some direct or indirect way. To collect information, these space vehicles must carry with them some means of collecting and distinguishing this data. Sensors are one type of instrument that can collect information.

Sensors, as applied to spacecraft, are instruments and devices that can detect alterations or variations in the space environment and send electrical, radio, or other types of signals or transmissions back to a main collection or recording device. Such a device can be aboard the spacecraft itself, on another spacecraft nearby or close by, or at a receiving station or receptacle on Earth, such as a radio antenna.

While some sensors gather data remotely about the conditions found in space or on another planetary body, other types of sensors can be used by

space vehicles to make determinations about the position or location of the vehicle itself or its condition while in flight. Such sensors, onboard the space vehicle and active during its flight, are essential elements in controlling the craft or flying it to a specific destination in space.

In their roles in remotely sensing the space environment, sensors can be of many different types and collect many different types of information. Radiometers aboard a probe can gather data on the temperature of a planet or Moon's surface, or the temperature of the gases contained in an atmosphere. A spectrometer can break down the composition of planetary gases or surface features across the visible or invisible spectrum of light. These instruments can also gather information on the environmental or weather conditions where they are located. Small **radar** units emitting radar signals can gather information about a planet's surface composition based on the radar's "return," or bounce, from the surface up to the sensor's instrument.

Satellite sensors may also include devices such as a thermocouple. This instrument, made from different types of metals, produces electrical voltage that can vary depending upon the temperature of the material through which the electrical current passes. In this way the device can chart changes in temperatures over a distance or across an altitude.

Sensors used to control satellites can include **gyroscopes** for attitude control or navigation equipment that can plot the craft's location based on sightings of stars and other space objects whose locations are fixed. Sensors are an essential part of a spacecraft, and they can contribute much to the mission and to the accuracy of the spacecraft's flight while in space. SEE ALSO GYROSCOPES (VOLUME 3); ROBOTIC EXPLORATION OF SPACE (VOLUME 2); SPACECRAFT BUSES (VOLUME 2).

Frank Sietzen, Jr.

radar a technique for detecting distant objects by emitting a pulse of radio-wavelength radiation and then recording echoes of the pulse off the distant objects

gyroscope a spinning disk mounted so that its axis can turn freely and maintain a constant orientation in space

Bibliography

Gatland, Kenneth. *Illustrated Encyclopedia of Space Technology.* New York: Harmony Books, 1981.

Plant, Malcolm. *Dictionary of Space.* New York: Longman Publishers, 1986.

Internet Resources

NASA Ames Research Center. <http://www.arc.nasa.gov/>.

SETI

Of all the scientific efforts to find life in space, none has potential consequences as profound as SETI, the Search for Extraterrestrial Intelligence. SETI researchers are trying to uncover other civilizations whose technical sophistication is at a human level or higher.

While science fiction routinely describes face-to-face encounters with intelligent aliens, it may be that we will never actually meet extraterrestrials. Building fast rockets capable of carrying living cargo to the stars is a formidable, perhaps even impossible, challenge. The amount of energy required to hurl a craft the size of the space shuttle at even half the speed of light is enormous—equivalent to the energy required to keep New York City running for 10,000 years. ✳ This is a problem of physics, not technology.

On the other hand, there are ways to reach other civilizations without **interstellar** travel. In 1959 Philip Morrison and Giuseppe Cocconi, two physicists at Cornell University, made a simple calculation to determine how far away a good radio receiver and a large antenna could detect our most powerful military radar transmitters. To their surprise, the answer turned out to be light-years—typical of the distances to the stars. Morrison and Cocconi realized that while interstellar rocketry was hard, interstellar communication by radio was easy. They suggested that other galactic civilizations might be discovered by simply eavesdropping on their radio traffic.

The Search for Extraterrestrial Intelligence

Within months, Frank Drake, a young radio astronomer at the National Radio Astronomy Observatory in Green Bank, West Virginia, tried to do just that. He was unaware of the work of the two Cornell physicists but had independently thought of the same idea. For several weeks in the spring of 1960, Drake pointed an 85-foot antenna (a radio telescope) at two Sun-like stars, Tau Ceti and Epsilon Eridani, tuning his receiver up and down the dial near 1,420 megahertz (MHz). This particular frequency was chosen because it is truly a universal radio channel. Hydrogen gas, which drifts and swirls through the immense spaces between the stars, naturally emits some radio noise at 1,420 MHz. Drake believed that every sophisticated society in the cosmos would know of this hydrogen hiss, and consequently it would make sense to broadcast interstellar hailing signals near this sweet spot on the dial.

Drake's Project Ozma was the first modern SETI search. By the early twenty-first century, about seventy others were undertaken. One of the most ambitious was the NASA SETI program, which ultimately became known as the High Resolution Microwave Survey. NASA got into SETI slowly, beginning in the 1970s with a technical study of the equipment and strat-

✳ **At half the speed of light, travel time to Alpha Centauri, the nearest star system, would take nine years.**

interstellar between the stars

Frank Drake stands under the Arecibo radio telescope on October 12, 1992. Drake is one of the leaders in the Search for Extraterrestrial Intelligence (SETI).

egy required for a serious search. In the fall of 1992, sufficient equipment had been built to start the listening. However, very shortly thereafter, the U.S. Congress stopped all NASA SETI efforts. The rationale for canceling this research was to reduce federal spending during an era of large budget deficits.

SETI work continued, however, funded in the United States by private donations. Most of these projects have been radio experiments, of the type pioneered by Drake. The SETI Institute, in California, runs the most sensitive search, known as Project Phoenix. Various large radio telescopes, including the king-sized 305-meter (1,000-foot) antenna at Arecibo, Puerto Rico, have been used by Project Phoenix to carefully examine the neighborhoods of nearby, Sun-like stars. Other projects, such as the University of California, Berkeley's SERENDIP experiment, sweep the sky in an attempt to survey greater amounts of cosmic real estate. While more of the heavens are examined, the sensitivity in any given direction is lower. Some of the SERENDIP data have been distributed on the Web for processing at home with a screen saver program known as SETI@home.

Additional radio SETI experiments are being carried out in Australia (Southern SERENDIP), Argentina (META II), and Italy. Starting in the late twentieth century, another approach to SETI has gained a number of adherents: so-called optical SETI. Rather than tuning the dial in search of persistent, artificial signals, ordinary telescopes (with mirrors and lens) are

The receiver for the Arecibo Radio Telescope in Puerto Rico requires three large poles to support its weight above the 305 meter (1,000 foot) dish. Scientists monitoring this telescope, which scrutinizes about 1,000 Sun-like stars less than 150 light-years distant, hope to find signs of intelligent extraterrestrial life.

outfitted with special detectors designed to find very short (less than a billionth of a second) laser pulses from distant worlds.

The Probability of Success

So far, no confirmed extraterrestrial signals—either radio or optical—have been found by SETI scientists. What are the chances that the aliens will *ever* be found? In 1961 Drake summarized the problem with a simple formula that predicts the number of galactic civilizations that are broadcasting now. Known as the Drake Equation, the computation is simply a product of factors bearing on the existence of intelligence. These factors include the number of galactic stars capable of supporting life, the fraction of such stars with planets, the number of planets in a solar system on which life evolves, the fraction of inhabited worlds where intelligence appears, and the lifetime of a broadcasting society. While we still do not know many of these factors, some scientists contend that the recent evidence for **extrasolar planets** and the growing suspicion that biology might be a common phenomenon have increased the chances for finding intelligence among the stars.

extrasolar planets planets orbiting stars other than the Sun

SETI scientists have made plans to greatly expand their search during the first two decades of the twenty-first century. The SETI Institute will build the Allen Telescope Array, a large grouping of small antennas that will be used for full-time searching. A world consortium of radio astronomers is considering the construction of a radio telescope a kilometer in size, a gargantuan instrument that could also be used for SETI. Optical SETI experiments are already increasing in number and sophistication.

In light of this rapid improvement in experimental technique, some scientists are optimistic that a signal will be found in the early decades of the twenty-first century. If so, the consequences would be dramatic. If we can ever successfully find and decode any message accompanying the signal, we might learn something of the knowledge and culture of other galactic beings, most likely from a society technologically far more advanced than our own. Even if we never understand or reply to an interstellar message, simply knowing that we are not the only "game" in town—let alone the most interesting game—would give us new and valuable perspective. SEE ALSO EXTRASOLAR PLANETS (VOLUME 2); LIFE IN THE UNIVERSE, SEARCH FOR (VOLUME 2); WHY HUMAN EXPLORATION? (VOLUME 3).

Seth Shostak

Bibliography

Cocconi, Giuseppe, and Philip Morrison. "Searching for Interstellar Communications." *Nature* 84 (1959):844.

Davies, Paul. *Are We Alone? Philosophical Implications of the Discovery of Extraterrestrial Life*. New York: Basic Books, 1996.

Dick, Steven. J. *Life on Other Worlds*. Cambridge, UK: Cambridge University Press, 1998.

Drake, Frank D., and Dava Sobel. *Is Anyone out There?* New York: Delacorte Press, 1992.

Goldsmith, Donald, and Tobias C. Owen. *The Search for Life in the Universe*. Reading, MA: Addison-Wesley, 1992.

Harrison, Albert. *After Contact: The Response to Extraterrestrial Life*. New York: Plenum, 1997.

Regis, Edward, Jr., ed. *Extraterrestrials: Science and Alien Intelligence*. Cambridge, UK: Cambridge University Press, 1985.

Shostak, Seth. *Sharing the Universe: Perspectives on Extraterrestrial Life.* Berkeley, CA: Berkeley Hills Books, 1998.

Shapley, Harlow

Astronomer
1885–1972

Harlow Shapley was born on November 2, 1885, in Nashville, Missouri. He worked for a newspaper in Kansas and later attended the University of Missouri, intending to study journalism, but taking up astronomy instead. In 1911 Shapley went to Princeton University, where he worked with Henry Norris Russell on eclipsing binary stars. After completing his doctoral thesis, Shapley began work at Mt. Wilson Observatory in California in 1914, where he studied **Cepheid variables**. Brighter Cepheids have longer periods, and Shapley was able to determine the distances to faint Cepheids and show that the Milky Way galaxy was far larger than previously believed.

Shapley's most important contribution to astronomy was to note that the **globular clusters** were concentrated toward the constellation Sagittarius, and he made the correct assumption that the center of this concentration marked the center of the Milky Way. He thus moved the universe from a Copernican Sun-centered system to a Sun located far from the galactic center in one of the spiral arms.

On April 20, 1920, a famous debate was held between Shapley and fellow astronomer Heber Curtis on the subject of "The Scale of the Universe." Curtis was correct in arguing that spiral nebulae were galaxies like our own but incorrect in placing the Sun at the center of the Milky Way. Shapley was correct in placing the Sun far from the center of the Milky Way but incorrect in saying the spiral nebulae were nearby gas clouds.

In 1920 Shapley was offered the directorship of the Harvard College Observatory, where he stayed for the rest of his career. He died in Boulder, Colorado, on October 20, 1972. Shapley's legacy as a popularizer of astronomy is maintained by the American Astronomical Society's Harlow Shapley Visiting Lectureships Program, which sends astronomers on two-day visits to universities and colleges in the United States, Canada, and Mexico. SEE ALSO ASTRONOMY, HISTORY OF (VOLUME 2); COPERNICUS, NICHOLAS (VOLUME 2); GALAXIES (VOLUME 2); STARS (VOLUME 2).

A. G. Davis Philip

Bibliography

Grindlay, Jonathan E. and A. G. Davis Philip, eds. *The Harlow Shapley Symposium on Globular Cluster Systems*, IAU Symposium No. 126. London, UK: Kluwer, 1988.

Shapley, Harlow. *The View from a Distant Star.* New York: Basic Books, 1963.

Shoemaker, Eugene

American Astrogeologist
1928–1997

Eugene Merle Shoemaker was instrumental in establishing the discipline of planetary geology. He founded the U.S. Geological Survey's Branch of

Harlow Shapley discovered that the solar system does not rest in the center of the Milky Way, but is actually located in the outer regions of the galaxy.

Cepheid variables a class of variable stars whose luminosity is related to their period; their periods can range from a few hours to about 100 days—the longer the period, the brighter the star

globular clusters roughly spherical collections of hundreds of thousands of old stars found in galactic haloes

Eugene Shoemaker, here outside the Palomar Observatory near San Diego, is credited with originating the field of astrogeology within the U.S. Geological Survey.

stratigraphy the study of rock layers known as strata, especially the age and distribution of various kinds of sedimentary rocks

impact craters bowl-shaped depressions on the surfaces of planets or satellites that result from the impact of space debris moving at high speeds

Astrogeology, which mapped the Moon and prepared astronauts for lunar exploration.

Born in 1928, Shoemaker's early fascination with the Grand Canyon led him to recognize that the powerful tool of **stratigraphy** could be applied to unraveling the history of the Moon. His research at Meteor Crater in Arizona led to an appreciation of the role of asteroid and comet impacts as a primal and fundamental process in the evolution of planets.

Shoemaker contributed greatly to space science exploration, particularly of the Moon. Although he hungered to become an Apollo astronaut himself, that aspiration was unfulfilled. Shoemaker was part of a leading comet-hunting team that discovered comet Shoemaker–Levy 9 and charted the object's breakup. Pieces of the comet slammed into Jupiter in July 1994—an unprecedented event in the history of astronomical observations. That same year, Shoemaker also led the U.S. Defense Department's Clementine mission, which first detected the possibility of pockets of water ice at the Moon's south pole.

While carrying out research on **impact craters** in the Australian outback in 1997, Shoemaker was killed in a car accident. A small vial of the astrogeologist's ashes were scattered on the lunar surface, deposited there by the National Aeronautics and Space Administration's Lunar Prospector spacecraft, which was purposely crashed on the Moon on July 31, 1999, after completing its mission. SEE ALSO APOLLO (VOLUME 3); ASTEROIDS (VOLUME 2); JUPITER (VOLUME 2).

Leonard David

Bibliography

Levy, David H. *Shoemaker: The Man Who Made An Impact.* Princeton, NJ: Princeton University Press, 2000.

Small Bodies

For much of history, the solar system was thought to consist of large objects: Earth, the Moon, the Sun, and other planets. However, within the last two centuries astronomers have discovered a menagerie of small objects throughout the solar system, including asteroids, comets, Kuiper belt objects, and many small moons orbiting planets. Studies of these objects, from both telescopes and spacecraft, have provided new insights into the formation of the solar system and its true nature.

There is no good definition for what constitutes a "small body" in the solar system. Nevertheless, one definition used by some planetary scientists considers objects less than 400 kilometers (250 miles) in diameter to be small bodies. When objects are larger than 400 kilometers, they have enough mass so that their gravity is powerful enough to shape the object into a roughly spherical form. For smaller objects, their gravity is not powerful enough to accomplish this, and as a result many have irregular shapes. Even this, though, is not a precise definition. In reality, there is a continuum of objects from very large to very small, with no distinct, obvious differences between "large" and "small" bodies.

Phobos is the larger of the two moons of Mars, both of which are less than 30 kilometers in diameter and are irregularly shaped.

The Discovery of Small Bodies

The first objects that would qualify as small bodies were asteroids, discovered in the early nineteenth century. While the largest asteroids, such as Ceres (the first asteroid to be discovered), may be somewhat larger than the above definition for a small body✳, within a few decades a number of small bodies were discovered. In 1877 American astronomer Asaph Hall discovered Phobos and Deimos, the two moons of Mars. Both moons are very small objects, each less than 30 kilometers (18.5 miles) in diameter and irregularly shaped. In 1892 Amalthea, a moon less than 250 kilometers (155 miles) in diameter, was found orbiting Jupiter. In the late nineteenth and early twentieth centuries, a number of small moons were found orbiting Jupiter and Saturn. In addition, during this time the rate of asteroid discoveries increased.

✳ **Ceres was initially considered to be a planet until other asteroids with similar orbits were found.**

The era of spacecraft exploration brought many more discoveries of small bodies. The Voyager 1 and Voyager 2 spacecraft discovered many small moons orbiting Jupiter, Saturn, Uranus, and Neptune. Telescopes on the ground, as well as the Hubble Space Telescope, have also discovered small moons around these planets. The rate of asteroid and comet discoveries increased dramatically starting in the 1990s. By October 2001, more than 30,000 asteroids had been discovered, compared to 20,000 just at the beginning of that year. Dozens of comets are discovered each year as well, many by automated telescopes and spacecraft.

This montage of the smaller satellites of Saturn was taken by Voyager 2 as it made its way through the Saturnian system. These satellites range in size from small on the asteroidal scales, to nearly as large as Saturn's moon Mimas.

Difficulties in Classification

These discoveries, and follow-up observations, have shown how difficult it can be to classify small bodies. Astronomers in the past attempted to fit these bodies into one of three classes: asteroids, comets, and moons. Now, however, there is evidence that many asteroids may be extinct comets, having exhausted their supply of ice that generates a tail when approaching the Sun. Some small moons orbiting Jupiter, as well as Phobos and Deimos, may have originally been asteroids captured into orbit by the gravity of Mars and Jupiter. In the outer solar system astronomers discovered in the early 1990s a family of icy bodies called Kuiper belt objects, some of which are larger than even the largest asteroids and are far larger than an ordinary comet. Spacecraft and ground-based telescopes have discovered several asteroids that have their own small moons.

Some planetary scientists have elected to classify objects in a different way, based on their composition and likely location in the solar system where they formed. Bodies that formed from the Sun out to a distance of about 2.5 **astronomical units** (AU) are primarily rocky and metallic. That close to the Sun, temperatures are too high for anything else to condense out of the protoplanetary nebula from which the solar system formed. This explains the composition of much of the asteroid belt as well as the inner planets. At around 2.5 to 2.7 AU is what some call the "soot line." At this distance temperatures are low enough for carbon-rich compounds, such as soot, to form. Asteroids in the outer portion of the belt, beyond the soot line, tend to be rich in these materials. At around 3 to 4 AU is the "frost line," beyond which water ice can form. Objects that formed beyond this distance tend to be rich in water ice and often in carbon dioxide, methane, and other ices.

astronomical units one AU is the average distance between Earth and the Sun (152 million kilometers [93 million miles])

This scheme allows scientists to understand where an object originated, regardless of where it is today. With this information, scientists can then try to understand how the object evolved over the history of the solar system from the location where it formed to where it is now located. The problem with this approach is that there is only limited information about the composition of many small bodies, including many of the moons of the giant planets. Even Phobos and Deimos, the two moons of nearby Mars, have not been examined enough to know their compositions well.

Sorting out the true nature of small bodies in the solar system will take many more years of research and observations by telescopes and spacecraft.

A number of missions by the National Aeronautics and Space Administration (NASA) and the European Space Agency will study asteroids and comets in detail in the early twenty-first century. In addition, NASA's Cassini spacecraft will arrive at Saturn in 2004 and spend several years studying the planet and its moons, which may uncover key clues about the origin of Saturn's small moons. Through these observations, it should be possible to learn not only about the origins of the small bodies in the solar system but also how the solar system itself formed. SEE ALSO ASTEROIDS (VOLUME 2); COMETS (VOLUME 2); EXPLORATION PROGRAMS (VOLUME 2); JUPITER (VOLUME 2); KUIPER BELT (VOLUME 2); MARS (VOLUME 2); METEORITES (VOLUME 2); NEPTUNE (VOLUME 2); OORT CLOUD (VOLUME 2); PLANETESIMALS (VOLUME 2); SATURN (VOLUME 2); URANUS (VOLUME 2).

Jeff Foust

Bibliography

Hartmann, William K. "Small Worlds: Patterns and Relationships." In *The New Solar System*, 4th ed., ed. J. Kelly Beatty, Carolyn Collins Petersen, and Andrew Chaikin. Cambridge, MA: Sky Publishing Corp., 1999.

Internet Resources

Arnett, Bill. "Small Bodies." <http://www.nineplanets.org/smallbodies.html>.

Small Bodies Node. University of Maryland. <http://pdssbn.astro.umd.edu/outreach/index.html>.

Solar Particle Radiation

The Sun radiates more than just life-sustaining light into the solar system. At irregular intervals, it also produces bursts of high-energy particles. These solar particles have energies that range from 30,000 **electron volts** to 30 billion electron volts per **nucleon** and consist primarily of protons (96% of the total number of nuclei) and helium nuclei (3%). The remaining particles are ions of elements that are common in the solar atmosphere, such as carbon, nitrogen, oxygen, neon, magnesium, silicon, and iron, as well as small numbers of even heavier elements. The processes that produce high-energy protons and ions also accelerate **electrons** to at least 20 million electron volts. Collisions between energetic particles and the solar atmosphere also produce **neutrons** and **gamma rays**. All these particles flow outward from the Sun into the **heliosphere**, where they can affect space systems and are a major concern for astronaut safety.

The Origins of Solar Particles

Until recently, it was thought that solar energetic particles came only from **flares**. Now solar physicists know that they are produced in both flares and coronal mass ejections. Flares occur when stressed magnetic fields in solar active regions release their energy. The energy appears as both heated plasma and energetic particles, some of which stream out along magnetic field lines into the heliosphere. Because they come from a small area on the Sun, the energetic particles follow a narrow set of field lines and affect only a small region of the heliosphere. Flare-generated energetic particle events tend to be impulsive, meaning that the flux of particles measured near Earth rises and decays rapidly, often within a day.

electron volt unit of energy equal to the energy gained by an electron when it passes through a potential difference of 1 volt in a vacuum

nucleon a proton or a neutron; one of the two particles found in a nucleus

electrons negatively charged subatomic particles

neutrons subatomic particles with no electrical charge

gamma rays a form of radiation with a shorter wavelength and more energy than X rays

heliosphere the volume of space extending outward from the Sun that is dominated by solar wind; it ends where the solar wind transitions into the interstellar medium, somewhere between 40 and 100 astronomical units from the Sun

flares intense, sudden releases of energy

Prominences that drift like clouds above the solar surface may suddenly erupt and break away from the Sun in a cataclysmic action.

solar corona the thin outer atmosphere of the Sun that gradually transitions into the solar wind

solar prominence cool material with temperatures typical of the solar photosphere or chromosphere; suspended in the corona above the visible surface layers

solar wind a continuous, but varying, stream of charged particles (mostly electrons and protons) generated by the Sun; it establishes and affects the interplanetary magnetic field; it also deforms the magnetic field about Earth and sends particles streaming toward Earth at its poles

magnetosphere the magnetic cavity that surrounds Earth or any other planet with a magnetic field. It is formed by the interaction of the solar wind with the planet's magnetic field

Coronal mass ejections are the result of a large-scale restructuring of the magnetic field in the **solar corona**. In this process significant amounts of plasma are ejected into the heliosphere. Usually a coronal mass ejection includes the eruption of a **solar prominence** and often is accompanied by a flare. The fastest coronal mass ejections travel at speeds above 800 kilometers per second (500 miles per second) and drive shock waves, which accelerate coronal plasma and **solar wind** into energetic particle events. Since the coronal mass ejection is a large-scale event, the accelerated particles cover a much broader region of the heliosphere than is the case for particles accelerated in flares alone. Coronal mass ejection-associated energetic particle events tend to be gradual, sometimes lasting for many days.

The Impact of Solar Particle Radiation

Because solar energetic particles have been stripped of some or all their electrons, they are positively charged and must follow the magnetic field lines away from the Sun. Near Earth, they are prevented from directly penetrating the near-Earth environment by the **magnetosphere** that surrounds the planet. Some particles can penetrate in the polar regions where Earth's magnetic field lines connect more directly with the space environment. There they produce fade-outs of radio communication at high latitudes and can bombard high-flying aircraft, including commercial flights. During a solar particle event, a passenger on a high-flying supersonic aircraft can receive a radiation dose equivalent to about one chest X ray an hour.

Energetic particles are stopped when they strike other matter. When this happens, they give up their energy to that material. On an orbiting satellite, energetic particle exposure degrades the efficiency of the solar-cell panels used to provide operating power. A large energetic particle event can also damage sensitive electronic components, leading to the failure of critical subsystems and loss of the satellite.

When energetic particles strike living tissue, the transfer of energy to the atoms and molecules in the cellular structure causes the atoms or molecules to become ionized or excited. These processes can break chemical bonds, produce highly reactive **free radicals**, and produce new chemical bonds and cross-linkage between **macromolecules**. Cells can repair small amounts of damage from low doses of particle radiation. Higher doses overwhelm this ability, resulting in cell death. If the dose of radiation is high enough, entire organs can fail to function properly and the organism dies.

Radiation doses are measured in rads or grays, where 1 gray equals 100 rads. One rad equals 100 ergs of absorbed energy per gram of target matter. The potential for radiation to cause biological damage is called the dose equivalent, which is measured in rems or sieverts, where 1 sievert equals 100 rems. The dose equivalent is simply the dose (in rads or grays) multiplied by the so-called radiation weighting factor, which depends on the type of radiation and other factors. The average American receives 360 millirems a year, and a typical X ray gives a patient 50 millirems (a millirem is a thousandth of a rem). The National Aeronautics and Space Administration (NASA) limits exposure to radiation absorbed by the skin to 600 rems for an astronaut's career, with additional limits of 300 rems per year and 150 rems for every thirty-day period.

A large solar particle event can produce enough radiation to kill an unprotected astronaut. For example, the large solar storm of August 1972 would have given an unshielded astronaut on the Moon a dose equivalent of 2,600 rems, probably resulting in death. **Shielding** can reduce the radiation levels, but the amount required for a large solar particle event is too large for an entire Mars-bound spacecraft or lunar surface base. Instead, a highly shielded storm shelter is necessary. This must be combined with a warning capability to give astronauts who are away from the shelter sufficient time to seek safety.

Solar activity is monitored continuously from specially designed ground-based observatories and from the National Oceanic and Atmospheric Administration's geostationary operational environmental satellites. These satellites continuously observe the solar flux in soft X rays and monitor energetic particles at the satellite location. A sudden increase in soft X rays signifies a solar flare. Coronal mass ejections are not currently monitored continuously, but ground-based observatories can often detect the disappearance of a solar filament, which is usually related to a coronal mass ejection. Significant solar particle events occur much less frequently than flares and coronal mass ejections, so many false alarms are possible. Thus, with current observing systems, astronauts must always be able to seek a sheltered environment within the roughly one-hour period that it takes for the particle radiation to rise to dangerous levels. This limits, for example, the distances away from a lunar base that an astronaut can safely explore. SEE ALSO LIVING ON OTHER WORLDS (VOLUME 4); SOLAR WIND (VOLUME 2); SPACE ENVIRONMENT, NATURE OF THE (VOLUME 2); SUN (VOLUME 2).

John T. Mariska

free radical a molecule with a high degree of chemical reactivity due to the presence of an unpaired electron

macromolecules large molecules such as proteins or DNA containing thousands or millions of individual atoms

shielding providing protection for humans and electronic equipment from cosmic rays, energetic particles from the Sun, and other radioactive materials

Bibliography

Odenwald, Sten. "Solar Storms: The Silent Menace." *Sky and Telescope* 99, no. 3 (2000):50–56.

Phillips, Kenneth J. H. *Guide to the Sun*. Cambridge, UK: Cambridge University Press, 1992.

Reames, Donald V. "Particle Acceleration at the Sun and the Heliosphere." *Space Science Reviews* 90 (1999):413–491.

Internet Resource

Space Environment Center Homepage. National Oceanic and Atmospheric Administration, Space Environment Center. <http://www.sec.noaa.gov/>.

Solar Wind

The area between the Sun and the planets, the interplanetary medium, is a turbulent area dominated by a constant stream of hot plasma that billows out from the Sun's corona. This hot plasma is called the solar wind.

The first indication that the Sun might be emitting a "wind" came in the seventeenth century from observations of comet tails. The tails were always seen to point away from the Sun, regardless of whether the comet was approaching the Sun or moving away from it.

Basic Characteristics

protons positively charged subatomic particles

electrons negatively charged subatomic particles

The solar wind is composed mostly of **protons** and **electrons** but also contains ions of almost every element in the periodic table. The temperature of the corona is so great that the Sun's gravity is unable to hold on to these accelerated and charged particles and they are ejected in a stream of coronal gases at speeds of about 400 kilometers per second (1 million miles per hour). Although the composition of the solar wind is known, the exact mechanism of formation is not known at this time.

sunspots dark, cooler areas on the solar surface consisting of transient, concentrated magnetic fields

flares intense, sudden releases of energy

The solar wind is not ejected uniformly from the Sun's corona but escapes primarily through holes in the honeycomb-like solar magnetic field. These gaps, located at the Sun's poles, are called coronal holes. In addition, massive disturbances associated with **sunspots**, called solar **flares**, can dramatically increase the strength and speed of the solar wind. These events occur during the peak of the Sun's eleven-year sunspot cycle.

The solar wind affects the magnetic fields of all planets in the solar system. The interaction of the solar wind, Earth's magnetic field, and Earth's upper atmosphere causes geomagnetic storms that produce the awe-inspiring Aurora Borealis (northern lights) and Aurora Australis (southern lights).

Undesirable Consequences

auroras atmospheric phenomena consisting of glowing bands or sheets of light in the sky caused by high-speed charged particles striking atoms in Earth's upper atmosphere

Although the solar wind produces beautiful **auroras**, it can also cause a variety of undesirable consequences. Electrical current surges in power lines; interference in broadcast of satellite radio, television, and telephone signals; and problems with defense communications are all associated with geomagnetic storms. Odd behavior in air and marine navigation instruments have also been observed, and geomagnetic storms are known to alter the atmospheric ozone layer and even increase the speed of pipeline corrosion in Alaska. For this reason, the U.S. government uses satellite measurements of the solar wind and observations of the Sun to predict space weather.

Major solar wind activity is also a very serious concern during spaceflight. Communications can be seriously disrupted. Large solar disturbances

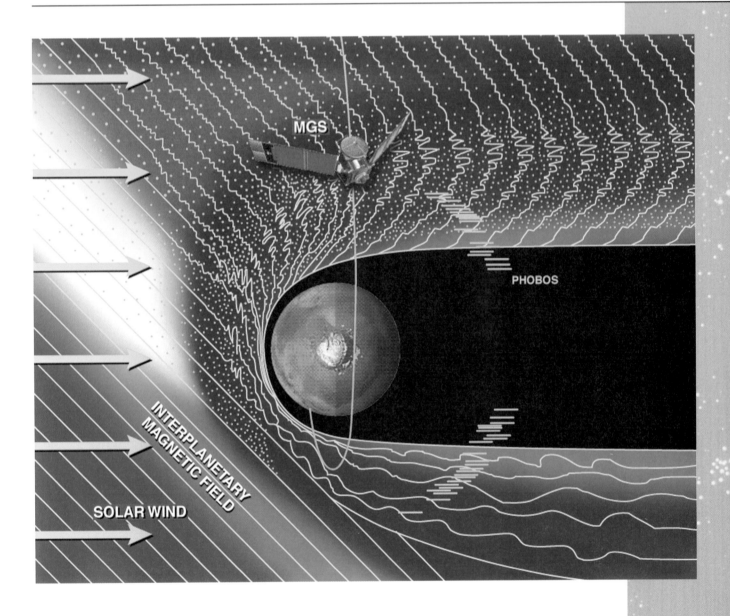

MGS

PHOBOS

INTERPLANETARY MAGNETIC FIELD

SOLAR WIND

heat Earth's upper atmosphere, causing it to expand. This creates increased atmospheric **drag** on spacecraft in low orbits, shortening their orbital lifetime. Intense solar flare events contain very high levels of radiation. On Earth humans are protected by Earth's **magnetosphere**, but beyond it astronauts could be subjected to lethal doses of radiation.

There have been a number of scientific missions that have enabled scientists to learn more about the Sun and the solar wind. Such missions have included Voyager, Ulysses, SOHO, Wind, and POLAR. The latest mission, Genesis, was launched in August 2001 and during its two years in orbit it will unfold its collectors and "sunbathe" before returning to Earth with its samples of solar wind particles. Scientists will study these solar wind samples for years to come. SEE ALSO SOLAR PARTICLE RADIATION (VOLUME 2); SPACE ENVIRONMENT, NATURE OF THE (VOLUME 2); SUN (VOLUME 2).

Alison Cridland Schutt

Depiction of the response of solar wind to an obstacle—Mars—in its path. (MGS identifies the Mars Surveyor spacecraft.)

drag a force that opposes the motion of an aircraft or spacecraft through the atmosphere

magnetosphere the magnetic cavity that surrounds Earth or any other planet with a magnetic field. It is formed by the interaction of the solar wind with the planet's magnetic field

Bibliography

Kaler, James B. *Extreme Stars.* Cambridge, UK: Cambridge University Press, 2001.

Space Debris

The term "space debris" in its largest sense includes all naturally occurring remains of solar and planetary processes: interplanetary dust, meteoroids, asteroids, and comets. Human-made space debris in orbit around Earth is commonly called orbital debris. Examples include dead satellites, spent rocket bodies, explosive bolt fragments, telescope lens covers, and the bits and pieces left over from satellite explosions and collisions.

Orbital debris is found wherever there are working satellites. Of the more than 9,000 objects larger than 10 centimeters (3.94 inches) in Earth orbit, 94 percent are debris. The densest regions are low Earth orbit (LEO), an altitude range from 400 to 2,000 kilometers (248.5 to 1,242.7 miles) above Earth; and geosynchronous Earth orbit (GEO) at 35,786 kilometers (22,300 miles), sometimes called the **Clarke orbit**, where the orbital period of a satellite is one day. More than 100,000 particles between 1 and 10 centimeters (0.39 and 3.94 inches) are thought to exist and probably tens of millions of times smaller than that can be found in space. By mass, there are more than 4 million kilograms (4,409.2 tons) in orbit.

The U.S. Air Force and Navy operate a network of **radar** sensors all over the world that can observe objects in space. These observations are combined to produce mathematical orbits that are maintained at U.S. Space Command as the Space Surveillance Catalog (SSC). Objects in LEO as small as 10 centimeters (3.94 inches) can be reliably tracked. GEO objects are harder to track because of their high altitude. Telescopes are used to observe GEO objects, and those observations are converted into orbits that are included in the SSC.

From 10 centimeters down to about 3 millimeters (0.12 inches), powerful ground radars like the Massachusetts Institute of Technology Haystack radar in Westford, Massachusetts are used to statistically sample the debris population. Analysis of **impact craters** on returned spacecraft surfaces, such as those from the Long Duration Exposure Facility, produce data concerning very small particles, those under 0.5 millimeters (0.02 inches) in size.

Orbital debris can severely damage or destroy a spacecraft. Due to the high average speed at impact, about 10 kilometers (6.21 miles) per second, a 3-millimeter (0.12 inch) fragment could penetrate the walls of a pressurized spacecraft. An unmanned satellite could be disabled by debris smaller than 1 millimeter (0.04 inches) if such particles were to disable critical power or data cables. The International Space Station (ISS) carries a variety of shields to protect it against space debris up to 1 centimeter in size. Debris objects larger than 10 centimeters will be avoided using orbital information from U.S. Space Command. Too large to shield against and too small to track with radar, objects of 1 to 10 centimeters pose a risk to the ISS that, while small, cannot be eliminated.

Because satellites can stay in orbit for more than 10,000 years, care must be taken in the world's policies concerning orbital debris. The Inter-Agency Space Debris Coordination Committee, composed of representatives from the world's leading space agencies, has developed agreements that minimize the creation of new debris. Most countries deplete their spent rocket bodies and payloads of stored energy, thereby decreasing their possibility of exploding and minimizing the largest historical source of debris. Other topics,

Clarke orbit geostationary orbit; named after science fiction writer Arthur C. Clarke, who first realized the usefulness of this type of orbit for communications and weather satellites

radar a technique for detecting distant objects by emitting a pulse of radio-wavelength radiation and then recording echoes of the pulse off the distant objects

impact craters bowl-shaped depressions on the surfaces of planets or satellites that result from the impact of space debris moving at high speeds

During this Space Debris Impact Experiment, changes can be seen in the prelaunch versus postlaunch color condition of Tray D06. Prelaunch, the center panel section has a pink tint and the end section panel is pale green. Postlaunch, the center panel has a green tint, and the end section panel has a pink tint. At the time of this experiment, there was no definitive explanation for the color changes in the panels.

such as forced de-orbiting of satellites, continue to be discussed in an international effort to control space debris. SEE ALSO LONG DURATION EXPOSURE FACILITY (LDEF) (VOLUME 2).

Jeffrey R. Theall

Bibliography

National Research Council. *Orbital Debris, A Technical Assessment.* Washington, DC: National Academy Press, 1995.

Office of Science and Technology Policy. *Interagency Report on Orbital Debris.* Washington, DC: National Science and Technology Council, November 1995.

Internet Resources

Orbital Debris Research at JSC. NASA Johnson Space Center. <http://sn-callisto.jsc.nasa.gov/>.

Space Environment, Nature of the

Near-Earth space is a complex, dynamic environment that affects not just objects in space, but our everyday lives as well. It exists as the interaction

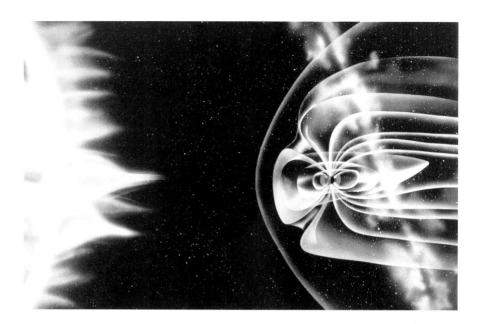

This illustration of the space environment shows the Earth, surrounded by Van Allen Radiation Belts (orange) and its outlying magnetosphere, and solar ejections from the Sun.

solar radiation total energy of any wavelength and all charged particles emitted by the Sun

infrared portion of the electromagnetic spectrum with wavelengths slightly longer than visible light

wavelength the distance from crest to crest on a wave at an instant in time

ultraviolet radiation electromagnetic radiation with a shorter wavelength and higher energy than visible light

electrons negatively charged subatomic particles

ionosphere a charged particle region of several layers in the upper atmosphere created by radiation interacting with upper atmospheric gases

of energy and mass from a variety of sources. Earth, with its magnetic field and its atmosphere, interacts with the Sun to form the solar-terrestrial system, which accounts for most of the effects of the near-Earth space environment. Deep space sources (e.g., other stars and galaxies) contribute particle and electromagnetic radiation that also interacts with Earth. Finally, there are solid bodies within and passing through the solar system that can and do interact with Earth. All of these systems affect orbiting artificial satellites and also have more direct effects on life on Earth, right down to the planet's surface.

The Sun's Interactions with Earth

The Sun is the greatest source of energy in the solar system. It drives most of the activity of the near-Earth space environment. Solar energy couples with Earth's atmosphere and surface, giving rise to terrestrial weather. In a similar way, **solar radiation** interacts with near-Earth space to give rise to space weather. The Sun continuously emits radiation in two primary forms: electromagnetic and particle.

The electromagnetic radiation emitted by the Sun spans the spectrum, from radio waves up through **infrared** and the visible light **wavelengths** to the ionizing energies of extreme ultraviolet, X-ray, and gamma radiation. **Ultraviolet radiation** is the familiar radiation that can burn human skin and fade curtains. Fortunately, the gases in Earth's atmosphere shield us from most ultraviolet radiation. It is the interaction of intense radiation, such as extreme ultraviolet radiation, that strips **electrons** from (or ionizes) the gases in the upper atmosphere, creating what is called the **ionosphere**. One example of how the ionosphere is affected by direct radiation by the Sun and by nighttime shielding by Earth is AM radio. At night, the thickness of the ionosphere shrinks. Radio waves then bounce off the bottom of the ionosphere at a higher altitude, giving these waves longer pathways to follow. This leads to the signals of certain AM stations reaching much larger areas at night than they do during the day.

The interaction of the Earth's magnetic field, Earth's upper atmosphere, and solar wind produces geomagnetic storms that produce the Aurora Australis (pictured here from space) and the Aurora Borealis.

Particle-type radiation from the Sun, referred to as the **solar wind**, consists primarily of electrons and protons that are thrust from the Sun's surface at speeds of hundreds of kilometers per second. These flowing charged particles constitute and interact with an interplanetary magnetic field. When these particles stream past Earth, they change the shape of Earth's magnetic field (called the geomagnetic field), creating a region called the magnetosphere, and affecting currents that flow about the planet. Charged particles are accelerated along the concentrated field lines at Earth's magnetic polls, generating eerie and beautiful auroral displays.

solar wind a continuous, but varying, stream of charged particles (mostly electrons and protons) generated by the Sun

Satellites are affected by the harsh radiation environment more directly. Penetrating charged particles can cause upsets in sensitive electronics components. Surface charge buildup and discharge can cause a wide variety of failures. Intense radiation can reduce the effectiveness of solar power arrays. And thermal expansion and contraction can cause mechanical failures.

Deep Space Contributions

Other sources that contribute to the near-Earth space environment include galactic cosmic ray particles, which originate from outside of the solar system. The higher the altitude, the less atmosphere there is to act as a shield, leading to greater exposure to these cosmic ray particles. A Geiger counter would detect a much higher number of such particles during an airplane flight than it would on the surface of Earth (away from radioactive sources, of course). For astronauts, the radiation hazards from all sources are serious. The National Aeronautics and Space Administration (NASA) monitors many different sources of information on the radiation environment to keep astronauts safe. Significant (though not complete) shielding can be afforded to spacewalking astronauts by simply having them go back inside their shuttle or space station.

Interactions with Comets and Asteroids

Comets and asteroids also contribute to the near-Earth space environment. Comets pass through the solar system, sometimes repeatedly because of their

THE IMPACT OF SOLAR FLARES

With fair frequency, the Sun's surface erupts with solar flares, which send intense bursts of electromagnetic radiation into space. If directed at Earth, this radiation heats the atmosphere, expanding gases into higher altitudes, which increases drag on low Earth orbiting satellites. This radiation also generates more charged particles in the ionosphere, affecting the reflection and transmission paths of radio frequencies used by satellites and ground-based communications. Since the space-based global positioning system uses radio frequency signals, navigation errors are also introduced.

★ **Solar wind is responsible for the direction of a comet's tail, which always points away from the Sun.**

meteoroid a piece of interplanetary material smaller than an asteroid or comet

meteor the physical manifestation of a meteoroid interacting with Earth's atmosphere; this includes visible light and radio frequency generation, and an ionized trail from which radar signals can be reflected

cover glass a thin sheet of glass used to cover the solid state device in a solar cell

perturbations term used in orbital mechanics to refer to changes in orbits due to "perturbing" forces, such as gravity

CORONAL MASS EJECTIONS

In addition to solar flares, another frequently occurring solar event is the coronal mass ejection, in which the Sun sends concentrated bursts of solar wind into space. These events typically generate the equivalent energy of several atomic bombs. If Earth is in the path of the increased solar wind from one of these events, a geomagnetic storm may occur as a result of the solar wind distorting Earth's magnetic field (the magnetosphere). Such storms have been responsible for power system failures on Earth, including a major blackout in Quebec, Canada, on March 13, 1989, which left six million people without power for nine hours.

orbits. Both forms of solar radiation act on these dirty snowballs in space. As comets come near the Sun, the absorbed heat and solar wind pressure cause particles to come loose from the comet. ★ Solid particles of ice and rock (**meteoroids**) blown off by the Sun stay suspended in the comet's orbital path. When that path is close to our own orbit around the Sun, Earth will collide with these particle streams, giving rise to our annual **meteor** showers. While most of these particles are very small and carry very little mass, they pose yet another hazard to our satellites. Though immediate failure from meteoroid impacts seldom occurs, the continuous bombardment of these grains of sand have a degrading effect on satellite surfaces. For example, they chip and crack the **cover glass** on solar panels, making them less efficient. Pitting allows atomic oxygen, present in low Earth orbits, to react with an exposed surface, causing corrosion and reducing the serviceable lifetimes of satellites.

Meteoroids can also come from outside the solar system (sporadics) or from other Sun-orbiting bodies, such as asteroids. The main asteroid belt lies between Mars and Jupiter, and may be the remnants of what would have been another planet that never formed in the solar system. While most of these stay in safe orbits away from Earth, some have made their way into Earth-crossing orbits, perhaps through collisions and gravitational **perturbations**. Such objects have been known to collide with Earth over time and are expected to do so in the future. An asteroid as small as 0.5 to 1 kilometer (0.3 to 0.6 mile) in diameter impacting Earth can cause significant immediate and long-lasting damage. It is believed that a somewhat larger event may be responsible for the extinction of the dinosaurs and the destruction of perhaps one-fourth of all life on Earth about 65 million years ago. Another major impact event occurred in Siberia in 1908. What may have been a small asteroid exploded over the Tunguska forestlands, laying flat hundreds of square kilometers of trees. Evidence of these large impact events exists in the form of the craters they have left on Earth.

Recognizing that such events are very rare but may occur and cause great catastrophe, in the early 1990s the U.S. Congress formalized a scientific effort called Planetary Defense to look for near Earth objects (NEOs) that might collide with Earth. Since that time, many NEOs have been discovered. Major impacts are certain to occur in the future, but it is hard to say when.

Space Debris: A Growing Concern

Finally, humankind itself contributes to our near-Earth space environment through launch and space activities over the years. Space debris from such activities is a growing concern. Windows on the space shuttle are replaced regularly because of chipping caused by collisions with small pieces of space junk or by natural meteoroid strikes. Collision risk during launch, orbit, and reentry operations continues to rise. Larger objects in low Earth orbits (rocket boosters, space stations, etc.) eventually fall back through Earth's atmosphere, posing a small, yet real risk to human life. Debris objects actually reenter the atmosphere quite frequently. Observers often mistake these reentering objects for meteors or UFOs. A woman named Lottie Williams may have the distinction of being the first person to be hit by reentering space debris. While her claim of being hit on the shoulder by a small piece

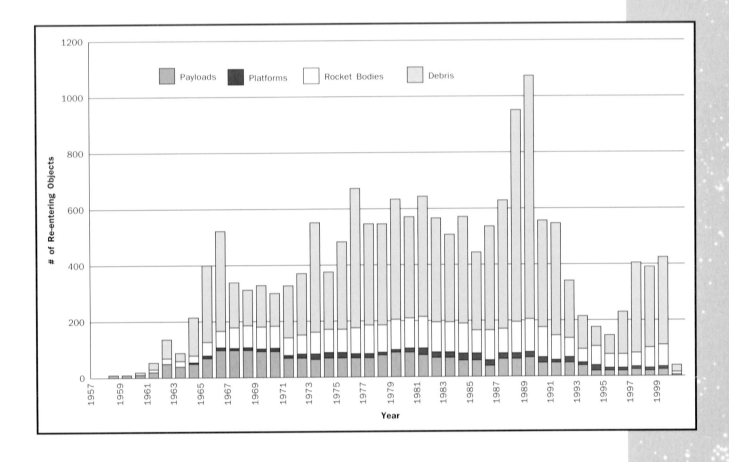

of a Delta II rocket may be difficult to verify, the piece of debris that she claims hit her has been verified to be from just such an object. SEE ALSO ASTEROIDS (VOLUME 2); CLOSE ENCOUNTERS (VOLUME 2); COMETS (VOLUME 2); COSMIC RAYS (VOLUME 2); SOLAR PARTICLE RADIATION (VOLUME 2); SPACE DEBRIS (VOLUME 2); SUN (VOLUME 2); WEATHER, SPACE (VOLUME 2).

David Desrocher

Bibliography

Gombosi, Tamas I. *Physics of the Space Environment.* Cambridge, UK: Cambridge University Press, 1998.

Johnson, Francis. *Satellite Environment Handbook.* Sanford, CA: Stanford University Press, 1999.

Steele, Duncan. *Rogue Asteroids and Doomsday Comets: The Search for the Million Megaton Menace that Threatens Life on Earth.* New York: John Wiley & Sons, 1997.

Tascione, Thomas F. *Introduction to the Space Environment.* Malabar, FL: Krieger Publishing Company, 1988.

Internet Resources

Center for Orbital and Reentry Debris Studies. Aerospace Corporation. <http://www.aero.org/cords/>.

Hamilton, Calvin J. "Terrestrial Impact Craters." <http://www.solarviews.com/eng/tercrate.htm>.

Living with a Star Program. National Aeronautics and Space Administration. <http://lws.gsfc.nasa.gov/lws.htm>.

Planetary Defense Workshop. Lawrence Livermore National Laboratory. <http://www.llnl.gov/planetary/>.

Primer on the Space Environment. National Oceanic and Atmospheric Administration, Space Environment Center. <http://sec.noaa.gov/primer/primer.html>.

SpaceWeather.com Daily Update Page. National Aeronautics and Space Administration. <http://www.spaceweather.com/>.

What is the Magnetosphere? Space Plasma Physics Branch, NASA Marshall Space Flight Center. <http://science.nasa.gov/ssl/pad/sppb/edu/magnetosphere/>.

Spacecraft Buses

payload any cargo launched aboard a rocket that is destined for space, including communications satellites or modules, supplies, equipment, and astronauts; does not include the vehicle used to move the cargo or the propellant that powers the vehicle

The structural body and primary system of a space vehicle is commonly referred to as a spacecraft bus. The spacecraft bus is used as a transport mechanism for a spacecraft **payload** much like an ordinary city bus is a transport vehicle for its passengers. Although each spacecraft payload may be quite different from another, all spacecraft buses are similar in their makeup. The spacecraft bus consists of several different subsystems, each with a unique purpose. The structural subsystem consists of the primary structure of the spacecraft, and supports all the spacecraft hardware, including the payload instruments. The structure, which can take various forms depending on the requirements of the particular mission, must be designed to minimize mass and still survive the severe forces exerted on it during launch and on its short trip to space.

The electrical power subsystem provides power for the payload, as well as the rest of the bus. This is usually achieved through the use of solar panels that convert solar radiation into electrical current. The solar panels sometimes must be quite large, so they are hinged and folded during launch then deployed once in orbit. The subsystem also may consist of batteries for storing energy to be used when the spacecraft is in Earth's shadow. Another major subsystem is command and data handling, which consists of the computer "brain" that runs the spacecraft, and all the electronics that control how data is transported from component to component. All other subsystems "talk" to this subsystem by sending data back and forth through hundreds of feet of wiring carefully routed throughout the spacecraft bus.

The communications subsystem contains components such as receivers and transmitters to communicate with controllers back on Earth. Many operations the spacecraft must perform are controlled through software commands sent from Earth by radio signals. Another important subsystem is the attitude control subsystem. This consists of specialized sensors able to look at the Earth, Sun, and stars to determine the exact position of the spacecraft and the direction in which it should point. Many operations spacecraft perform require very precise pointing, such as positioning imaging satellites that must point at specific spots on Earth.

In order to adjust the orbit to maintain the spacecraft in orbit for many years, a propulsion subsystem is sometimes required. There are many types of propulsion systems, but most consist of various types of rocket thrusters, which are small engines that burn special fuel to produce thrust. One additional crucial subsystem worth discussion is the thermal control subsystem, which maintains the proper temperatures for the entire spacecraft bus and all its components. This is achieved through the use of small heater strips,

Typical spacecraft bus (High Energy Solar Spectroscopic Imager [HESSI] built by Spectrum Astro, Inc., for NASA).

special paints and coatings that either reflect or absorb heat from Earth and the Sun, and multi-layered insulation blankets to protect from the extreme cold of space. SEE ALSO EXPLORATION PROGRAMS (VOLUME 2); GOVERNMENT SPACE PROGRAMS (VOLUME 2); ROBOTIC EXPLORATION OF SPACE (VOLUME 2); SENSORS (VOLUME 2); SOLID ROCKET BOOSTERS (VOLUME 3).

Kevin Jardine

Bibliography

Fortescue, Peter W., and Stark, John P. W. *Spacecraft Systems Engineering.* Chichester, West Sussex, UK: John Wiley & Sons, 1991.

Spaceflight, History of

On October 4, 1957, the Union of Soviet Socialist Republics (USSR) launched a rocket that inserted a small satellite into orbit around Earth. Three months later, on January 31, 1958, the United States launched a satellite into a higher Earth orbit. Most historians consider these two events as denoting the beginning of the space age. This new age historically marked the first time that humans had been able to send objects—and, later, themselves—into outer space, that is, the region beyond the detectable atmosphere. The flight of machines into and through space, while a product of mid-twentieth-century technology, was a dream held by scientists, engineers, political leaders, and visionaries for many centuries before the means existed to convert these ideas into reality. And while the USSR and United States first created the enabling space technologies, the ideas that shaped these machines spanned other continents and the peoples of many other nations. The idea of spaceflight, like the capabilities that today's spaceships and rockets make possible, belongs to humanity without the limitations of any single nation or people.

The Origins of Spaceflight

The ideas that gave birth to spaceflight are ancient in origin and international in scope. Like many such revolutionary concepts, spaceflight was first expressed in myth and later in the writings of fiction authors and academicians. The Chinese developed rudimentary forms of rockets, adapted from solid gunpowder, as devices for celebrations of religious anniversaries. In 1232 China used rockets for the first time as weapons against invading Mongols. A decade later, Roger Bacon, an English monk, developed a formula for mixing gunpowder into controlled explosive devices. In the eighteenth century, British Captain Thomas Desaguliers conducted studies of rockets obtained from India in an attempt to determine their range and capabilities. In the nineteenth century William Congrieve, a British colonel, developed a series of rockets that extended the range of the rockets developed by India and that were adapted for use by armies. Congrieve's rockets were used in the Napoleonic wars of 1806. The nineteenth century would see the growth of the technology of solid-fueled rockets as weapons and the wider application of their use.

In 1857 a self-taught Russian mathematics teacher, Konstantin Eduardovich Tsiolkovsky, was born. Over the next eight decades, his writings and teachings would form the basis of modern spaceflight goals and systems, including multistage rockets, winged shuttlecraft, space stations, and interplanetary missions. Upon his death in Kaluga, Russia, on September 19, 1935, Tsiolkovsky would be considered one of the major influences upon space technology and later became known as the "Father" of the Soviet Union's space exploration program.

During the same period, the idea of space travel received attention in the form of fiction writings. The French science fiction author Jules Verne penned several novels with spaceflight themes. In his 1865 novel *From the Earth to the Moon,* Verne constructed a scenario for a piloted flight to the Moon that contained elements of the future space missions a century later, including a launching site on the Florida coast and a spaceship named Co-

During a test of the Vanguard rocket for the United States Geophysical Year, a first stage malfunction causes a loss of thrust, resulting in the destruction of the vehicle on December 6, 1957.

lumbia, the same name chosen for the Apollo 11 spaceship that made the first lunar landing mission in July 1969. In his 1869 novel *The Brick Moon* and an 1870 sequel *Life in the Brick Moon*, American writer Edward Everett Hale predicted the first uses for an orbiting space station, including military and navigation functions. These novels, first published in the *Atlantic Monthly* magazine, address issues related to permanent spaceflight and satellite observations of Earth.

Twentieth Century Development of the Liquid-Fueled Rocket

In the early years of the twentieth century, American academician Robert Goddard developed the first controlled liquid-fueled rocket. Launching from a rudimentary test laboratory in Auburn, Massachusetts, on March 16, 1926, his rocket flights and test stand firings advanced the technology of rocketry. In Europe, rocket enthusiasts formed the Society for Space Travel to better promote rocket development and space exploration themes. Members of the group included Hermann Oberth, whose writings and space advocacy would include engineering and mathematical models for interplanetary rocket flights, and Wernher von Braun, who designed the Saturn V booster that carried Apollo spacecraft to the Moon. Germany also paid for rocketry research conducted by Austrian engineer Eugen Sänger. Sänger and his research assistant (and later his wife) Irene Brendt contributed studies on advanced winged cargo rockets that were the forerunners of today's space shuttles. In the Soviet Union, academician Valentin P. Glushko developed the USSR's first liquid propellant rockets.

Although there were many others as well whose works detailed different types of space vehicles, space missions, and space utilization, the basis of many of the space launch vehicles of the twentieth century arose from the work of von Braun, who was subsidized by the German government during World War II (1939–1945). Working at a laboratory and launching complex called Peenemünde on the Baltic coast, von Braun and his associates developed the first **ballistic** missiles capable of exiting Earth's atmosphere during their brief flights. The most advanced of these designs was called the V-2. On October 3, 1942, the first of the V-2 rockets were successfully launched to an altitude of 93 kilometers (58 miles) and a range of 190 kilometers (118 miles). The successful test was referred to by German Captain Walter Dornberger, von Braun's superior at the Peenemünde complex, as the "birth of the Space Age," for it marked the first flight of a missile out of the atmosphere, in essence the world's first spaceship. While von Braun's task was to develop military weapons, he and his staff stole away as much time as possible to work on rocket-powered spaceship designs, a fact that was discovered by the German military. This discovery led von Braun to be briefly imprisoned until he was able to assure the Nazi military that the energy of his workers was directed toward weapons and not planetary rocket flights.

After the war ended in 1945, von Braun, his engineers and technicians, his unfired inventory of V-2 rockets, and his research data formed a treasure trove of space and rocketry concepts for both the United States and the Soviet Union. Von Braun himself and much of his team came to the United States, bringing along a good portion of the German rocketry archive and many V-2 rockets and rocket parts. Others of the von Braun group and some of the V-2 missiles and data were captured by the Soviet government. These two elements of the former German rocketeers led to major advances for the space enthusiasts of the United States and USSR. Beginning in January 1947, at a site located at White Sands, New Mexico, von Braun modified his V-2 rockets for scientific flights to the upper atmosphere. In the Soviet Union, Sergei Korolev undertook similar testing, using the captured V-2s. Over the next decade, data gained from firings of the V-2s led each nation to develop its own rocket and space vehicle designs.

Space Program Development

In the Soviet Union, a ballistic missile called the R-7 was the first design to emerge from the early Soviet rocket programs that was powerful enough to strike targets in the United States or to insert satellites into Earth orbits. In the United States, a series of intermediate, medium, and intercontinental missiles emerged from the drawing boards. These had names such as Thor, Redstone, Atlas, and Titan. Along with the R-7, these missiles became the foundation of space-launching vehicles used by both nations to send the first

ballistic the path of an object in unpowered flight; the path of a spacecraft after the engines have shut down

satellites, probes, and human beings into space. Once begun on October 4, 1957, this so-called space race for dominance of the space environment was a defining element of the Cold War between the two superpowers. Rocketry gave each nation both a means to carry destructive nuclear weapons to the soil of the other country and a means of gaining scientific exploration of space. This race eventually formed up around four major elements: humans in space, advanced space exploration, reusable spaceflight, and permanent spaceflight.

The early humans in space efforts saw leadership by the Soviet Union. On April 12, 1961, using a version of the R-7 missile, the Soviet Union launched the first human, Air Force Major Yuri Gagarin, into orbital flight around Earth. Sealed inside a single seat in the space capsule Vostok 1, Gagarin completed a single orbit before descending under parachute for a landing in the Soviet Union (Gagarin himself actually ejected from the capsule's cabin before it landed in a field about ninety minutes after liftoff). The United States followed with more limited suborbital flights of astronauts Alan B. Shepard Jr. and Virgil I. "Gus" Grissom aboard single-seat Mercury spacecraft named Freedom 7 and Liberty Bell 7 on May 5 and July 19, 1961. Throughout 1961, 1962, and 1963 the United States and the USSR launched astronauts and cosmonauts into Earth orbit aboard these limited craft.

Beginning in 1965, the United States launched a two-seat space capsule called Gemini using larger Titan II missiles. The Soviet Union continued to launch Vostok capsules, modified to carry two and three persons. But the American Gemini craft were more capable, performing rendezvous and docking and long-duration space missions. This era of advanced human spaceflight now centered on a race between the superpowers to send a human expedition to the Moon's surface.

The Americans announced a program to land astronauts on the Moon called Project Apollo. The United States initiated a series of advanced space vehicles, including a new three-seat capsule capable of maneuvering between Earth and the Moon, a lunar landing craft that could carry two astronauts to the Moon's surface, and a family of advanced space rockets not based on earlier missile designs. The Soviets began a series of advanced rocket designs and a series of advanced Earth-orbiting space capsules called Soyuz. A lunar landing program was also underway in secret in the Soviet Union. But from 1965 to 1969 the Americans maintained a lead in human space missions that included the first space rendezvous and docking of two craft in orbit and long-duration spaceflights of one and two weeks in duration. Following the first walk in space performed during the Soviet Voshkod 2 mission in March 1965, American spacewalkers achieved extensive data on working outside space vehicles, considered key learning steps before astronauts could walk on the Moon's surface. But both nations suffered casualties during this peaceful scientific race. In January 1967, the first crew of an Apollo flight, Apollo 1, was killed in a launch pad fire. In April 1967 the first cosmonaut testing the Soyuz capsule was killed during a reentry mishap. Gradually, however, the United States was pulling ahead in the lunar race.

Using the lifting power of the Saturn rockets, the United States sent the first astronauts beyond Earth to lunar orbit in December 1968. The following summer, Apollo 11 and its crew of astronauts Neil A. Armstrong, Edwin E. "Buzz" Aldrin Jr., and Michael Collins were launched toward the

Moon and on July 20, 1969, accomplished the first of six piloted landings. The Soviets were forced to abandon their lunar landing program because of continued malfunctions of the large N-1 lunar booster. No Russian cosmonauts ever made the attempt.

Instead, the Soviet space program redefined itself by the development of semipermanent space stations. The first in this series, called Salyut, was launched in 1971. Eventually the experience gained in the Salyut space station series led the Soviets to develop a larger and more expandable station complex called Mir. The Mir space station, resupplied by both Soyuz rockets and U.S. space shuttles, provided valuable long-duration space experience from 1986 to the spring of 2001 when the Mir complex was successfully and safely deorbited.

Both the U.S. and Soviet programs explored space with robotic probes. The Soviets were successful in accomplishing landings on Venus with an unmanned probe called Venera. Soviet robots also landed on the Moon and returned lunar soils to Earth for analysis by Russian scientists. The United States successfully accomplished robotic landings on Mars in 1976 and 1997 in the Viking and Mars Pathfinder programs.

An era of reusable space vehicles began in April 1981 with the first launch of the partially reusable space shuttle. From 1981 through 2001 more than 100 flights of the shuttles were accomplished. Only one, the launch of space shuttle Challenger on January 28, 1986, was unsuccessful and resulted in the loss of the spacecraft and the entire crew of seven astronauts. Following the accident, the shuttles were redesigned and returned to safe spaceflight. A Soviet shuttle project called Buran was abandoned in 1993 because of the collapse of the Russian economy. Construction of a permanent space station began in 1998. The project brought together sixteen international partners, including Russia and the United States.

As the twenty-first century began, space activities assumed more of an international and commercial flavor, begetting a process of evolution and change as old as the idea of spaceflight itself. SEE ALSO APOLLO (VOLUME 3); GOVERNMENT SPACE PROGRAMS (VOLUME 2); INTERNATIONAL COOPERATION (VOLUME 3); INTERNATIONAL SPACE STATION (VOLUME 3); MIR (VOLUME 3); NASA (VOLUME 3).

Frank Sietzen, Jr.

Bibliography

Emme, Eugene M. *A History of Space Flight.* New York: Holt, Rinehart and Winston, 1965.

Gatland, Kenneth. *The Illustrated Encyclopedia of Space Technology.* New York: Harmony Books, 1981.

Ley, Willy. *Events in Space.* New York: David McKay Co., 1969.

Neal, Valerie, Cathleen S. Lewis, and Frank H. Winter. *Spaceflight: A Smithsonian Guide.* New York: Macmillan, 1995.

microgravity the condition experienced in free-fall as a spacecraft orbits Earth or another body; commonly called weightlessness; only very small forces are perceived in freefall, on the order of one-millionth the force of gravity on Earth's surface

Spacelab

Spacelab was a cylindrically shaped reusable laboratory carried aboard the space shuttle that was designed to allow scientists to perform experiments in **microgravity** conditions while orbiting Earth. It was designed and de-

An access tunnel joins with the Spacelab 1 module in the cargo bay of the space shuttle Columbia at the Kennedy Space Center. Spacelab was a reusable laboratory carried aboard the space shuttle, which was designed to allow scientists to perform experiments in microgravity conditions.

veloped by the European Space Agency (ESA) in cooperation with the National Aeronautics and Space Administration's (NASA) George C. Marshall Space Flight Center. Cradled in the shuttle's spacious cargo bay, Spacelab was used on numerous shuttle missions between 1983 and 1997. In addition to the United States, countries like Germany and Japan also conducted dedicated Spacelab missions.

Spacelab Components

Spacelab was developed as a modular structure with several components that could be connected and installed to meet specific mission requirements. For each mission, Spacelab components were assembled and placed into the shuttle's cargo bay at Kennedy Space Center (KSC) in Florida. Its four principal components consisted of the pressurized module, which contained a laboratory with a shirt-sleeve working environment; one or more open pallets that exposed materials and equipment to space; a tunnel to gain access to the module from the shuttle; and an instrument pointing subsystem.

An interior view of an empty Spacelab module, prior to its move to the National Air and Space Museum in Washington, D.C., September 22, 1998.

The pressurized module, or laboratory, provided a habitable environment for the crew. It was available in several configurations that included either one segment (core) or two segments (core and experiment) that could be reused for up to fifty missions. The core segment contained supporting systems, such as data processing equipment and utilities for the pressurized module and pallets. Inside the module, laboratory equipment was mounted in racks and other areas. The laboratory also had fixtures, such as floor-mounted racks and a workbench. The so-called "experiment segment," provided more working laboratory space and contained only floor-mounted racks.

Together, the core and experiment segments were approximately 7 meters long. Added to this assembly were U-shaped pallets located outside the pressurized module. The pallets were often used on Spacelab missions for mounting instrumentation, large instruments, experiments needing exposure to space, and instruments requiring a large field of view, such as telescopes.

Conducting Experiments on Spacelab

Scientists onboard Spacelab included mission specialists and **payload specialists**, who were primarily scientists, not career astronauts. To make working in space easier, handrails were mounted on the racks and overhead. Foot restraints were also provided on the floor and on rack platforms. During some missions, the crew split into two twelve-hour shifts, allowing research to continue around the clock.

Each Spacelab mission required years of planning. Before each mission, support personnel developed a timeline for conducting experiments, and worked closely with the principal investigators to make sure the resources for each experiment were available. In addition, scientists on the ground could follow the progress of experiments aboard Spacelab using television and computer displays from orbit. Earth-bound scientists also could command experiments, and talk with the crew.

Spacelab Research

Research into many fields of science was performed aboard Spacelab. Experiments on Earth's atmosphere included research into atmospheric chemistry, energy, and dynamics. In addition, Spacelab was used to correlate atmospheric data from satellites. The space-based laboratory also provided an ideal platform to conduct experiments on space plasma. From orbit, scientists could closely observe the electrified gases in the **ionosphere** layer of the atmosphere.

Studies of the Sun were a major focus of Spacelab activities. Crews onboard Spacelab were able to observe all of the Sun's radiant energy. Using the instruments on Spacelab missions, astronomers obtained some of the best images of the Sun in both still photographs and videos. Additionally, sensitive **spectrometers** collected information on the chemistry and the physics of our nearest star. The images and spectral analysis contributed to the modeling of the Sun's dynamics and structure.

Spacelab also allowed scientists to look farther into space and conduct sophisticated astronomical research. While in orbit, scientists had the opportunity to select targets, fine-tune their instruments based on the current conditions, and look at interesting events, just like an astronomer at a ground-based observatory. In addition, astronomers were able to view the universe at various **wavelengths**, including cosmic rays, **X rays**, **ultraviolet**, and **infrared**. Increasingly complex astronomical instruments were deployed with succeeding Spacelab missions, allowing scientists to increase the quality and quantity of data collected.

Progress in materials science on Earth has been limited in some areas due to the effects of gravity. However, the microgravity condition on Spacelab provided scientists an opportunity to study how materials behave outside

payload specialists scientists or engineers selected by a company or a government employer for their expertise in conducting a specific experiment or commercial venture on a space shuttle mission

ionosphere a charged particle region of several layers in the upper atmosphere created by radiation interacting with upper atmospheric gases

spectrometer an instrument with a scale for measuring the wavelength of light

wavelength the distance from crest to crest on a wave at an instant in time

X rays a form of high-energy radiation just beyond the ultraviolet portion of the electromagnetic spectrum

ultraviolet the portion of the electromagnetic spectrum just beyond (having shorter wavelengths than) violet

infrared portion of the electromagnetic spectrum with wavelengths slightly longer than visible light

of the influence of Earth's gravity. Experiments conducted on Spacelab significantly advanced the science of material processing by providing a sustained microgravity environment for melting, combining or separating raw materials into useful products, and creating defect-free crystals.

In addition, the microgravity environment in the Spacelab module allowed scientists to test basic theories and to develop new processing techniques. This research advanced the study of new metals and alloys, as well as protein crystals for drug research, electronics and **semiconductors**, and fluid physics. Improvements in processing developed on Spacelab might lead to the development of valuable drugs; high-strength, temperature resistant ceramics and alloys; and other improved materials.

Many life sciences experiments were also conducted aboard Spacelab. Scientists were able to study life from the simplest, one-celled forms such as bacteria, to larger, more complex systems such as animals and humans. The Spacelab module provided habitats for plants and animals, and most importantly, the trained scientists to perform experiments. Basic biology questions were investigated, as well as practical questions related human adaptation to space and the phenomena of "space sickness." Commercial and pharmacologic products were also produced in purer forms than ever before onboard Spacelab. At the same time, biological materials could be studied with great precision because crystals could be grown both larger and purer.

Spacelab's Contribution to Space Exploration

Over a 15-year period, Spacelab served as an orbiting laboratory that allowed scientists to study the universe, the Sun and Earth, and conduct materials and biologic experiments. During that time, Spacelab served as both a laboratory and an observatory as scientists could both stimulate the environment with active experiments and observe the effects. Work on Spacelab also provided a "dress rehearsal" into the types of activities that are currently performed on the International Space Station. SEE ALSO CRYSTAL GROWTH (VOLUME 3); INTERNATIONAL SPACE STATION (VOLUMES 1 AND 3); MADE IN SPACE (VOLUME 1); MICROGRAVITY (VOLUME 2); SPACE SHUTTLE (VOLUME 3).

John F. Kross

Bibliography

Yenne, Bill. *The Encyclopedia of US Spacecraft*. New York: Exeter Books, 1988.

Internet Resources

"Spacelab Science." Kennedy Space Center. <http://science.ksc.nasa.gov/shuttle/technology/sts-newsref/spacelab.html#spacelab>.

"Spacelab." Manned Spaceflight Center. <http://liftoff.msfc.nasa.gov/Shuttle/spacelab>.

Stars

Stars are huge balls of very hot, mostly ionized, gas (plasma) that are held together by gravity. They form when vast agglomerations of gas and dust known as molecular clouds (typically 10 to 100 **light years** across) fragment into denser cores (tenths of a light-year across) that can collapse inward under their own gravity. Matter falling inward forms one or more dense, hot, central objects known as protostars. Rotation forces some of the matter to

Optical image of the Pleiades open star cluster in the constellation of Taurus. Stars are huge balls of very hot gas (plasma) that are held together by gravity.

accumulate in a disk rotating around the protostar(s). As gravity pulls rotating material inward, it spins faster, akin to what happens to figure skaters when they pull their initially outstretched arms in toward their bodies.

In order for material to fall onto a protostar from a rapidly spinning disk, it must slow down. Recent theoretical work suggests that this is accomplished through the interaction of the material with magnetic fields that thread the disks of protostars. Near the disk, the magnetic field is bent into an hourglass shape. Gas particles are flung off the rotating disk by **centrifugal** force, slowing the rotation of the disk. The ejected material is channeled into narrow jets perpendicular to the disk, while material from the disk falls onto the protostar. Planets may eventually form within the disk. The jets plow into the surrounding medium, sweeping up a **bipolar outflow** on opposite sides of the protostar. It is not yet known whether the final mass of a star is determined by the initial mass of the core in which it was born or from the clearing of material by bipolar outflows. In any case, the final mass of the star determines how it will evolve from this point on.

centrifugal directed away from the center through spinning

bipolar outflow jets of material (gas and dust) flowing away from a central object (e.g., a protostar) in opposite directions

Main Sequence Stars

When the star has accumulated enough material so that the temperature and pressure are high enough, nuclear **fusion** reactions, which convert hydrogen into helium, begin deep within the core of the star. The energy from the reactions makes its way to the surface of the star in about a million years, causing the star to shine. The pressure from these nuclear reactions at the star's core balances the pull of gravity, and the star is now called a main sequence star.

This name is derived from the relationship between a star's intrinsic brightness and its temperature, which was discovered independently by Danish astronomer Ejnar Hertzsprung (in 1911) and American astronomer Henry Norris Russell (in 1913). This relationship is displayed in a Hertzsprung-Russell diagram. A star's color depends on its surface temperature; red stars

fusion releasing nuclear energy by combining lighter elements such as hydrogen into heavier elements

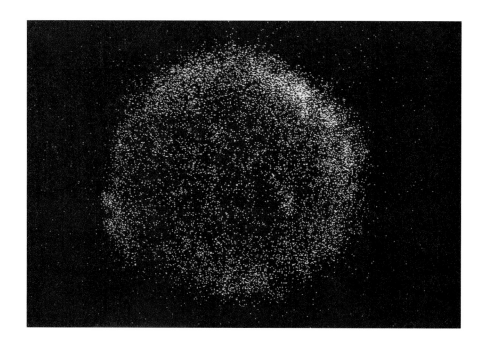

Massive stars may explode in an event known as a supernova. This supernova in the constellation Cassiopeia was observed by astronomer Tycho Brahe in 1572.

are the coolest and blue stars are the hottest. The temperature, brightness, and longevity of a star on the main sequence are determined by its mass; the least massive main sequence stars are the coolest and dimmest, and the most massive stars are the hottest and brightest. Objects less than about one-thirteenth the mass of the Sun can never sustain fusion reactions. These objects are known as brown dwarfs.

Red Giants and Red Supergiants

Counterintuitively, the more massive a star is, the more rapidly it uses up the hydrogen at its core. The most massive stars deplete their central hydrogen supply in a million years, whereas stars that are only about one-tenth the mass of the Sun remain on the main sequence for hundreds of billions of years. When hydrogen becomes depleted in the core, the core starts to collapse, and the temperature and pressure rise, so that fusion reactions can begin in a shell around the helium core. This new heat supply causes the outer layers of the star to expand and cool, and the star becomes a red giant, or a red supergiant if it is very massive.

Planetary Nebulae, White Dwarfs, and Black Dwarfs

Once stars up to a few times the mass of the Sun reach the red giant phase, the core continues to contract and temperatures and pressures in the core become high enough for helium nuclei to fuse together to form carbon. This process occurs rapidly (only a few minutes in a star like the Sun), and the star begins to shed the outer layers of its atmosphere as a **diffuse** cloud called a planetary nebula. Eventually, only about 20 percent of the star's initial mass remains in a very dense core, about the size of Earth, called a white dwarf. White dwarfs are stable because the pressure of **electrons** repulsing each other balances the pull of gravity. There is no fuel left to burn, so the star slowly cools over billions of years, eventually becoming a cold, dark object known as a black dwarf.

diffuse spread out; not concentrated

electrons negatively charged subatomic particles

Supernovae, Neutron Stars, and Black Holes

After a star more than about five times the mass of the Sun has become a red supergiant, its core goes through several contractions, becoming hotter and denser each time, initiating a new series of nuclear reactions that release energy and temporarily halt the collapse. Once the core has become primarily iron, however, energy can no longer be released through fusion reactions, because energy is required to fuse iron into heavier elements. The core then collapses violently in less than a tenth of a second.

The energy released from this collapse sends a shock wave through the star's outer layers, compressing the material and fusing new elements and radioactive isotopes, which are propelled into space in a spectacular explosion known as a supernova. This material seeds space with heavy elements and may collide with other clouds of gas and dust, compressing them and initiating the formation of new stars. The core that remains behind after the explosion may become either a neutron star, as the intense pressure forces electrons to combine with **protons**, or a black hole, if the original star was massive enough so that not even the pressure of the neutrons can overcome gravity. Black holes are stars that have literally collapsed out of existence, leaving behind only an intense gravitational pull. SEE ALSO ASTRONOMER (VOLUME 2); ASTRONOMY, KINDS OF (VOLUME 2); BLACK HOLES (VOLUME 2); GALAXIES (VOLUME 2); GRAVITY (VOLUME 2); PULSARS (VOLUME 2); SUN (VOLUME 2); SUPERNOVA (VOLUME 2).

Grace Wolf-Chase

protons positively charged subatomic particles

Bibliography

Bennett, Jeffrey, Megan Donahue, Nicholas Schneider, and Mark Voit. *The Cosmic Perspective.* Menlo Park, CA: Addison Wesley Longman, 1999.

Kaler, James B. *Stars.* New York: Scientific American Library and W. H. Freeman, 1992.

Seeds, Michael A. *Horizons: Exploring the Universe,* 6th ed. Pacific Grove, CA: Brooks/Cole, 2000.

Internet Resources

Imagine the Universe! NASA Goddard Space Flight Center. <http://imagine.gsfc.nasa .gov/docs/science/know_l2/stars.html>.

Sun

Of all of the astronomical objects, the Sun is the most important to human beings. Since the dawn of civilization, knowing the daily and annual behavior of the Sun has meant the difference between life and death for people learning when to plant crops and when to harvest. Ancient mythologies preserved this knowledge in story form. These were often picturesque descriptions of the Sun's behavior—for example, the Chinese interpretation of a solar eclipse as a dragon chasing and eating the Sun. Sometimes the stories included precise enough details for predicting solar behavior—for instance, in the version from India, the dragon is sliced into two invisible halves. When the position in the sky of one of these halves is lined up with the Sun and the Moon, an eclipse occurs.

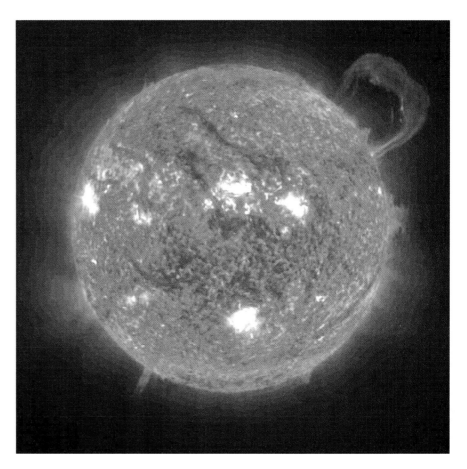

A large, handle-shaped prominence is visible in this extreme ultraviolet imaging telescope (EIT) shot of the Sun. Prominences are giant clouds of relatively cool plasma suspended in the Sun's hot corona, which occasionally can erupt. The cooler temperatures are indicated by darker red areas, while the hottest areas appear almost white.

Solar Eclipses

Over centuries of observations and study, a scientific understanding of the Sun has grown out of these myths. The invisible dragon halves were a way of describing the serendipitous arrangement of the relative locations and sizes of Earth, the Moon, and the Sun. In order for a solar eclipse to happen, the Moon not only has to be in new phase (between the Sun and Earth) but also has to line up exactly with the disk of the Sun. Since the Moon's orbit around Earth is tilted with respect to Earth's orbit around the Sun, this happens about twice a year instead of once a month. Solar eclipses are not visible all over Earth, but only under the moving shadow of the Moon. In areas not completely covered by the Moon's shadow, observers see a "partial eclipse," which looks like a bite has been taken out of the Sun. Or, if the Moon is in the far reaches of its orbit it might not be quite big enough to cover the Sun's disk. Then observers would see the Sun shining in a thin, bright ring around the Moon in what is known as an "annular eclipse," even if they are perfectly lined up. Total eclipses of the Sun are rarely seen, because the timing and geometry have to be just right to position a large enough Moon-shadow right over a particular location. When this happens, observers in that location have an opportunity to observe parts of the Sun that are usually impossible to see.

Solar Corona

It is when the Sun is totally eclipsed that the solar corona is visible. "Corona" means "crown," and indeed the outer atmosphere of the Sun appears to encircle its blacked-out disk in an extended pearly crown. Ordinarily, the corona is so much dimmer than the bright disk of the Sun that it cannot be seen—even during a partial or annular eclipse. There is another way to see the corona, however, even without an eclipse. Although the part of the Sun seen with the naked eye normally outshines it, the corona is actually the brightest part of the Sun when observed with an X-ray telescope. The Sun emits light at a wide range of **frequencies**, or colors. Most of the light it emits is in the range visible to human eyes—the colors that make up a rainbow. Human eyes have actually adapted to be sensitive to the frequencies at which the majority of the sunlight shines. X rays are light emitted at much higher frequencies than humans can see, in the same way as a dog whistle blows at a frequency that is beyond the sensitivity of the human ear. An X-ray telescope filters out all the light from the Sun except X rays, and what is left is mostly the solar corona.

Because the corona shines in X rays we know it is very hot. This is strange. It means that although the temperature of the Sun decreases from its center out to its surface (from several million degrees Celsius down to several thousand), it increases again in the corona (up again to several million degrees). How and why the corona gets heated is one of the big mysteries of solar physics. It probably has to do with the energy that comes from magnetic fields generated inside the Sun, which is dumped into the corona, heating it up.

Sunspots and Magnetic Fields

Besides the more obvious daily and annual variations of the Sun, an approximately eleven-year cycle was discovered once people started observing with telescopes. This was first seen by counting the number of sunspots on the Sun. Sunspots are dark regions on the solar surface that are fairly infrequent during the minimum phase of the eleven-year solar cycle, but that become more and more common during the maximum phase. They are dark because they are cooler than their surroundings, and they are cool because they are regions of very strong magnetic field where less heat escapes the solar surface.

Sunspots are not the only solar features that are most abundant at solar cycle maximum. Explosive flashes known as "solar flares," and massive eruptions of material out from the Sun known as coronal mass ejections also become more and more frequent. The material that is hurled outward in a coronal mass ejection can affect us here on Earth, damaging satellites and even power stations, and potentially causing blackouts or disrupting satellite TV or cell phone transmissions. Like sunspots, flares and coronal mass ejections are related to solar magnetic fields. In general, magnetic activity increases at solar cycle maximum.

Magnetic fields are an important part of almost everything that is observed about the Sun. So where do they come from? The motions of sunspots provide a clue. Like Earth, the Sun is spinning so it has its own north pole, south pole, and equator. As they move around as the Sun spins, sunspots

An eclipse of the Sun is photographed from the Apollo 12 spacecraft during its journey back to Earth from the Moon.

frequency the number of oscillations or vibrations per second of an electromagnetic wave or any wave

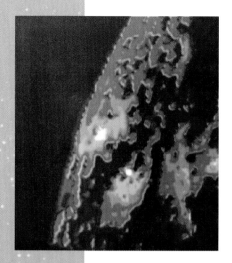

The Sun is photographed here by the Apollo Telescope Mount through a spectroheliometer. The reds, yellows, and white areas are the Sun's corona, approximately 70,000 kilometers above its surface, while the black regions are the surface of the star.

near the solar equator return to their starting point in about twenty-five days. Sunspots near the north and south pole of the Sun, however, take about thirty-five days to spin all the way around. The reason for this difference is that the Sun is not solid like a baseball, but fluid—more like a water balloon. Just below the surface this fluid is vigorously boiling and churning around, and this motion causes different parts of the Sun to spin around at different speeds. Furthermore, all this churning and spinning creates a magnetic field that is pointing one way near the north pole of the Sun and the opposite way near the south pole, like a giant bar magnet. Every eleven years, this magnet flips upside down so that in twenty-two years it has flipped over twice and is back where it started. Solar minimum happens when the magnet is pointing either due north or due south, and solar maximum occurs while it is in the process of flipping over.

Inside the Sun

When we look at the Sun, we see only the outside; how do we know what is happening below the surface? It turns out we can use techniques that are similar to those used in studying earthquakes. The surface of the Sun is continuously vibrating like a never-ending earthquake or a bell that is constantly being rung. By looking at the pattern of these vibrations and their frequency (like the tone of the bell), we can figure out what the inside of the Sun must be like. Thanks in part to these vibrations, we can confidently say that the churning motions below the surface not only create magnetic fields and make the Sun spin at different speeds, but they also move heat from the center of the Sun to the surface, where it is radiated away as light.

Near the center of the Sun the churning motions stop and the fluid becomes very dense and hot. Hydrogen atoms fly around at incredible speeds and when they collide they can stick together, creating helium atoms. This process, which is called fusion, provides the energy that causes stars to shine. In some stars, fusion can convert hydrogen and helium into heavier elements, such as carbon, oxygen, and nitrogen, which can in turn be combined to make still heavier elements, such as iron, lead, and even gold! In fact, everything on Earth—air, water, dirt, rocks, buildings, cars, trees, dogs, and even people—is made of elements that were created in stars by fusion.

The Evolution of the Sun

As exciting as it is, the Sun is often referred to as an "ordinary" star. This means that the information gained from the vast array of solar observations can be applied to understanding many of the stars in the sky. Furthermore by studying similar stars at various stages of their lifetimes, astronomers can tell how the Sun formed and how it will eventually die.

The Sun and the solar system began as a huge clump of gas in space, mostly made of hydrogen with some helium and only a relatively small amount of everything else (carbon, oxygen, iron, etc.). This clump slowly condensed and heated up due to gravity, and eventually it became dense and hot enough that fusion began and it started to shine. Not all of the gas fell into the young Sun; some of it stayed behind and was flattened into a pancake-like disk because it was spinning (just as a skilled pizza cook can flatten a clump of dough by tossing and spinning it). This disk then broke up into smaller clumps, which eventually became Earth and the other planets.

Meanwhile, the Sun settled down to a quieter life, slowly converting hydrogen into helium by fusion and shining the energy away into space. That was about 5 billion years ago and the Sun is still going strong.

The Sun's Future

But that is not the end of the story. Eventually, there will not be any hydrogen left in the center of the Sun to make helium. Gravity will then cause the center part of the Sun to collapse in on itself, and the energy given off by this implosion will cause the outer part to inflate. So, while the inner part of the Sun shrinks, the outer part will expand, and it will become so big that it will envelop Mercury, Venus, and even Earth.

The Sun will then continue its life as a red giant star, but not for long. As its last hydrogen is used up, the center of the Sun will heat up and start to convert helium into other elements in a last-ditch effort to keep fusion going and to keep shining. The available helium will be used up relatively quickly, however, and before long all fusion in the center will stop. The outer part of the Sun will then slowly expand and dissipate into space while the inner part will become a white dwarf, a relatively small, inactive lump of matter, which will slowly cool down as it radiates all its remaining energy into space. Life on Earth would not survive these events—but as this terrible fate is not due to happen for another 5 billion years, we have plenty more time to study the Sun in all its splendor! SEE ALSO COSMIC RAYS (VOLUME 2); SOLAR PARTICLE RADIATION (VOLUME 2); SOLAR WIND (VOLUME 2); SPACE ENVIRONMENT, NATURE OF (VOLUME 2); STARS (VOLUME 2); WEATHER, SPACE (VOLUME 2).

Sarah Gibson and Mark Miesch

Bibliography

Golub, Leon, and Jay M. Pasachoff. *The Solar Corona*. Cambridge, UK: Cambridge University Press, 1997.

Krupp, Edwin. C. *Echoes of Ancient Skies: Astronomy of Lost Civilizations*. New York: Harper and Row, 1983.

Phillips, Kenneth J. H. *Guide to the Sun*. Cambridge, UK: Cambridge University Press, 1992.

Strong, Keith. T., et al., eds. *The Many Faces of the Sun: A Summary of the Results from NASA's Solar Maximum Mission*. New York: Springer-Verlag, 1999.

Taylor, Peter O. and Nancy L. Hendrickson. *Beginner's Guide to the Sun*. Waukesha, WI: Kalmbach Publishing Company, 1995.

Taylor, Roger. J. *The Sun as a Star*. Cambridge, UK: Cambridge University Press, 1997.

Internet Resources

Mr. Eclipse. <http://www.MrEclipse.com/Special/SEprimer.html>.

Solar and Heliospheric Observatory. <http://sohowww.nascom.nasa.gov/>.

The Stanford Solar Center. <http://solar-center.stanford.edu>.

Supernova

As stars age, many use up their fuel and fade away to oblivion. Others, however, go out with a bang as supernovae, releasing energies of up to 10^{44} joules—an amount of energy equivalent to 30 times the power of a typical

The center of the Crab Nebula as viewed by the Hubble Space Telescope. The Crab was created by a supernova explosion on July 4, 1054, and is located approximately 6,500 light years from Earth.

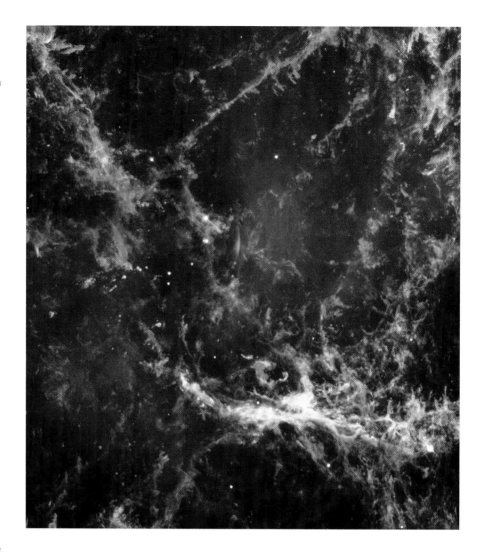

accretion the growth of a star or planet by accumulation of material from a companion star or the surrounding interstellar matter

spectra representations of the brightness of objects as a function of the wavelength of the emitted radiation

nuclear fusion the combining of low-mass atoms to create heavier ones; the heavier atom's mass is slightly less than the sum of the mass of its constituents, with the remaining mass converted to energy

galaxy a system of as many as hundreds of billions of stars that have a common gravitational attraction

nuclear bomb. The explosions of low-mass stars can be triggered by the **accretion** of mass from a companion star in a binary system to create classical, or Type Ia, supernovae. These supernovae show no hydrogen in their **spectra**. Massive stars, on the other hand, proceed through normal **nuclear fusion** but then, when their energy supply runs out, there is no outward pressure to hold them up and they rapidly collapse. The core is crushed into a neutron star or black hole, and the outer layers bounce and are then hurled outward into the surroundings at many million kilometers per hour. These are Type Ib and II supernovae. The Type II supernovae still eject some hydrogen from the unprocessed atmosphere of the star. During a supernova explosion, temperatures are so high that all the known elements can be produced by nuclear fusion.

The most recent supernova that was close enough to be seen without a telescope occurred in early 1987 within a nearby **galaxy**, the Large Magellanic Cloud. Known as 1987A, it is the only supernova for which there is accurate data on the **progenitor star** before it exploded. It has been a tremendous help in understanding how stars explode and expand.

The rapidly growing surface of the star can brighten by up to 100 billion times. Then, as the material gets diluted, it becomes transparent and the brightness fades on time scales of a few years. The **ejecta** are still moving rapidly, however, and quickly sweep up surrounding matter to form a shell that slows down as mass gets accumulated, an action similar to that of a snowplow. This is the beginning of the supernova remnant that can be visible for tens of thousands of years. 1987A is starting to show such interaction with its surroundings.

Supernova remnants emit various forms of radiation. The material is moving highly supersonically and creates a shock wave ahead of it. The shock heats the material in the shell to temperatures over 1 million degrees, producing bright **X rays**. In the presence of **interstellar** magnetism, shocks also accelerate some **electrons** to almost the speed of light, to produce strong **synchrotron radiation** at radio **wavelengths**. Sometimes, even high-energy **gamma rays** can be produced. Dense areas can also cool quickly and we observe filaments of cool gas, at about 10,000 degrees, in various **spectral lines** at optical wavelengths.

In 1054 astronomers in China and New Mexico observed a famous example of the explosion of a massive star. What remains is a large volume of material that, with a lot of imagination, looks like a crab and, hence, is named the Crab Nebula. The object is being stimulated by jets from a rapidly spinning (about thirty times a second) neutron star called a pulsar. In most supernova remnants, this pulsar wind nebula is surrounded by the shell discussed above, but remarkably, no one has yet detected the shell around the Crab Nebula. Oppositely, the young supernova remnant Cassiopeia A has a shell and a neutron star but no pulsar wind nebula. Astronomers hope to explain these and many other mysteries about supernovae and their remnants using more multiwavelength observations with new telescopes. SEE ALSO BLACK HOLES (VOLUME 2); COSMIC RAYS (VOLUME 2); GALAXIES (VOLUME 2); PULSARS (VOLUME 2); STARS (VOLUME 2).

John R. Dickel

Bibliography

Robinson, Leif. "Supernovae, Neutrinos, and Amateur Astronomers." *Sky and Telescope* 98, no. 2 (1999):31–37.

Wheeler, J. Craig. *Cosmic Catastrophes.* Cambridge, UK: Cambridge University Press, 2000.

Zimmerman, Robert. "Into the Maelstrom." *Astronomy* 26, no. 11 (1998):44–49.

Tombaugh, Clyde

American Astronomer
1906–1997

Clyde W. Tombaugh, famous for his discovery of Pluto, the solar system's ninth major planet, was born in 1906 in Streator, Illinois. His family moved to Burdett, Kansas, when he was young. In 1928 he sent planetary drawings done using his homemade nine-inch telescope to Lowell Observatory. Its director, Vesto Slipher, was so impressed with the young astronomer's ability to sketch what he saw in a telescope that he offered him a job to conduct the search for a suspected ninth planet. On February 18, 1930, Tombaugh discovered Pluto on two photographs he had taken of the

progenitor star the star that existed before a dramatic change, such as a supernova, occurred

ejecta the pieces of material thrown off by a star when it explodes

X rays a form of high-energy radiation just beyond the ultraviolet portion of the electromagnetic spectrum

interstellar between the stars

electrons negatively charged subatomic particles

synchrotron radiation the radiation from electrons moving at almost the speed of light inside giant magnetic accelerators of particles, called synchrotrons, either on Earth or in space

wavelengths the distance from crest to crest on a wave at an instant in time

gamma rays a form of radiation with a shorter wavelength and more energy than X rays

spectral lines the unique pattern of radiation at discrete wavelengths that every material produces

Astronomer Clyde Tombaugh with his home-made, backyard telescope in Las Cruces, New Mexico. Tombaugh's most famous discovery, the planet Pluto, was photographed during a scan of the Delta Geminorum star region.

variable star a star whose light output varies over time

region centered on the star Delta Geminorum. During his fifteen-year sky search, he also discovered the cataclysmic **variable star** TV Corvi, six star clusters, and a supercluster of galaxies.

After World War II, at the fledgling New Mexico missile site called White Sands, Tombaugh developed the optical telescopes used to track the first rockets of the U.S. space program. In the 1950s he conducted the first and only search for small natural Earth satellites, a contribution to science that will, thanks to modern artificial satellites, be forevermore impossible to replicate.

Tombaugh died just short of his ninety-first birthday at his home in Las Cruces, New Mexico, where he had lived the second half of his productive and interesting life as a professor, writer, and observer. SEE ALSO ASTRONOMER (VOLUME 2); PLUTO (VOLUME 2).

David H. Levy

Bibliography

Levy, David H. *Clyde Tombaugh: Discoverer of Planet Pluto.* Tucson, AZ: University of Arizona Press, 1991.

Stern, S. Alan, and Jacqueline Mitton. *Pluto and Charon: Icy Worlds at the Ragged Edge of the Solar System.* New York: John Wiley & Sons, 1999.

Trajectories

Trajectories are the paths followed by spacecraft as they travel from one point to another. They are governed by two key factors: the spacecraft's own propulsion system and the gravity of the Sun, Earth, and other planets and moons. Because even the most powerful rockets have only a limited amount of thrust, engineers must carefully develop trajectories for spacecraft that will allow them to reach their intended destination. In some cases this can lead to complicated trajectories that get a boost from the gravity of other worlds.

The trajectory needed for a spacecraft to go into orbit around Earth is relatively straightforward. The spacecraft needs to gain enough altitude—typically at least 200 kilometers (124 miles)—to clear Earth's atmosphere and enough speed to keep from falling back to Earth. This minimum **orbital velocity** around Earth is about 28,000 kilometers per hour (17,360 miles per hour) for **low Earth orbits** and slower than that for higher orbits as Earth's gravitational pull weakens. Other parameters of the orbit, such as the inclination of the orbit to Earth's equator, can be altered by changing the direction of the spacecraft's launch.

Launching a spacecraft beyond Earth, such as on a mission to Mars or another planet, is more complicated. Because of the great distances between planets and the limited power of modern rockets, one cannot simply aim a spacecraft directly at its destination and launch it. Instead, trajectories must

orbital velocity velocity at which an object needs to travel so that its flight path matches the curve of the planet it is circling; approximately 8 kilometers (5 miles) per second for low-altitude orbit around Earth

low Earth orbits orbits between 300 and 800 kilometers above Earth's surface

Mission Control flight controllers at work confirming flight trajectories prior to the launch of the STS-101 shuttle mission on May 19, 2000.

be carefully calculated to allow a spacecraft to travel to its destination given the limited amount of rocket power available. A common way to do this is to use a Hohmann transfer orbit, a type of orbit that minimizes the amount of propellant needed to send a spacecraft to its destination. A Hohmann transfer orbit is an elliptical orbit with its **perihelion** at Earth and **aphelion** at the destination planet (or the reverse if traveling towards the Sun). If launched at the proper time a spacecraft will spend only half an orbit in a Hohmann orbit, catching up with the destination world at the opposite point of its orbit from Earth. To do this, the spacecraft much be launched during a relatively short launch window. For a mission to Mars, such launch windows are available every twenty-six months, for only a couple months at a time.

Even a Hohmann orbit, however, may require more energy than a rocket can provide. Another technique, known as gravity assist, can allow spacecraft to reach more distant destinations by taking advantage of the gravity of other worlds. A spacecraft is launched on a Hohmann trajectory toward an intermediate destination, usually another planet. The spacecraft flies by this planet, gaining velocity by taking, though gravitational interaction, an infinitesimally small amount of the planet's **angular momentum**. This added velocity allows the spacecraft to continue on to its destination. Gravity assists allowed the Voyager 2 spacecraft, launched from Earth with only enough velocity to reach Jupiter, to travel on to Saturn, Uranus, and Neptune. Gravity assist **flybys** of Venus, Earth, and Jupiter will also allow the Cassini spacecraft to reach Saturn in 2004. SEE ALSO ORBITS (VOLUME 2).

Jeff Foust

Bibliography

Wertz, James R., and Wiley J. Larson, eds. *Space Mission Analysis and Design.* Dordrecht, Netherlands: Kluwer, 1991.

Internet Resources

Basics of Space Flight. NASA Jet Propulsion Laboratory, California Institute of Technology. <http://www.jpl.nasa.gov/basics/>.

perihelion the point in an object's orbit that is closest to the Sun

aphelion the point in an object's orbit that is farthest from the Sun

angular momentum the angular equivalent of linear momentum; the product of angular velocity and moment of inertia (moment of inertia = mass × radius²)

flyby flight path that takes the spacecraft close enough to a planet to obtain good observations; the spacecraft then continues on a path away from the planet but may make multiple passes

Uranus

Uranus was the first planet to be discovered that had not been known since antiquity. Although Uranus is just bright enough to be seen with the naked eye, and in fact had appeared in some early star charts as an unidentified star, English astronomer William Herschel was the first to recognize it as a planet in 1781.

The planet's benign appearance gives no hint of a history fraught with catastrophe: Sometime in Uranus's past, a huge collision wrenched the young planet. As a result, the rotation pole of Uranus is now tilted more than 90 degrees from the plane of the planet's orbit. Uranus travels in a nearly circular orbit at an average distance of almost 3 billion kilometers (1.9 billion miles) from the Sun (about nineteen times the distance from Earth to the Sun).

A Somewhat Small Gas Planet

The composition of Uranus is similar to that of the other giant planets✳ and the Sun, consisting predominantly of hydrogen (about 80 percent) and helium (15 percent). The remainder of Uranus's atmosphere is methane (less than 3 percent), hydrocarbons (mixtures of carbon, nitrogen, hydrogen, and oxygen), and other trace elements. Uranus's color is caused by the methane, which preferentially absorbs red light, rendering the remaining reflected light a greenish-blue color.

✳ **There are four giant planets in the solar system: Jupiter, Saturn, Uranus, and Neptune**

Like Jupiter and Saturn, Uranus is a gas planet, although a somewhat small one (at its equator, its radius is about 25,559 kilometers [15,847 miles]). We see the outermost layers of clouds, which are probably composed of icy crystals of methane. Below this layer of clouds, the atmosphere gets thicker and warmer. Deep within the center of Uranus, at extremely high pressure, a core of rocky material is hypothesized to exist, with a mass almost five times that of Earth.

One of the more puzzling aspects of Uranus is the lack of excess heat radiating from its interior. In comparison, the other three giant planets radiate significant excess heat. Astronomers believe that this excess heat is left over from the time of the planets' formation and from continuing **gravitational contraction**. Why then does Uranus have none? Scientists theorize that perhaps the heat is there but is trapped by layers in the atmosphere, or perhaps the event that knocked Uranus over on its side somehow caused much of the heat to be released early in the planet's history.

gravitational contraction the collapse of a cloud of gas and dust due to the mutual gravitational attraction of the parts of the cloud; a possible source of excess heat radiated by some Jovian planets

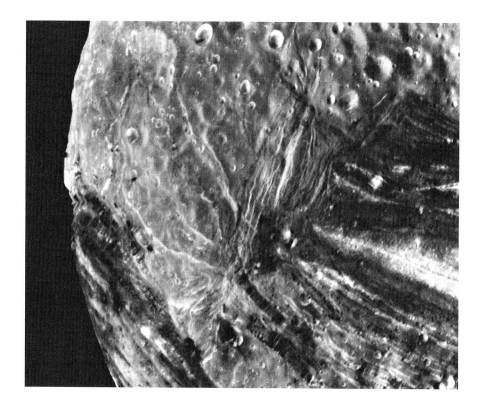

This Voyager 2 image of Miranda reveals a significant variety of fractures and troughs, with varying densities of impact craters on them. These differences suggest that the moon had a long, complicated geologic evolution.

Magnetic Field

When the Voyager 2 spacecraft flew by Uranus in 1986, it detected a magnetic field about fifty times stronger than that of Earth. In a surprising twist, the magnetic field's source was not only offset from the center of the planet to the outer edge of the rocky core, but it was also tilted nearly 60 degrees from the planet's rotation axis. From variations in the magnetic field strength detected by Voyager 2, scientists determined that the planet's internal rotation period was 17.2 hours. The winds in the visible cloud layers have rotation periods ranging from about 16 to 18 hours depending on latitude, implying that wind speeds reach 300 meters per second (670 miles per hour) for some regions.

The Moons of Uranus

Within six years of the discovery of Uranus, two moons were discovered. They were subsequently named Titania and Oberon. It was more than sixty years before the next two Uranian moons, Ariel and Umbriel, were discovered. Nearly a century elapsed before Miranda was discovered in 1948, bringing the total of Uranus's large moons to five. Little was known about their surface structure or history until the Voyager 2 spacecraft returned detailed images of the surfaces of these moons.

On Miranda, huge geologic features dominate the small moon's landscape, indicating that some kind of intense heating must have occurred in the past. It is not yet clear whether a massive collision disturbed the small moon, which then reassembled into the current jumble, or whether, in the past, tidal interactions with other moons produced heat to melt and modify the surface, as is the case for Jupiter's moon Io. Oberon, the outermost major moon, shows many large craters, some with **bright rays**. Titania has fewer large craters, indicating that its surface has been "wiped clean" by resurfacing sometime in the moon's past. Ariel has the youngest surface of the major moons, based on cratering rates. Umbriel is much darker and smoother. Its heavily cratered surface is probably the oldest of the satellites.

In 1986, Voyager 2 discovered ten additional moons, with Puck being the largest. Voyager 2 images of Puck showed it to be an irregularly shaped body with a mottled surface. Voyager 2 did not venture close enough to the other small moons to learn much about them. Since 1986, six more tiny moons have been discovered around Uranus, bringing its total to twenty-one. Little is known about these moons other than their sizes and orbits.

Rings and Seasons

In 1977, astronomers discovered that Uranus has a ring system. Voyager 2 studied the rings in detail when it flew by Uranus in 1986. There are nine well-defined rings, plus a fainter ring and a wider fuzzy ring. Unlike the broad system of Saturnian rings, the main Uranian rings are narrow. The rings are not perfectly circular and also vary in width. Like the rings of Saturn, the Uranian rings are thought to be composed mainly of rocky material (ranging in size from dust particles to house-sized boulders) mixed with small amounts of ice.

The atmosphere of Uranus has often been called bland, and even boring. These epithets are a consequence of fate and unfortunate timing. It was fate that caused the early collision of Uranus with a large body, creating the planet's extreme axial tilt, which in turn created extreme seasons. It was unfortunate timing that the Voyager 2 encounter (which gave us our highest resolution pictures) occurred at peak southern summer, when we had a view of only the southern half of the planet. Historically, this season is when Uranus has appeared blandest in the past.

As Uranus continues its eighty-four-year-long progression around the Sun, its equatorial region is now receiving sunlight again, and parts of its northern hemisphere are being bathed in **solar radiation** for the first time in decades. Today, images from the Hubble Space Telescope are revealing multiple bright cloud features and stunning banded structures on Uranus. It is fascinating to speculate how Uranus will appear to us by the time it reaches equinox in 2007. SEE ALSO EXPLORATION PROGRAMS (VOLUME 2); HERSCHEL FAMILY (VOLUME 2); NASA (VOLUME 3); NEPTUNE (VOLUME 2); ROBOTIC EXPLORATION OF SPACE (VOLUME 2).

Heidi B. Hammel

bright rays lines of lighter material visible on the surface of a body, which are caused by relatively recent impacts

solar radiation total energy of any wavelength and all charged particles emitted by the Sun

Bibliography

Bergstralh, Jay T., Ellis D. Miner, and Mildred Shapley Matthews, eds. *Uranus.* Tucson, AZ: University of Arizona Press, 1991.

Miner, Ellis D. *Uranus: The Planet, Rings, and Satellites,* 2nd ed. Chichester, UK: Praxis Publishing, 1998.

Standage, Tom. *The Neptune File: A Story of Astronomical Rivalry and the Pioneers of Planet Hunting.* New York: Walker and Company, 2000.

Venus

Venus was one of the last planets to be explored, despite its position as the closest planet to Earth. This is largely because it is perpetually shrouded in a uniformly bland covering of clouds. The cloud cover made looking at Venus through a telescope about as exciting as staring at a billiard ball. While Mars and the Moon were objects of much attention by early telescopic observation, the surface of Venus remained a mystery. It was even easier to say something about the outer planets, such as Jupiter and Saturn, than it was to make meaningful observations of Venus.

The absence of information about Venus was particularly ironic because Venus is the most like Earth in size and position within the solar system, thus suggesting that it could be more like Earth than any of the other planets. Venus's diameter is only 651 kilometers (404 miles) smaller than Earth's diameter of 12,755 kilometers (7,908 miles). Venus's density is 0.9 times that on Earth, and its surface gravity is 0.8 that of Earth. Venus orbits the Sun in just under one Earth year (224.7 days). When compared to Earth, all of the planets except Venus are much larger or smaller, higher or lower in density, located at much greater or lesser distances from the Sun, or enveloped in atmospheres much thinner or colder. Thus Venus was a cornerstone in scientists' survey of the solar system and offered the chance to see how an Earth-sized planet might have evolved similarly or differently. Planetary geologists now know that it is very different. This fact has revealed that the details of how a planet geologically evolves are probably as important in planetary evolution as differences in fundamental characteristics. Venus and Earth are truly twins separated at birth.

Atmospheric Characteristics

Because of the cloud cover, one of the first things that could be determined about Venus in the early days of planetary astronomy was the characteristics of the visible atmosphere. This was done first through telescopic measurements and early spacecraft **flybys**. In the nineteenth century, rare transits of Venus across the surface of the Sun were used to prove that Venus was enveloped in an atmosphere. This led to all sorts of early speculation that the clouds were, like clouds on Earth, water vapor clouds, and that the surface was a teeming **primordial swamp** filled with plants and animals similar to the **Paleozoic** coal swamps of Earth. This speculation withered under results of early spectroscopic observations, which were able to determine that the atmosphere was largely carbon dioxide, not oxygen and nitrogen as on Earth, and later on, that the clouds appeared to be sulfuric acid fog, not water vapor.

By the 1960s little was still known about Venus, but modern instruments were beginning to reveal more. Early surface temperature estimates were made by observing **infrared** wavelengths to better determine the temperature. Such observations, using radio telescopes and the first U.S. interplanetary flyby spacecraft, Mariner 2, in 1962, implied that the surface temperatures were high. Over the next decade, several U.S. atmospheric entry probes (Pioneer-Venus 1 and 2) and Soviet landers (Venera 9, 10, 13, and 14) directly measured the temperature and pressure within the atmosphere. These measurements revealed a surface temperature of 450°C

flyby flight path that takes the spacecraft close enough to a planet to obtain good observations; the spacecraft then continues on a path away from the planet but may make multiple passes

primordial swamp warm, wet conditions postulated to have occurred early in Earth's history as life was beginning to develop

Paleozoic relating to the first appearance of animal life on Earth

infrared portion of the electromagnetic spectrum with wavelengths slightly longer than visible light

A global view of Venus, created with data obtained during the Magellan mission. At image center is a bright feature, the mountainous region of Ovda Regio, located in the western portion of the great Aphrodite equatorial highland.

(842°F), or about as hot as the surface of a catalytic converter on an automobile. The surface pressure on Venus was found to be ninety-two times that of Earth (92 bars or 9.2 million pascals). This is equivalent to pressures at about 1 kilometer (0.6 mile) of depth in the sea, or about fifty times greater than a pressure cooker.

The atmosphere is so dense that pressures and temperatures similar to the surface of Earth occur at about 60 to 70 kilometers (37 to 43 miles) of altitude. Most of the atmosphere on Earth lies below 10 kilometers (6.2 miles), and it pretty much peters out before 30 kilometers (18.6 miles). On Venus, a daring future adventurer in a balloon with an oxygen mask (and protection from sulfuric acid clouds) could float in the upper atmosphere at an altitude of 60 to 70 kilometers (37 to 43 miles) in relative comfort. The global trip would be rapid since the atmosphere super-rotates, meaning that it flows from west to east faster than the underlying surface rotates. A balloon traveler in the upper atmosphere of Venus could circumnavigate the planet in only four days, especially near the equator where the speeds are greatest. This is unlike Earth, where the surface spins beneath a relatively sluggishly moving atmosphere that takes several weeks for a complete circuit. The balloon traveler's view would be boring, however, because the surface of Venus would be obscured by the main cloud layer, which occurs at 45 to 70 kilometers (28 to 43 miles) of altitude.

These conditions are the result of early development of a thick atmosphere consisting mostly of carbon dioxide (about 97 percent) through a

THE GREENHOUSE EFFECT ON VENUS

The greenhouse effect refers to a condition on Venus in which solar heating of the upper atmosphere at the short wavelengths of visible light radiation results in warming of the atmosphere, but the longer wavelengths of thermal radiation in the lower atmosphere cannot penetrate the atmosphere and reradiate to space. As a result, the temperature of the atmosphere continually increased.

radar a technique for detecting distant objects by emitting a pulse of radio-wavelength radiation and then recording echoes of the pulse off the distant objects

radar altimetry using radar signals bounced off the surface of a planet to map its variations in elevation

fault a fracture in rock in the upper crust of a planet along which there has been movement

rift valley a linear depression in the surface, several hundred to thousand kilometers long, along which part of the surface has been stretched, faulted, and dropped down along many normal faults

impact craters bowl-shaped depressions on the surfaces of planets or satellites that result from the impact of space debris moving at high speeds

orthogonal composed of right angles or relating to right angles

so-called runaway greenhouse effect. First discovered from the study of Venus, the greenhouse effect is now discussed for Earth, where it is recognized that industrial additions of carbon dioxide to the atmosphere pose potential environmental problems of similar global magnitude.

Surface Features and Geologic Findings

Although the surface of Venus has been seen locally around a few Russian landers with optical cameras, a true picture of the global surface was obtained only with the advent of **radar** that could penetrate the dense obscuring clouds and create radar images. Early results were obtained through Earth-based radio telescopes at Goldstone (in California) and Arecibo (in Puerto Rico), which emitted tight beams of radar and built up images showing differences in the radar reflecting properties of the surfaces. These images were low in resolution, but they enabled large areas of unusually radar-reflective terrain to be detected. These also allowed the first estimates of the rotation to be made, and they showed that Venus rotates backwards, or west to east, and slowly. It takes about 243 days to do so. Oddly, this is a little longer than its year (224.7 days). Stranger still, at closest approach to Earth (a distance of just under 40 million kilometers [24.8 million miles]), or opposition, Venus presents the same hemisphere to Earth. The origin of this unusual set of rotation conditions is not known.

The first truly global maps of Venus was made by the Pioneer-Venus orbiter using **radar altimetry**. This showed the surface elevations over the globe resolved at scales of about 100 kilometers (62 miles) and revealed a relatively flat surface, with the absolute range of elevations much less than that found on Earth. More than 80 percent of the surface area lies within a kilometer of the mean planetary radius (6,051.8 kilometers [3,752.1 miles]). A few highland regions rise from one to several kilometers above the mean planetary radius, but these cover only about 15 percent of the surface of Venus. Whereas Earth has two common elevations, seafloors and continents, Venus has one most common elevation, broad plains.

Radar images of the surface, somewhat similar to photographs, were made later for large areas of the globe by the Russian Venera 15 and 16 orbiters, for the northern quarter of Venus, and, several years later, by the U.S. Magellan orbiter, for about 98 percent of the surface. These efforts obtained images of the surface at scales of 2 kilometers (1.2 miles) and 0.3 kilometers (0.2 miles), respectively, thus generating the first true images of what the surface of Venus really looks like and permitting the first geologic analysis.

The radar images show that the surface of Venus is a complex of plains, mountains, **faults**, ridges, **rift valleys**, volcanoes, and a few **impact craters**, a surface more complex and geologically modified than any of the other planets seen previously. The highland regions seen first in low-resolution Earth-based radar images and Pioneer-Venus radar altimetry are among the most complex surfaces and consist of a terrain that is complexly faulted in **orthogonal** patterns. These regions are known as tessera after the Greek word for mosaics of tiles. The sequence of geologic surfaces suggests that tesserae (plural of "tessera") also represent the oldest preserved surfaces on Venus. One of the highlands is surrounded by ridgelike mountain belts that rise from 6 to 11 kilometers (3.7 to 6.8 miles) above the mean planetary radius and appear to have formed from compression and buckling of the surface, similar to mountain belts on Earth. Low ridges of possibly similar origin, in a range of sizes, occur singly or in belts throughout the plains areas.

Faults, **fractures**, and immense rift valleys are present in abundance. One rift valley, Diana Chasma, is similar in size to the great East African rift valley and Rio Grande rift valley of Earth. Like those on Earth, it probably formed from the stretching and pulling apart of the surface. On Earth, erosion and **sedimentation** quickly obscure all but the latest structures associated with such rifts. But on Venus the absence of erosion means that all of the structural details are perfectly preserved as a complex mass of faults.

Large volcanoes up to several hundreds of kilometers across but only a few kilometers high are common, as are long lava flow fields, extensive low-lying regions of lava plains, and lava channels. One lava channel is longer than the largest rivers on Earth. Low-relief domical volcanoes, many less than several kilometers in diameter, are globally abundant, numbering in the hundreds of thousands. Some volcanoes appear similar to those formed from eruption of thick, viscous lavas on Earth. Additional volcanic features include **calderas** similar to those on Earth, although generally much larger; complex topographic **annular** spider-and-web-shaped features known as arachnoids; and circular structural patterns up to several hundred kilometers across with associated volcanism known as coronae. Many of these are generally thought to represent local formation of large and deep magma reservoirs. Radial patterns of fractures associated with volcanoes are common and may represent the surface deformation associated with radial-dike-like magma intrusions.

Impact craters are about as numerous on Venus as they are on continental areas of Earth, and are thus not as common as they are on most other planets. Only about 900 have been identified on Venus. Meteors smaller than a certain size disintegrate on entering the atmosphere. As a result, impact craters smaller than 2 kilometers (1.2 miles) are infrequent. Morphologically, craters on Venus resemble those on other planets with several exceptions related to the interaction of the crater **ejecta** with the dense atmosphere. These include extensive parabola-shaped halos much like fallout from plumes associated with volcanic eruptions on Earth. These open to the west and possibly record the interaction of the upward expanding cloud of crater ejecta with the strong global easterly winds. Many craters are characterized by large lava-flow-like features that may represent molten ejecta flowing outward from the crater after the impact.

Impact craters also appear nearly uniformly distributed, unlike most planets where large areas of different crater abundance indicate variations in age of large areas of their surfaces. Based on estimates of their rates of formation on surfaces in the inner solar system, impact crater statistics indicate an average surface age on Venus of about 500 million years. Either most of the surface was formed over 500 million years ago in a catastrophic resurfacing event and volcanism has been much reduced since that time, or continual, widespread, and evenly spaced volcanism and **tectonism** remove craters with a rate that yields an average lifetime of the surface of 500 million years. The rate of volcanism on Venus is estimated to be less than 1 cubic kilometer per year, somewhat less than the 20 cubic kilometers associated largely with seafloor spreading on Earth. The surface of Venus

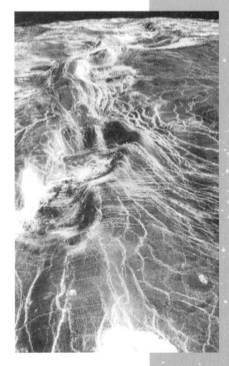

Venus has a complex surface, with plains, mountains, volcanoes, ridges, rift valleys, and a few impact craters.

fracture any break in rock, ranging from small "joints" that divide rocks into planar blocks (such as that seen in road cuts) to vast breaks in the crusts of unspecified movement

sedimentation process of depositing sediments, which result in a thick accumulation of rock debris eroded from high areas and deposited in low areas

calderas the bowl-shaped crater at the top of a volcano caused by the collapse of the central part of the volcano

annular ring-like

ejecta the material thrown out of an impact crater during its formation

tectonism process of deformation in a planetary surface as a result of geological forces acting on the crust; includes faulting, folding, uplift, and down-warping of the surface and crust

subduction the process by which one edge of a crustal plate is forced to move under another plate

lithosphere the rocky outer crust of a planet

convective processes processes that are driven by the movement of heated fluids resulting from a variation in density

asthenosphere the weaker portion of a planet's interior just below the rocky crust, over which tectonic plates slide

basalts dark, volcanic rock with abundant iron and magnesium and relatively low silica common on all of the terrestrial planets

appears to be dominated by volcanic hot spots rather than spreading and **subduction** associated with plate tectonics.

Another spacecraft observation method allowed something to be determined about the interior of Venus. By carefully tracking spacecraft orbits, variations in gravitational acceleration associated with differences in mass on and beneath the surface can be detected. On Venus, this technique reveals that the strength of gravity is mostly proportional to the surface topography, in contrast to Earth, where mass associated with topography is generally compensated underneath by lower density roots. This means that many large topographic features on Venus are supported either by strong **lithosphere** without a low-density root, or by topography originating from the dynamical uplift of the surface through **convective processes** in the deep interior. If the first type is assumed, it may indicate that the lithosphere is strong and that a low-strength layer at the base of the lithosphere (called the **asthenosphere** on Earth) is not present. The second type may be attributed to upwelling associated with volcanic hot spots.

Several Venera landers of the Russian space program returned both optical images of the surface and chemical information about the rocks at several sites. Early landers had searchlights in case the cloud cover made it too dark to see anything. Despite the dense cloud cover, enough light gets through that the surface is illuminated to the equivalent of a cloudy day on Earth. But the sky as seen from the surface is probably a bland fluorescent yellow-white, rather than mottled gray. The relatively rocky surroundings appeared to be volcanic lava flow surfaces or associated rubble. The measured chemical compositions are indistinguishable for the most part from tholeiitic and alkali **basalts** typical of ocean basins and hot spots on Earth.

The low number of impact craters scattered over the surface implies that only the last 20 percent of the history of Venus appears to be preserved, and little is known about the earlier surface geologic history. The geological complexity and young surface ages of both Venus and Earth relative to smaller terrestrial planets can be attributed to their larger sizes and correspondingly warmer and more mobile interiors, extensive surface deformation (tectonism), and mantle melts (volcanism) over a greater period of geological time. SEE ALSO EXPLORATION PROGRAMS (VOLUME 2); GOVERNMENT SPACE PROGRAMS (VOLUME 2); NASA (VOLUME 3); ROBOTIC EXPLORATION OF SPACE (VOLUME 2); PLANETARY EXPLORATION, FUTURE OF (VOLUME 2).

Larry S. Crumpler

Bibliography

Cattermole, Peter. *Venus: The Geological Story*. Baltimore, MD: Johns Hopkins University Press, 1994.

Cooper, Henry S. F., Jr. *The Evening Star*. New York: HarperCollins, 1993.

"*Magellan* at Venus" (special issues on results of *Magellan* mission to Venus). *Journal of Geophysical Research* 97, no. E8 (1992):13,063–13,675; 97, no. E10 (1992): 15,921–16,382.

Weather, Space

Space weather describes the conditions in space that affect Earth and its technological systems. Space weather is a consequence of the behavior of

GEOMAGNETIC STORMS

Scale	Category Descriptor	Effect	Physical measure	Average Frequency (1 cycle = 11 years)
			Kp values* determined every 3 hours	Number of storm events when Kp level was met; (number of storm days)
		Duration of event will influence severity of effects		
G 5	Extreme	*Power systems*: widespread voltage control problems and protective system problems can occur, some grid systems may experience complete collapse or blackouts. Transformers may experience damage. *Spacecraft operations*: may experience extensive surface charging, problems with orientation, uplink/downlink and tracking satellites. *Other systems*: pipeline currents can reach hundreds of amps, HF (high frequency) radio propagation may be impossible in many areas for one to two days, satellite navigation may be degraded for days, low-frequency radio navigation can be out for hours, and aurora has been seen as low as Florida and southern Texas (typically 40° geomagnetic lat.)**	Kp=9	4 per cycle (4 days per cycle)
G 4	Severe	*Power systems*: possible widespread voltage control problems and some protective systems will mistakenly trip out key assets from the grid. *Spacecraft operations*: may experience surface charging and tracking problems, corrections may be needed for orientation problems. *Other systems*: induced pipeline currents affect preventive measures, HF radio propagation sporadic, satellite navigation degraded for hours, low-frequency radio navigation disrupted, and aurora has been seen as low as Alabama and northern California (typically 45° geomagnetic lat.)**	Kp=8, including a 9-	100 per cycle (60 days per cycle)
G 3	Strong	*Power systems*: voltage corrections may be required, false alarms triggered on some protection devices. *Spacecraft operations*: surface charging may occur on satellite components, drag may increase on low-Earth-orbit satellites, and corrections may be needed for orientation problems. *Other systems*: intermittent satellite navigation and low-frequency radio navigation problems may occur, HF radio may be intermittent, and aurora has been seen as low as Illinois and Oregon (typically 50° geomagnetic lat.)**	Kp=7	200 per cycle (130 days per cycle)
G 2	Moderate	*Power systems*: high-latitude power systems may experience voltage alarms, long-duration storms may cause transformer damage. *Spacecraft operations*: corrective actions to orientation may be required by ground control; possible changes in drag affect orbit predictions. *Other systems*: HF radio propagation can fade at higher latitudes, and aurora has been seen as low as New York and Idaho (typically 55° geomagnetic lat.)**	Kp=6	600 per cycle (360 days per cycle)
G 1	Minor	*Power systems*: weak power grid fluctuations can occur. *Spacecraft operations*: minor impact on satellite operations possible. *Other systems*: migratory animals are affected at this and higher levels; aurora is commonly visible at high latitudes (northern Michigan and Maine)**	Kp=5	1700 per cycle (900 days per cycle)

* Based on this measure, but other physical measures are also considered.

** For specific locations around the globe, use geomagnetic latitude to determine likely sightings (see www.sec.noaa.gov/Aurora)

SOLAR RADIATION STORMS

Scale	Category Descriptor	Effect	Physical measure	Average Frequency (1 cycle = 11 years)
			Flux level of ≥ 10 MeV (ions)*	Number of events when flux level was met**
		Duration of event will influence severity of effects		
S 5	Extreme	*Biological*: unavoidable high radiation hazard to astronauts on EVA (extra-vehicular activity); high radiation exposure to passengers and crew in commercial jets at high latitudes (approximately 100 chest X-rays) is possible. *Satellite operations*: satellites may be rendered useless, memory impacts can cause loss of control, may cause serious noise in image data, star-trackers may be unable to locate sources; permanent damage to solar panels possible. *Other systems*: complete blackout of HF (high frequency) communications possible through the polar regions, and position errors make navigation operations extremely difficult.	10^5	Fewer than 1 per cycle
S 4	Severe	*Biological*: unavoidable radiation hazard to astronauts on EVA; elevated radiation exposure to passengers and crew in commercial jets at high latitudes (approximately 10 chest X-rays) is possible. *Satellite operations*: may experience memory device problems and noise on imaging systems; star-tracker problems may cause orientation problems, and solar panel efficiency can be degraded. *Other systems*: blackout of HF radio communications through the polar regions and increased navigation errors over several days are likely.	10^4	3 per cycle
S 3	Strong	*Biological*: radiation hazard avoidance recommended for astronauts on EVA; passengers and crew in commercial jets at high latitudes may receive low-level radiation exposure (approximately 1 chest X-ray). *Satellite operations*: single-event upsets, noise in imaging systems, and slight reduction of efficiency in solar panel are likely. *Other systems*: degraded HF radio propagation through the polar regions and navigation position errors likely.	10^3	10 per cycle
S 2	Moderate	*Biological*: none. *Satellite operations*: infrequent single-event upsets possible. *Other systems*: small effects on HF propagation through the polar regions and navigation at polar cap locations possibly affected.	10^2	25 per cycle
S 1	Minor	*Biological*: none. *Satellite operations*: none. *Other systems*: minor impacts on HF radio in the polar regions.	10	50 per cycle

* Flux levels are 5 minute averages. Flux in particles·s⁻¹·ster⁻¹·cm⁻² based on this measure, but other physical measures are also considered.

** These events can last more than one day.

RADIO BLACKOUTS

Category		Effect	Physical measure	Average Frequency (1 cycle = 11 years)
Scale	Descriptor		GOES X-ray peak brightness by class and by flux*	Number of events when flux level was met; (number of storm days)
		Duration of event will influence severity of effects		
R 5	Extreme	HF Radio: Complete HF (high frequency**) radio blackout on the entire sunlit side of the Earth lasting for a number of hours. This results in no HF radio contact with mariners and en route aviators in this sector. Navigation: Low-frequency navigation signals used by maritime and general aviation systems experience outages on the sunlit side of the Earth for many hours, causing loss in positioning. Increased satellite navigation errors in positioning for several hours on the sunlit side of Earth, which may spread into the night side.	X20 (2×10^{-3})	Fewer than 1 per cycle
R 4	Severe	HF Radio: HF radio communication blackout on most of the sunlit side of Earth for one to two hours. HF radio contact lost during this time. Navigation: Outages of low-frequency navigation signals cause increased error in positioning for one to two hours. Minor disruptions of satellite navigation possible on the sunlit side of Earth.	X10 (10^{-3})	8 per cycle (8 days per cycle)
R 3	Strong	HF Radio: Wide area blackout of HF radio communication, loss of radio contact for about an hour on sunlit side of Earth. Navigation: Low-frequency navigation signals degraded for about an hour.	X1 (10^{-4})	175 per cycle (140 days per cycle)
R 2	Moderate	HF Radio: Limited blackout of HF radio communication on sunlit side, loss of radio contact for tens of minutes. Navigation: Degradation of low-frequency navigation signals for tens of minutes.	M5 (5×10^{-5})	350 per cycle (300 days per cycle)
R 1	Minor	HF Radio: Weak or minor degradation of HF radio communication on sunlit side, occasional loss of radio contact. Navigation: Low-frequency navigation signals degraded for brief intervals.	M1 (10^{-5})	2000 per cycle (950 days per cycle)

* Flux, measured in the 0.1-0.8 nm range, in $W \cdot m^{-2}$. Based on this measure, but other physical measures are also considered.
** Other frequencies may also be affected by these conditions.

sunspots dark, cooler areas on the solar surface consisting of transient, concentrated magnetic fields

coronal holes large, dark holes seen when the Sun is viewed in X-ray or ultraviolet wavelengths; solar wind emanates from the coronal holes

prominences inactive "clouds" of solar material held above the solar surface by magnetic fields

flares intense, sudden releases of energy

coronal mass ejections large quantities of solar plasma and magnetic field launched from the Sun into space

vacuum a space where air and all other molecules and atoms of matter have been removed

the Sun, the nature of Earth's magnetic field and atmosphere, and our location in the solar system.

While most people know that the Sun is overwhelmingly important to life on Earth, few of us know about the effects caused by this star and its variations. Scientists can observe variations such as **sunspots**, **coronal holes**, **prominences**, **flares**, and **coronal mass ejections**. These dramatic changes to the Sun send material and energy hurtling towards Earth.

Space is sometimes considered a perfect **vacuum**, but between the Sun and the planets is a turbulent area dominated by the fast-moving **solar wind**. The solar wind flows around Earth and distorts the **geomagnetic field** lines. During solar storms, the solar wind can gust wildly, causing geomagnetic storms.

Systems affected by space weather include satellites, navigation, radio transmissions, and power grids. Space weather also produces harmful radiation to humans in space. The NOAA Space Weather Scales list the likely effects of various storms. The list of consequences has grown in proportion to humankind's dependence on technological systems, and will continue to do so. SEE ALSO Solar Wind (VOLUME 2); Space Environment, Nature of (VOLUME 2); Sun (VOLUME 2).

Barbara Poppe

Internet Resources

NOAA Space Weather Scales. <http://www.sec.noaa.gov/NOAAscales/>.

What is Space?

Space, most generally, might be described as the boundless container of the universe. Its contents are all physical things that we know of, and more. To describe the contents of space, we use terms of distance, mass, force, motion, energy, and time. The units we use depend on the scale we are considering. Units useful to us on the scale of human life become difficult to use for much smaller domains (such as atoms), and as we describe space beyond our planet Earth.

Consider a distribution of mass at very large distances throughout space and the motion and energy transformation processes going on all the time. While these masses are very distant from each other, they do interact in various ways. These include gravitational force attraction; emitting, absorbing, or reflecting energy; and sometimes, though statistically very seldom, colliding.

Interplanetary Space

Interplanetary space refers to that region of our container that holds the Sun, the nine major planets that revolve about the Sun, and all other mass, distances, force interactions, motions, and transformations of energy within that realm. Distances are often given in terms of the average distance separating the Sun and Earth, in units called **astronomical units** (AU). Pluto, our most distant planet, is 39 AU from the Sun. Each body in the solar system exerts a gravitational pull on every other body, proportional to their masses but reduced by the separation between them. These forces keep the planets in orbit about the Sun, and moons in orbit about planets. Earth's moon, though relatively small in mass, is close enough to cause tidal changes in Earth's oceans with its pull. The massive planet Jupiter, however, affects orbits throughout the solar system.

Between Mars and Jupiter lies a ring of debris called the main asteroid belt, consisting of fragments of material that never became a planet. Gravitational forces (primarily those of the Sun and Jupiter) pull the asteroids into more defined orbits within this doughnut-shaped region. Some of these asteroids, because of collisions or by gravitational **perturbations**, leave the main belt and fall into Earth-crossing orbits. These are the ones that are the subject of disaster films and to which craters on Earth and the extinction of the dinosaurs are attributed. In addition to these asteroids, comets (which are essentially large, dirty snowballs) pass through the solar system, leaving dust trails. As Earth travels about the Sun, it collides with some of these dust trails, giving rise to meteor showers.

As we peer into space from our home planet, whether with our naked eye or with the most powerful telescope, we are looking back through time. Light arriving at our eyes carries information about how the source looked at a time equal to the travel time of the photons of light. For example, our

solar wind a continuous, but varying, stream of charged particles (mostly electrons and protons) generated by the Sun

geomagnetic field Earth's magnetic field; under the influence of solar wind, the magnetic field is compressed in the Sun-ward direction and stretched out in the downwind direction, creating the magnetosphere, a complex, teardrop-shaped cavity around Earth.

astronomical units one AU is the average distance between Earth and the Sun (152 million kilometers [93 million miles])

perturbations term used in orbital mechanics to refer to changes in orbits due to "perturbing" forces, such as gravity

APPROXIMATE DISTANCE CONVERSIONS

Distance	Miles	Kilometers	Astronomical Units (AU)	Light Years*
1 mile	1	1.609	0.0000000108	0.0000000000001701
1 kilometer	0.62150404	1	0.0000000067	0.0000000000001057
1 AU	92,977,004	149,600,000	1	0.0000158129
1 light year*	5,879,833,998,757	9,460,652,904,000	63,239.66	1

*A light year is the distance that light travels through a vacuum in the period of one year.

light year the distance that light in a vacuum would travel in one year, or about 5.9 trillion miles (9.5 trillion kilometers)

nearest star neighbor is a three-star system called Alpha Centauri, which is 4.3 **light years** away. This means that, as we look at this star system, we can know only how it looked 4.3 years ago, and never as it looks right now. We see the light from more distant stars that may have died and vanished many, many years ago. SEE ALSO SPACE ENVIRONMENT, NATURE OF (VOLUME 2).

David Desrocher

Bibliography

National Geographic Atlas of the World, 7th ed. Washington, DC: National Geographic, 1999.

Shirley, James H. *Encyclopedia of Planetary Sciences.* New York: Kluwer Academic Publishers, 2001.

Weissman, Paul. *Encyclopedia of the Solar System.* San Diego, CA: Academic Press, 1998.

Internet Resources

Solar System Dynamics. NASA Jet Propulsion Laboratory, California Institute of Technology. <http://ssd.jpl.nasa.gov/>.

Solar System Exploration Home Page. National Aeronautics and Space Administration. <http://solarsystem.nasa.gov/>.

Solar System Simulator. NASA Jet Propulsion Laboratory, California Institute of Technology. <http://space.jpl.nasa.gov/>.

Virtual Solar System. National Geographic. <http://www.nationalgeographic.com/solarsystem/splash.html>.

Photo and
Illustration Credits

Volume 1

AP/Wide World Photos, Inc.: **2, 5, 25, 26, 41, 56, 77, 82, 84, 113, 115, 117, 143, 148, 151, 153, 156, 161, 180, 208, 211;** Associated Press: **7;** NASA: **13, 21, 28, 30, 32, 34, 38, 43, 54, 67, 71, 79, 80, 90, 94, 96, 101, 102, 107, 121, 123, 158, 169, 178, 182, 193, 195, 200, 203, 207;** MSFC–NASA: **15;** Photograph by Kipp Teague. NASA: **16;** Illustration by Bonestell Space Art. © Bonestell Space Art: **18, 19;** Reuters/Mike Theiler/Archive Photos: **20;** © Roger Ressmeyer/Corbis: **46, 144, 199;** The Kobal Collection: **49, 73;** AP Photo/ Lenin Kumarasiri: **53;** © Bettmann/CORBIS: **58, 64;** © CORBIS: **60, 98;** © Reuters NewMedia Inc./CORBIS: **63;** © AFP/Corbis: **86, 165, 185;** Courtesy NASA/JPL/Caltech: **92;** International Space University: **106;** European Space Agency/Photo Researchers, Inc.: **109;** Photograph by David Parker. ESA/National Audubon Society Collection/ Photo Researchers, Inc.: **119;** © Joseph Sohm, ChromoSohm Inc./Corbis: **126;** Archive Photos, Inc.: **129, 134;** Illustration by Pat Rawlings. NASA: **136, 190;** Photograph © Dr. Dennis Morrison and Dr. Benjamin Mosier. Instrumentation Technology Associates, Inc.: **139;** © Dr. Allen Edmunsen. Instrumentation Technology Associates, Inc.: **140;** ©NASA/Roger Ressmeyer/Corbis: **157;** The Dwight D. Eisenhower Library: **162;** Landsat 7 Science Team/USDA Agricultural Research Service/NASA: **172;** Richard T. Nowitz/Corbis: **186;** Courtesy of Walter A. Lyons: **189;** UPI/Corbis-Bettmann: **206;** The Library of Congress: **213.**

Volume 2

NASA and The Hubble Heritage Team (STScI/AURA): **2, 51, 53, 206;** NASA/ JHUAPL: **4, 5;** AP/Wide World Photos/ Johns Hopkins University/NASA: **6;** © Roger Ressmeyer/Corbis: **8, 21, 24, 122, 147, 170, 171, 174;** NASA: **11, 29, 33, 37, 54, 65, 69, 72, 91, 96, 99, 100, 102, 107, 113, 117, 127, 128, 130, 135, 143, 149, 155, 157, 162, 165, 166, 178, 183, 191, 195, 196, 200, 203, 204, 209, 217;** NASA/STScI: **15;** © Royal Observatory, Edinburgh/Science Photo Library, National Audubon Society Collection/Photo Researchers, Inc.: **17;** Illustration by Don Davis. NASA: **26;** The Library of Congress: **28, 30, 68;** Photograph by Robert E. Wallace. U.S. Geological Survey: **36;** The Bettmann Archive/Corbis-Bettmann: **39;** Photograph by Kipp Teague. NASA: **41;** Courtesy of NASA/JPL/Caltech: **43, 44, 79, 88, 94, 110, 111, 116, 136, 150, 161, 176, 181, 215;** © AFP/Corbis: **49, 58, 60;** © Corbis: **57, 185;** © Bettmann/Corbis: **75, 81, 84, 119;** Courtesy of NASA/JPL/ University of Arizona: **77;** Courtesy of NASA/JPL: **78, 144, 167, 175, 212;** AP/Wide World Photos: **85, 139, 164, 173, 208;** David Crisp and the WFPC2 Science Team (JPL/Caltech): **93;** Courtesy of Robert G. Strom and the Lunar and Planetary Lab: **102;** © Rick Kujawa: **105;** © Richard and Dorothy Norton, Science Graphics: **106;** © Reiters NewMedia Inc./Corbis: **108;**

Stephen and Donna O'Meara/Photo Researchers, Inc.: **120;** Photograph by Seth Shostak. Photo Researchers, Inc.: **133;** © Sandy Felsenthal/Corbis: **140;** MSFC–NASA: **169;** National Oceanic and Atmospheric Administration/Department of Commerce: **184;** Spectrum Astro, Inc., 2000: **189;** Photo Researchers, Inc.: **199;** Courtesy of SOHO/EIT Consortium: **202;** Kenneth Seidelmann, U.S. Naval Observatory, and NASA: **211.**

Volume 3

© Bettmann/Corbis: **2, 83;** NASA: **4, 6, 10, 14, 16, 17, 18, 20, 25, 26, 35, 38, 41, 43, 45, 48, 51, 53, 55, 57, 59, 60, 64, 65, 66, 70, 72, 74, 76, 78, 80, 85, 90, 94, 97, 101, 103, 107, 108, 111, 113, 120, 123, 126, 128, 131, 133, 134, 139, 140, 142, 144, 148, 150, 152, 158, 160, 161, 169, 172, 176, 178, 181, 184, 185, 188, 192, 193, 196, 200, 202, 207, 209, 211, 214, 219, 223, 225, 226, 233, 235;** Courtesy of Brad Joliff: **13;** © Metropolitan Tucson Convention & Visitors Bureau: **22;** AP/Wide World Photos: **28, 32, 155, 164, 203, 217;** MSFC–NASA: **49, 106, 118, 130, 154, 163, 175;** © NASA/Roger Ressmeyer/Corbis: **61;** Archive Photos, Inc.: **65, 137, 171;** Courtesy of NASA/JPL/Caltech: **73;** Illustration by Pat Rawlings. NASA: **88;** Hulton Getty Collection/Archive Photos: **116;** © Roger Ressmeyer/Corbis: **167, 222;** Photo

Researchers, Inc.: **189;** The Library of Congress: **216;** Getty Images: **228.**

Volume 4

NASA: **2, 8, 10, 14, 19, 22, 26, 35, 46, 54, 59, 67, 72, 82, 88, 90, 93, 99, 104, 112, 113, 119, 120, 122, 130, 144, 148, 153, 157, 164, 172, 189, 200, 202;** Denise Watt/NASA: **4;** Courtesy of NASA/JPL/Caltech: **5, 62, 100, 116;** © Corbis: **7, 126;** © Ted Streshinsky/Corbis: **17;** Photo courtesy of NASA/Viking Orbiter : **9;** Royal Observatory, Edinburgh/Science Photo Library/Photo Researchers, Inc.: **23;** © Paul A. Souders/Corbis: **25;** AP/Wide World Photos: **32, 155, 181, 182, 186;** © Charles O'Rear/Corbis: **39, 139;** Photograph by Detlev Van Ravenswaay. Picture Press/Corbis–Bettmann: **41;** Kobal Collection/Lucasfilm/20th Century Fox: **43;** Kobal Collection/Universal: **44;** Courtesy of NASA/JPL: **65, 94;** MSFC–NASA: **69, 80;** © Reuters NewMedia Inc./Corbis: **74;** © Bettmann/Corbis: **78, 151;** © AFP/Corbis: **102;** Agence France Presse/Corbis–Bettmann: **107;** Archive Photos, Inc.: **109, 135;** Dennis Davidson/NASA: **86;** Paramount Pictures/Archive Photos: **110;** John Frassanito and Associates/NASA: **141;** Illustration by Pat Rawlings. NASA: **168, 174;** Painting by Michael Carroll: **191;** Tethers Unlimited, Inc.: **195, 196;** Kobal Collection/Amblin/Universal: **198.**

Glossary

ablation removal of the outer layers of an object by erosion, melting, or vaporization

abort-to-orbit emergency procedure planned for the space shuttle and other spacecraft if the spacecraft reaches a lower than planned orbit

accretion the growth of a star or planet through the accumulation of material from a companion star or the surrounding interstellar matter

adaptive optics the use of computers to adjust the shape of a telescope's optical system to compensate for gravity or temperature variations

aeroballistic describes the combined aerodynamics and ballistics of an object, such as a spacecraft, in flight

aerobraking the technique of using a planet's atmosphere to slow down an incoming spacecraft; its use requires the spacecraft to have a heat shield, because the friction that slows the craft is turned into intense heat

aerodynamic heating heating of the exterior skin of a spacecraft, aircraft, or other object moving at high speed through the atmosphere

Agena a multipurpose rocket designed to perform ascent, precision orbit injection, and missions from low Earth orbit to interplanetary space; also served as a docking target for the Gemini spacecraft

algae simple photosynthetic organisms, often aquatic

alpha proton X-ray analytical instrument that bombards a sample with alpha particles (consisting of two protons and two neutrons); the X rays are generated through the interaction of the alpha particles and the sample

altimeter an instrument designed to measure altitude above sea level

amplitude the height of a wave or other oscillation; the range or extent of a process or phenomenon

angular momentum the angular equivalent of linear momentum; the product of angular velocity and moment of inertia (moment of inertia = mass $\times$ radius^2)

angular velocity the rotational speed of an object, usually measured in radians per second

anisotropy a quantity that is different when measured in different directions or along different axes

annular ring-like

anomalies phenomena that are different from what is expected

anorthosite a light-colored rock composed mainly of the mineral feldspar (an aluminum silicate); commonly occurs in the crusts of Earth and the Moon

anthropocentrism valuing humans above all else

antimatter matter composed of antiparticles, such as positrons and antiprotons

antipodal at the opposite pole; two points on a planet that are diametrically opposite

aperture an opening, door, or hatch

aphelion the point in an object's orbit that is farthest from the Sun

Apollo American program to land men on the Moon; Apollo 11, 12, 14, 15, 16, and 17 delivered twelve men to the lunar surface between 1969 and 1972 and returned them safely back to Earth

asthenosphere the weaker portion of a planet's interior just below the rocky crust

astrometry the measurement of the positions of stars on the sky

astronomical unit the average distance between Earth and the Sun (152 million kilometers [93 million miles])

atmospheric probe a separate piece of a spacecraft that is launched from it and separately enters the atmosphere of a planet on a one-way trip, making measurements until it hits a surface, burns up, or otherwise ends its mission

atmospheric refraction the bending of sunlight or other light caused by the varying optical density of the atmosphere

atomic nucleus the protons and neutrons that make up the core of an atom

atrophy condition that involves withering, shrinking, or wasting away

auroras atmospheric phenomena consisting of glowing bands or sheets of light in the sky caused by high-speed charged particles striking atoms in Earth's upper atmosphere

avionics electronic equipment designed for use on aircraft, spacecraft, and missiles

azimuth horizontal angular distance from true north measured clockwise from true north (e.g., if North = 0 degrees; East = 90 degrees; South = 180 degrees; West = 270 degrees)

ballast heavy substance used to increase the stability of a vehicle

ballistic the path of an object in unpowered flight; the path of a spacecraft after the engines have shut down

basalt a dark, volcanic rock with abundant iron and magnesium and relatively low silica common on all of the terrestrial planets

base load the minimum amount of energy needed for a power grid

beacon signal generator a radio transmitter emitting signals for guidance or for showing location

berth space the human accommodations needed by a space station, cargo ship, or other vessel

Big Bang name given by astronomers to the event marking the beginning of the universe when all matter and energy came into being

biocentric notion that all living organisms have intrinsic value

biogenic resulting from the actions of living organisms; or, necessary for life

bioregenerative referring to a life support system in which biological processes are used; physiochemical and/or nonregenerative processes may also be used

biosignatures the unique traces left in the geological record by living organisms

biosphere the interaction of living organisms on a global scale

bipolar outflow jets of material (gas and dust) flowing away from a central object (e.g., a protostar) in opposite directions

bitumen a thick, almost solid form of hydrocarbons, often mixed with other minerals

black holes objects so massive for their size that their gravitational pull prevents everything, even light, from escaping

bone mineral density the mass of minerals, mostly calcium, in a given volume of bone

breccia mixed rock composed of fragments of different rock types; formed by the shock and heat of meteorite impacts

bright rays lines of lighter material visible on the surface of a body and caused by relatively recent impacts

brown dwarf star-like object less massive than 0.08 times the mass of the Sun, which cannot undergo thermonuclear process to generate its own luminosity

calderas the bowl-shaped crater at the top of a volcano caused by the collapse of the central part of the volcano

Callisto one of the four large moons of Jupiter; named for one of the Greek nymphs

Caloris basin the largest (1,300 kilometers [806 miles] in diameter) well-preserved impact basin on Mercury viewed by Mariner 10

capsule a closed compartment designed to hold and protect humans, instruments, and/or equipment, as in a spacecraft

carbon-fiber composites combinations of carbon fibers with other materials such as resins or ceramics; carbon fiber composites are strong and lightweight

carbonaceous meteorites the rarest kind of meteorites, they contain a high percentage of carbon and carbon-rich compounds

carbonate a class of minerals, such as chalk and limestone, formed by carbon dioxide reacting in water

cartographic relating to the making of maps

Cassini mission a robotic spacecraft mission to the planet Saturn scheduled to arrive in July 2004 when the Huygens probe will be dropped into Titan's atmosphere while the Cassini spacecraft studies the planet

catalyst a chemical compound that accelerates a chemical reaction without itself being used up; any process that acts to accelerate change in a system

catalyze to change by the use of a catalyst

cell culture a means of growing mammalian (including human) cells in the research laboratory under defined experimental conditions

cellular array the three-dimensional placement of cells within a tissue

centrifugal directed away from the center through spinning

centrifuge a device that uses centrifugal force caused by spinning to simulate gravity

Cepheid variables a class of variable stars whose luminosity is related to their period. Their periods can range from a few hours to about 100 days and the longer the period, the brighter the star

Čerenkov light light emitted by a charged particle moving through a medium, such as air or water, at a velocity greater than the phase velocity of light in that medium; usually a faint, eerie, bluish, optical glow

chassis frame on which a vehicle is constructed

chondrite meteorites a type of meteorite that contains spherical clumps of loosely consolidated minerals

cinder field an area dominated by volcanic rock, especially the cinders ejected from explosive volcanoes

circadian rhythm activities and bodily functions that recur every twenty-four hours, such as sleeping and eating

Clarke orbit geostationary orbit; named after science fiction writer Arthur C. Clarke, who first realized the usefulness of this type of orbit for communication and weather satellites

coagulate to cause to come together into a coherent mass

comet matrix material the substances that form the nucleus of a comet; dust grains embedded in frozen methane, ammonia, carbon dioxide, and water

cometary outgassing vaporization of the frozen gases that form a comet nucleus as the comet approaches the Sun and warms

communications infrastructure the physical structures that support a network of telephone, Internet, mobile phones, and other communication systems

convection the movement of heated fluid caused by a variation in density; hot fluid rises while cool fluid sinks

convection currents mechanism by which thermal energy moves because its density differs from that of surrounding material. Convection current is the movement pattern of thermal energy transferring within a medium

convective processes processes that are driven by the movement of heated fluids resulting from a variation in density

coronal holes large, dark holes seen when the Sun is viewed in X-ray or ultraviolet wavelengths; solar wind emanates from the coronal holes

coronal mass ejections large quantities of solar plasma and magnetic field launched from the Sun into space

cosmic microwave background ubiquitous, diffuse, uniform, thermal radiation created during the earliest hot phases of the universe

cosmic radiation high energy particles that enter Earth's atmosphere from outer space causing cascades of mesons and other particles

cosmocentric ethic an ethical position that establishes the universe as the priority in a value system or appeals to something characteristic of the universe that provides justification of value

cover glass a sheet of glass used to cover the solid state device in a solar cell

crash-landers or hard-lander; a spacecraft that collides with the planet, making no—or little—attempt to slow down; after collision, the spacecraft ceases to function because of the (intentional) catastrophic failure

crawler transporter large, tracked vehicles used to move the assembled Apollo/Saturn from the VAB to the launch pad

cryogenic related to extremely low temperatures; the temperature of liquid nitrogen or lower

cryptocometary another name for carbonaceous asteroids—asteroids that contain a high percentage of carbon compounds mixed with frozen gases

cryptoendolithic microbial microbial ecosystems that live inside sandstone in extreme environments such as Antarctica

crystal lattice the arrangement of atoms inside a crystal

crystallography the study of the internal structure of crystals

dark matter matter that interacts with ordinary matter by gravity but does not emit electromagnetic radiation; its composition is unknown

density-separation jigs a form of gravity separation of materials with different densities that uses a pulsating fluid

desiccation the process of drying up

detruents microorganisms that act as decomposers in a controlled environmental life support system

diffuse spread out; not concentrated

DNA deoxyribonucleic acid; the molecule used by all living things on Earth to transmit genetic information

docking system mechanical and electronic devices that work jointly to bring together and physically link two spacecraft in space

doped semiconductor such as silicon with an addition of small amounts of an impurity such as phosphorous to generate more charge carriers (such as electrons)

dormant comet a comet whose volatile gases have all been vaporized, leaving behind only the heavy materials

downlink the radio dish and receiver through which a satellite or spacecraft transmits information back to Earth

drag a force that opposes the motion of an aircraft or spacecraft through the atmosphere

dunites rock type composed almost entirely of the mineral olivine, crystallized from magma beneath the Moon's surface

dynamic isotope power the decay of isotopes such as plutonium-238, and polonium-210 produces heat, which can be transformed into electricity by radioisotopic thermoelectric generators

Earth-Moon LaGrange five points in space relative to Earth and the Moon where the gravitational forces on an object balance; two points, 60 degrees from the Moon in orbit, are candidate points for a permanent space settlement due to their gravitational stability

eccentric the term that describes how oval the orbit of a planet is

ecliptic the plane of Earth's orbit

EH condrites a rare form of meteorite containing a high concentration of the mineral enstatite (a type of pyroxene) and over 30 percent iron

ejecta the pieces of material thrown off by a star when it explodes; or, material thrown out of an impact crater during its formation

ejector ramjet engine design that uses a small rocket mounted in front of the ramjet to provide a flow of heated air, allowing the ramjet to provide thrust when stationary

electrodynamic pertaining to the interaction of moving electric charges with magnetic and electric fields

electrolytes a substance that when dissolved in water creates an electrically conducting solution

electromagnetic spectrum the entire range of wavelengths of electromagnetic radiation

electron a negatively charged subatomic particle

electron volts units of energy equal to the energy gained by an electron when it passes through a potential difference of 1 volt in a vacuum

electrostatic separation separation of substances by the use of electrically charged plates

elliptical having an oval shape

encapsulation enclosing within a capsule

endocrine system in the body that creates and secretes substances called hormones into the blood

equatorial orbit an orbit parallel to a body's geographic equator

equilibruim point the point where forces are in balance

Europa one of the large satellites of Jupiter

eV an electron volt is the energy gained by an electron when moved across a potential of one volt. Ordinary molecules, such as air, have an energy of about $3x10^{-2}$ eV

event horizon the imaginary spherical shell surrounding a black hole that marks the boundary where no light or any other information can escape

excavation a hole formed by mining or digging

expendable launch vehicles launch vehicles, such as a rocket, not intended to be reused

extrasolar planets planets orbiting stars other than the Sun

extravehicular activity a space walk conducted outside a spacecraft cabin, with the crew member protected from the environment by a pressurized space suit

extremophiles microorganisms that can survive in extreme environments such as high salinity or near boiling water

extruded forced through an opening

failsafe a system designed to be failure resistant through robust construction and redundant functions

fairing a structure designed to provide low aerodynamic drag for an aircraft or spacecraft in flight

fault a fracture in rock in the upper crust of a planet along which there has been movement

feedstock the raw materials introduced into an industrial process from which a finished product is made

feldspathic rock containing a high proportion of the mineral feldspar

fiber-optic cable a thin strand of ultrapure glass that carries information in the form of light, with the light turned on and off rapidly to represent the information sent

fission act of splitting a heavy atomic nucleus into two lighter ones, releasing tremendous energy

flares intense, sudden releases of energy

flybys flight path that takes the spacecraft close enough to a planet to obtain good observations; the spacecraft then continues on a path away from the planet but may make multiple passes

fracture any break in rock, from small "joints" that divide rocks into planar blocks (such as that seen in road cuts) to vast breaks in the crusts of unspecified movement

freefall the motion of a body acted on by no forces other than gravity, usually in orbit around Earth or another celestial body

free radical a molecule with a high degree of chemical reactivity due to the presence of an unpaired electron

frequencies the number of oscillations or vibrations per second of an electromagnetic wave or any wave

fuel cells cells that react a fuel (such as hydrogen) and an oxidizer (such as oxygen) together; the chemical energy of the initial reactants is released by the fuel cell in the form of electricity

fusion the act of releasing nuclear energy by combining lighter elements such as hydrogen into heavier elements

fusion fuel fuel suitable for use in a nuclear fusion reactor

G force the force an astronaut or pilot experiences when undergoing large accelerations

galaxy a system of as many as hundreds of billions of stars that have a common gravitational attraction

Galilean satellite one of the four large moons of Jupiter first discovered by Galileo

Galileo mission succesful robot exploration of the outer solar system; this mission used gravity assists from Venus and Earth to reach Jupiter, where it dropped a probe into the atmosphere and studied the planet for nearly seven years

gamma rays a form of radiation with a shorter wavelength and more energy than X rays

Ganymede one of the four large moons of Jupiter; the largest moon in the solar system

Gemini the second series of American-piloted spacecraft, crewed by two astronauts; the Gemini missions were rehearsals of the spaceflight techniques needed to go to the Moon

general relativity a branch of science first described by Albert Einstein showing the relationship between gravity and acceleration

geocentric a model that places Earth at the center of the universe

geodetic survey determination of the exact position of points on Earth's surface and measurement of the size and shape of Earth and of Earth's gravitational and magnetic fields

geomagnetic field Earth's magnetic field; under the influence of solar wind, the magnetic field is compressed in the Sunward direction and stretched out in the downwind direction, creating the magnetosphere, a complex, teardrop-shaped cavity around Earth

geospatial relating to measurement of Earth's surface as well as positions on its surface

geostationary remaining above a fixed point above Earth's equator

geostationary orbit a specific altitude of an equatorial orbit where the time required to circle the planet matches the time it takes the planet to rotate on its axis. An object in geostationary orbit will always remain over the same geographic location on the equator of the planet it orbits

geosynchronous remaining fixed in an orbit 35,786 kilometers (22,300 miles) above Earth's surface

geosynchronous orbit a specific altitude of an equatorial orbit where the time required to circle the planet matches the time it takes the planet to rotate on its axis. An object in geostationary orbit will always remain over the same geographic location on the equator of the planet it orbits

gimbal motors motors that direct the nozzle of a rocket engine to provide steering

global change a change, such as average ocean temperature, affecting the entire planet

global positioning systems a system of satellites and receivers that provide direct determination of the geographical location of the receiver

globular clusters roughly spherical collections of hundreds of thousands of old stars found in galactic haloes

grand unified theory (GUT) states that, at a high enough energy level (about 10^{25} eV), the electromagnetic force, strong force, and weak force all merge into a single force

gravitational assist the technique of flying by a planet to use its energy to "catapult" a spacecraft on its way—this saves fuel and thus mass and cost of a mission; gravitational assists typically make the total mission duration longer, but they also make things possible that otherwise would not be possible

gravitational contraction the collapse of a cloud of gas and dust due to the mutual gravitational attraction of the parts of the cloud; a possible source of excess heat radiated by some Jovian planets

gravitational lenses two or more images of a distant object formed by the bending of light around an intervening massive object

gravity assist using the gravity of a planet during a close encounter to add energy to the motion of a spacecraft

gravity gradient the difference in the acceleration of gravity at different points on Earth and at different distances from Earth

gravity waves waves that propagate through space and are caused by the movement of large massive bodies, such as black holes and exploding stars

greenhouse effect process by which short wavelength energy (e.g., visible light) penetrates an object's atmosphere and is absorbed by the surface, which reradiates this energy as longer wavelength infrared (thermal) energy; this energy is blocked from escaping to space by molecules (e.g., H_2O and CO_2) in the atmosphere; and as a result, the surface warms

gyroscope a spinning disk mounted so that its axis can turn freely and maintain a constant orientation in space

hard-lander spacecraft that collides with the planet or satellite, making no attempt to slow its descent; also called crash-landers

heliosphere the volume of space extending outward from the Sun that is dominated by solar wind; it ends where the solar wind transitions into the interstellar medium, somewhere between 40 and 100 astronomical units from the Sun

helium-3 a stable isotope of helium whose nucleus contains two protons and one neutron

hertz unit of frequency equal to one cycle per second

high-power klystron tubes a type of electron tube used to generate high frequency electromagnetic waves

hilly and lineated terrain the broken-up surface of Mercury at the antipode of the Caloris impact basin

hydrazine a dangerous and corrosive compound of nitrogen and hydrogen commonly used in high powered rockets and jet engines

hydroponics growing plants using water and nutrients in solution instead of soil as the root medium

hydrothermal relating to high temperature water

hyperbaric chamber compartment where air pressure can be carefully controlled; used to gradually acclimate divers, astronauts, and others to changes in pressure and air composition

hypergolic fuels and oxidizers that ignite on contact with each other and need no ignition source

hypersonic capable of speeds over five times the speed of sound

hyperspectral imaging technique in remote sensing that uses at least sixteen contiguous bands of high spectral resolution over a region of the electromagnetic spectrum; used in NASA spacecraft Lewis' payload

ilmenite an important ore of titanium

Imbrium Basin impact largest and latest of the giant impact events that formed the mare-filled basins on the lunar near side

impact craters bowl-shaped depressions on the surfaces of planets or satellites that result from the impact of space debris moving at high speeds

impact winter the period following a large asteroidal or cometary impact when the Sun is dimmed by stratospheric dust and the climate becomes cold worldwide

impact-melt molten material produced by the shock and heat transfer from an impacting asteroid or meteorite

in situ in the natural or original location

incandescence glowing due to high temperature

indurated rocks rocks that have been hardened by natural processes

information age the era of our time when many businesses and persons are involved in creating, transmitting, sharing, using, and selling information, particularly through the use of computers

infrared portion of the electromagnetic spectrum with waves slightly longer than visible light

infrared radiation radiation whose wavelength is slightly longer than the wavelength of light

infrastructure the physical structures, such as roads and bridges, necessary to the functioning of a complex system

intercrater plains the oldest plains on Mercury that occur in the highlands and that formed during the period of heavy meteoroid bombardment

interferometers devices that use two or more telescopes to observe the same object at the same time in the same wavelength to increase angular resolution

interplanetary trajectories the solar orbits followed by spacecraft moving from one planet in the solar system to another

interstellar between the stars

interstellar medium the gas and dust found in the space between the stars

ion propulsion a propulsion system that uses charged particles accelerated by electric fields to provide thrust

ionization removing one or more electrons from an atom or molecule

ionosphere a charged particle region of several layers in the upper atmosphere created by radiation interacting with upper atmospheric gases

isotopic ratios the naturally occurring ratios between different isotopes of an element

jettison to eject, throw overboard, or get rid of

Jovian relating to the planet Jupiter

Kevlar® a tough aramid fiber resistant to penetration

kinetic energy the energy an object has due to its motion

KREEP acronym for material rich in potassium (K), rare earth elements (REE), and phosphorus (P)

L-4 the gravitationally stable Lagrange point 60 degrees ahead of the orbiting planet

L-5 the gravitationally stable Lagrange point 60 degrees behind the orbiting planet

Lagrangian point one of five gravitationally stable points related to two orbiting masses; three points are metastable, but L4 and L5 are stable

laser-pulsing firing periodic pulses from a powerful laser at a surface and measuring the length of time for return in order to determine topography

libration point one of five gravitationally stable points related to two orbiting masses; three points are metastable, but L4 and L5 are stable

lichen fungus that grows symbiotically with algae

light year the distance that light in a vacuum would travel in one year, or about 9.5 trillion kilometers (5.9 trillion miles)

lithosphere the rocky outer crust of a body

littoral the region along a coast or beach between high and low tides

lobate scarps a long sinuous cliff

low Earth orbit an orbit between 300 and 800 kilometers above Earth's surface

lunar maria the large, dark, lava-filled impact basins on the Moon thought by early astronomers to resemble seas

Lunar Orbiter a series of five unmanned missions in 1966 and 1967 that photographed much of the Moon at medium to high resolution from orbit

macromolecules large molecules such as proteins or DNA containing thousands or millions of individual atoms

magnetohydrodynamic waves a low frequency oscillation in a plasma in the presence of a magnetic field

magnetometer an instrument used to measure the strength and direction of a magnetic field

magnetosphere the magnetic cavity that surrounds Earth or any other planet with a magnetic field. It is formed by the interaction of the solar wind with the planet's magnetic field

majority carriers the more abundant charge carriers in semiconductors; the less abundant are called minority carriers; for n-type semiconductors, electrons are the majority carriers

malady a disorder or disease of the body

many-bodied problem in celestial mechanics, the problem of finding solutions to the equations for more than two orbiting bodies

mare dark-colored plains of solidified lava that mainly fill the large impact basins and other low-lying regions on the Moon

Mercury the first American piloted spacecraft, which carried a single astronaut into space; six Mercury missions took place between 1961 and 1963

mesons any of a family of subatomic particle that have masses between electrons and protons and that respond to the strong nuclear force; produced in the upper atmosphere by cosmic rays

meteor the physical manifestation of a meteoroid interacting with Earth's atmosphere; this includes visible light and radio frequency generation, and an ionized trail from which radar signals can be reflected. Also called a "shooting star"

meteorites any part of a meteoroid that survives passage through Earth's atmosphere

meteoroid a piece of interplanetary material smaller than an asteroid or comet

meteorology the study of atmospheric phenomena or weather

meteorology satellites satellites designed to take measurements of the atmosphere for determining weather and climate change

microgravity the condition experienced in freefall as a spacecraft orbits Earth or another body; commonly called weightlessness; only very small forces are perceived in freefall, on the order of one-millionth the force of gravity on Earth's surface

micrometeoroid flux the total mass of micrometeoroids falling into an atmosphere or on a surface per unit of time

micrometeoroid any meteoroid ranging in size from a speck of dust to a pebble

microwave link a connection between two radio towers that each transmit and receive microwave (radio) signals as a method of carrying information (similar to radio communications)

minerals crystalline arrangements of atoms and molecules of specified proportions that make up rocks

missing matter the mass of the universe that cannot be accounted for but is necessary to produce a universe whose overall curvature is "flat"

monolithic massive, solid, and uniform; an asteroid that is formed of one kind of material fused or melted into a single mass

multi-bandgap photovoltaic photovoltaic cells designed to respond to several different wavelengths of electromagnetic radiation

multispectral referring to several different parts of the electromagnetic spectrum, such as visible, infrared, and radar

muons the decay product of the mesons produced by cosmic rays; muons are about 100 times more massive than electrons but are still considered leptons that do not respond to the strong nuclear force

near-Earth asteroids asteroids whose orbits cross the orbit of Earth; collisions between Earth and near Earth asteroids happen a few times every million years

nebulae clouds of interstellar gas and/or dust

neutron a subatomic particle with no electrical charge

neutron star the dense core of matter composed almost entirely of neutrons that remain after a supernova explosion has ended the life of a massive star

New Millennium a NASA program to identify, develop and validate key instrument and spacecraft technologies that can lower cost and increase performance of science missions in the twenty-first century

Next Generation Space Telescope the telescope scheduled to be launched in 2009 that will replace the Hubble Space Telescope

nuclear black holes black holes that are in the centers of galaxies; they range in mass from a thousand to a billion times the mass of the Sun

nuclear fusion the combining of low-mass atoms to create heavier ones; the heavier atom's mass is slightly less than the sum of the mass of its constituents, with the remaining mass converted to energy

nucleon a proton or a neutron; one of the two particles found in a nucleus

occultations a phenomena that occurs when one astronomical object passes in front of another

optical interferometry a branch of optical physics that uses the wavelength of visible light to measure very small changes within the environment

optical-interferometry based the use of two or more telescopes observing the same object at the same time at the same visible wavelength to increase angular resolution

optical radar a method of determining the speed of moving bodies by sending a pulse of light and measuring how long it takes for the reflected light to return to the sender

orbit the circular or elliptical path of an object around a much larger object, governed by the gravitational field of the larger object

orbital dynamics the mathematical study of the nature of the forces governing the movement of one object in the gravitational field of another object

orbital velocity velocity at which an object needs to travel so that its flight path matches the curve of the planet it is circling; approximately 8 kilometers (5 miles) per second for low-altitude orbit around Earth

orbiter spacecraft that uses engines and/or aerobraking, and is captured into circling a planet indefinitely

orthogonal composed of right angles or relating to right angles

oscillation energy that varies between alternate extremes with a definable period

osteoporosis the loss of bone density; can occur after extended stays in space

oxidizer a substance mixed with fuel to provide the oxygen needed for combustion

paleolake depression that shows geologic evidence of having contained a lake at some previous time

Paleozoic relating to the first appearance of animal life on Earth

parabolic trajectory trajectory followed by an object with velocity equal to escape velocity

parking orbit placing a spacecraft temporarily into Earth orbit, with the engines shut down, until it has been checked out or is in the correct location for the main burn that sends it away from Earth

payload any cargo launched aboard a rocket that is destined for space, including communications satellites or modules, supplies, equipment, and astronauts; does not include the vehicle used to move the cargo or the propellant that powers the vehicle

payload bay the area in the shuttle or other spacecraft designed to carry cargo

payload fairing structure surrounding a payload; it is designed to reduce drag

payload operations experiments or procedures involving cargo or "payload" carried into orbit

payload specialists scientists or engineers selected by a company or a government employer for their expertise in conducting a specific experiment or commercial venture on a space shuttle mission

perihelion the point in an object's orbit that is closest to the Sun

period of heavy meteoroid the earliest period in solar system history (more than 3.8 billion years ago) when the rate of meteoroid impact was very high compared to the present

perturbations term used in orbital mechanics to refer to changes in orbits due to "perturbing" forces, such as gravity

phased array a radar antenna design that allows rapid scanning of an area without the need to move the antenna; a computer controls the phase of each dipole in the antenna array

phased-array antennas radar antenna designs that allow rapid scanning of an area without the need to move the antenna; a computer controls the phase of each dipole in the antenna array

photolithography printing that uses a photographic process to create the printing plates

photometer instrument to measure intensity of light

photosynthesis a process performed by plants and algae whereby light is transformed into energy and sugars

photovoltaic pertaining to the direct generation of electricity from electromagnetic radiation (light)

photovoltaic arrays sets of solar panels grouped together in big sheets; these arrays collect light from the Sun and use it to make electricity to power the equipment and machines

photovoltaic cells cells consisting of a thin wafer of a semiconductor material that incorporates a p-n junction, which converts incident light into electrical power; a number of photovoltaic cells connected in series makes a solar array

plagioclase most common mineral of the light-colored lunar highlands

planetesimals objects in the early solar system that were the size of large asteroids or small moons, large enough to begin to gravitationally influence each other

pn single junction in a transistor or other solid state device, the boundary between the two different kinds of semiconductor material

point of presence an access point to the Internet with a unique Internet Protocol (IP) address; Internet service providers (ISP) like AOL generally have multiple POPs on the Internet

polar orbits orbits that carry a satellite over the poles of a planet

polarization state degree to which a beam of electromagnetic radiation has all of the vibrations in the same plane or direction

porous allowing the passage of a fluid or gas through holes or passages in the substance

power law energy spectrum spectrum in which the distribution of energies appears to follow a power law

primary the body (planet) about which a satellite orbits

primordial swamp warm, wet conditions postulated to have occurred early in Earth's history as life was beginning to develop

procurement the process of obtaining

progenitor star the star that existed before a dramatic change, such as a supernova, occurred

prograde having the same general sense of motion or rotation as the rest of the solar system, that is, counterclockwise as seen from above Earth's north pole

prominences inactive "clouds" of solar material held above the solar surface by magnetic fields

propagate to cause to move, to multiply, or to extend to a broader area

proton a positively charged subatomic particle

pseudoscience a system of theories that assumes the form of science but fails to give reproducible results under conditions of controlled experiments

pyroclastic pertaining to clastic (broken) rock material expelled from a volcanic vent

pyrotechnics fireworks display; the art of building fireworks

quantum foam the notion that there is a smallest distance scale at which space itself is not a continuous medium, but breaks up into a seething foam of wormholes and tiny black holes far smaller than a proton

quantum gravity an attempt to replace the inherently incompatible theories of quantum physics and Einstein gravity with some deeper theory that would have features of both, but be identical to neither

quantum physics branch of physics that uses quantum mechanics to explain physical systems

quantum vacuum consistent with the Heisenberg uncertainty principle, vacuum is not empty but is filled with zero-point energy and particle-antiparticle pairs constantly being created and then mutually annihilating each other

quasars luminous objects that appear star-like but are highly redshifted and radiate more energy than an entire ordinary galaxy; likely powered by black holes in the centers of distant galaxies

quiescent inactive

radar a technique for detecting distant objects by emitting a pulse of radio-wavelength radiation and then recording echoes of the pulse off the distant objects

radar altimetry using radar signals bounced off the surface of a planet to map its variations in elevation

radar images images made with radar illumination instead of visible light that show differences in radar brightness of the surface material or differences in brightness associated with surface slopes

radiation belts two wide bands of charged particles trapped in a planet's magnetic field

radio lobes active galaxies show two regions of radio emission above and below the plane of the galaxy, and are thought to originate from powerful jets being emitted from the accretion disk surrounding the massive black hole at the center of active galaxies

radiogenic isotope techniques use of the ratio between various isotopes produced by radioactive decay to determine age or place of origin of an object in geology, archaeology, and other areas

radioisotope a naturally or artificially produced radioactive isotope of an element

radioisotope thermoelectric device using solid state electronics and the heat produced by radioactive decay to generate electricity

range safety destruct systems system of procedures and equipment designed to safely abort a mission when a spacecraft malfunctions, and destroy the rocket in such a way as to create no risk of injury or property damage

Ranger series of spacecraft sent to the Moon to investigate lunar landing sites; designed to hard-land on the lunar surface after sending back television pictures of the lunar surface; Rangers 7, 8, and 9 (1964–1965) returned data

rarefaction decreased pressure and density in a material caused by the passage of a sound wave

reconnaissance a survey or preliminary exploration of a region of interest

reflex motion the orbital motion of one body, such as a star, in reaction to the gravitational tug of a second orbiting body, such as a planet

regolith upper few meters of a body's surface, composed of inorganic matter, such as unconsolidated rocks and fine soil

relative zero velocity two objects having the same speed and direction of movement, usually so that spacecraft can rendezvous

relativistic time dilation effect predicted by the theory of relativity that causes clocks on objects in strong gravitational fields or moving near the speed of light to run slower when viewed by a stationary observer

remote manipulator system a system, such as the external Canada2 arm on the International Space Station, designed to be operated from a remote location inside the space station

remote sensing the act of observing from orbit what may be seen or sensed below on Earth

retrograde having the opposite general sense of motion or rotation as the rest of the solar system, clockwise as seen from above Earth's north pole

reusable launch vehicles launch vehicles, such as the space shuttle, designed to be recovered and reused many times

reusables launches that can be used many times before discarding

rift valley a linear depression in the surface, several hundred to thousand kilometers long, along which part of the surface has been stretched, faulted, and dropped down along many normal faults

rille lava channels in regions of maria, typically beginning at a volcanic vent and extending downslope into a smooth mare surface

rocket vehicle or device that is especially designed to travel through space, and is propelled by one or more engines

"rocky" planets nickname given to inner or solid-surface planets of the solar system, including Mercury, Venus, Mars, and Earth

rover vehicle used to move about on a surface

rutile a red, brown, or black mineral, primarily titanium dioxide, used as a gemstone and also a commercially important ore of titanium

satellite any object launched by a rocket for the purpose of orbiting the Earth or another celestial body

scoria fragments of lava resembling cinders

secondary crater crater formed by the impact of blocks of rock blasted out of the initial crater formed by an asteroid or large meteorite

sedentary lifestyle a lifestyle characterized by little movement or exercise

sedimentation process of depositing sediments, which result in a thick accumulation of rock debris eroded from high areas and deposited in low areas

semiconductor one of the groups of elements with properties intermediate between the metals and nonmetals

semimajor axis one half of the major axis of an ellipse, equal to the average distance of a planet from the Sun

shepherding small satellites exerting their gravitational influence to cause or maintain structure in the rings of the outer planets

shield volcanoes volcanoes that form broad, low-relief cones, characterized by lava that flows freely

shielding providing protection for humans and electronic equipment from cosmic rays, energetic particles from the Sun, and other radioactive materials

sine wave a wave whose amplitude smoothly varies with time; a wave form that can be mathematically described by a sine function

smooth plains the youngest plains on Mercury with a relatively low impact crater abundance

soft-landers spacecraft that uses braking by engines or other techniques (e.g., parachutes, airbags) such that its landing is gentle enough that the spacecraft and its instruments are not damaged, and observations at the surface can be made

solar arrays groups of solar cells or other solar power collectors arranged to capture energy from the Sun and use it to generate electrical power

solar corona the thin outer atmosphere of the Sun that gradually transitions into the solar wind

solar flares explosions on the Sun that release bursts of electromagnetic radiation, such as light, ultraviolet waves, and X rays, along with high speed protons and other particles

solar nebula the cloud of gas and dust out of which the solar system formed

solar prominence cool material with temperatures typical of the solar photosphere or chromosphere suspended in the corona above the visible surface layers

solar radiation total energy of any wavelength and all charged particles emitted by the Sun

solar wind a continuous, but varying, stream of charged particles (mostly electrons and protons) generated by the Sun; it establishes and affects the interplanetary magnetic field; it also deforms the magnetic field about Earth and sends particles streaming toward Earth at its poles

sounding rocket a vehicle designed to fly straight up and then parachute back to Earth, usually designed to take measurements of the upper atmosphere

space station large orbital outpost equipped to support a human crew and designed to remain in orbit for an extended period; to date, only Earth-orbiting space stations have been launched

space-time in relativity, the four-dimensional space through which objects move and in which events happen

spacecraft bus the primary structure and subsystems of a spacecraft

spacewalking moving around outside a spaceship or space station, also known as extravehicular activity

special theory of relativity the fundamental idea of Einstein's theories, which demonstrated that measurements of certain physical quantities such as mass, length, and time depended on the relative motion of the object and observer

specific power amount of electric power generated by a solar cell per unit mass; for example watts per kilogram

spectra representations of the brightness of objects as a function of the wavelength of the emitted radiation

spectral lines the unique pattern of radiation at discrete wavelengths that many materials produce

spectrograph an instrument that can permanently record a spectra

spectrographic studies studies of the nature of matter and composition of substances by examining the light they emit

spectrometers an instrument with a scale for measuring the wavelength of light

spherules tiny glass spheres found in and among lunar rocks

spot beam technology narrow, pencil-like satellite beam that focuses highly radiated energy on a limited area of Earth's surface (about 100 to 500 miles in diameter) using steerable or directed antennas

stratigraphy the study of rock layers known as strata, especially the age and distribution of various kinds of sedimentary rocks

stratosphere a middle portion of a planet's atmosphere above the tropopause (the highest place where convection and "weather" occurs)

subduction the process by which one edge of a crustal plate is forced to move under another plate

sublimate to pass directly from a solid phase to a gas phase

suborbital trajectory the trajectory of a rocket or ballistic missile that has insufficient energy to reach orbit

subsolar point the point on a planet that receives direct rays from the Sun

substrate the surface, such as glass, metallic foil, or plastic sheet, on which a thin film of photovoltaic material is deposited

sunspots dark, cooler areas on the solar surface consisting of transient, concentrated magnetic fields

supercarbonaceous term given to P- and D-type meteorites that are richer in carbon than any other meteorites and are thought to come from the primitive asteroids in the outer part of the asteroid belt

supernova an explosion ending the life of a massive star

supernovae ejecta the mix of gas enriched by heavy metals that is launched into space by a supernova explosion

superstring theory the best candidate for a "theory of everything" unifying quantum mechanics and gravity, proposes that all particles are oscillations in tiny loops of matter only 10^{-35} meters long and moving in a space of ten dimensions

superstrings supersymmetric strings are tiny, one dimensional objects that are about 10^{-33} cm long, in a 10-dimensional spacetime. Their different vibration modes and shapes account for the elementary particles we see in our 4-dimensional spacetime

Surveyor a series of spacecraft designed to soft-land robotic laboratories to analyze and photograph the lunar surface; Surveyors 1, 3, and 5–7 landed between May 1966 and January 1968

synchrotron radiation the radiation from electrons moving at almost the speed of light inside giant magnetic accelerators of particles, called synchrotrons, either on Earth or in space

synthesis the act of combining different things so as to form new and different products or ideas

technology transfer the acquisition by one country or firm of the capability to develop a particular technology through its interactions with the existing technological capability of another country or firm, rather than through its own research efforts

tectonism process of deformation in a planetary surface as a result of geological forces acting on the crust; includes faulting, folding, uplift, and downwarping of the surface and crust

telescience the act of operation and monitoring of research equipment located in space by a scientist or engineer from their offices or laboratories on Earth

terrestrial planet a small rocky planet with high density orbiting close to the Sun; Mercury, Venus, Earth, and Mars

thermodynamically referring to the behavior of energy

thermostabilized designed to maintain a constant temperature

thrust fault a fault where the block on one side of the fault plane has been thrust up and over the opposite block by horizontal compressive forces

toxicological related to the study of the nature and effects on humans of poisons and the treatment of victims of poisoning

trajectories paths followed through space by missiles and spacecraft moving under the influence of gravity

transonic barrier the aerodynamic behavior of an aircraft moving near the speed of sound changes dramatically and, for early pioneers of transonic flight, dangerously, leading some to hypothesize there was a "sound barrier" where drag became infinite

transpiration process whereby water evaporates from the surface of leaves, allowing the plant to lose heat and to draw water up through the roots

transponder bandwidth-specific transmitter-receiver units

troctolite rock type composed of the minerals plagioclase and olivine, crystallized from magma

tunnelborer a mining machine designed to dig a tunnel using rotating cutting disks

Tycho event the impact of a large meteoroid into the lunar surface as recently as 100 million years ago, leaving a distinct set of bright rays across the lunar surface including a ray through the Apollo 17 landing site

ultramafic lavas dark, heavy lavas with a high percentage of magnesium and iron; usually found as boulders mixed in other lava rocks

ultraviolet the portion of the electromagnetic spectrum just beyond (having shorter wavelengths than) violet

ultraviolet radiation electromagnetic radiation with a shorter wavelength and higher energy than light

uncompressed density the lower density a planet would have if it did not have the force of gravity compressing it

Universal time current time in Greenwich, England, which is recognized as the standard time that Earth's time zones are based

vacuum an environment where air and all other molecules and atoms of matter have been removed

vacuum conditions the almost complete lack of atmosphere found on the surface of the Moon and in space

Van Allen radiation belts two belts of high energy charged particles captured from the solar wind by Earth's magnetic field

variable star a star whose light output varies over time

vector sum sum of two vector quantities taking both size and direction into consideration

velocity speed and direction of a moving object; a vector quantity

virtual-reality simulations a simulation used in training by pilots and astronauts to safely reproduce various conditions that can occur on board a real aircraft or spacecraft

visible spectrum the part of the electromagnetic spectrum with wavelengths between 400 nanometers and 700 nanometers; the part of the electromagnetic spectrum to which human eyes are sensitive

volatile ices (e.g., H_2O and CO_2) that are solids inside a comet nucleus but turn into gases when heated by sunlight

volatile materials materials that easily pass into the vapor phase when heated

wavelength the distance from crest to crest on a wave at an instant in time

X ray form of high-energy radiation just beyond the ultraviolet portion of the spectrum

X-ray diffraction analysis a method to determine the three-dimensional structure of molecules

Volume 2 Index

*Page numbers in **boldface type** indicate article titles; those in italic type indicate illustrations. A cumulative index, which combines the terms in all volumes of Space Sciences, can be found in volume 4 of this series.*

A

Acceleration, law of, 119

Action and reaction, law of, 119

Active galaxies, 53–54, *54*

Adams, John, 116

Adaptive optics, in telescopes, 122

Adler Planetarium, 140

Advanced degrees, in astronomy, 17–18

Advanced Satellite for Cosmology and Astrophysics (ASCA), 13

Aerobraking, 45–46

Air resistance, gravity and, 63

Aldrin, Edwin E. Jr. "Buzz," 42, 193–194

Allen telescope array, 178

ALMA (Atacama Large Millimeter Array), 126

Alpha Centauri, 170

Alpha proton X-ray spectrometer, 162

Alvarez, Louis, 25

Alvarez, Walter, 25

Amalthea (Moon), 92, 175

Amateur astronomers, 8

American Astronomical Society, 19

Andromeda, 54, 56

Antimatter, 129

Antiope (Asteroid), 4

Apollo, 142, 191–192, 193–194
    humans on Moon, 42
    robotics and, 160

Arecibo radio telescope, 171

Ares Vallis, 45

Ariane rocket, 59

Arianespace, 59

Ariel (Moon), 212

Aristotle
    geocentric system, 9
    on gravity, 63

Armstrong, Neil, 42, 193–194

Arrays, 121, 121*t*, 124*t*
    Atacama Large Millimeter Array, 126
    SETI use of, 178

Artemis satellites, 59

ASCA (Advanced Satellite for Cosmology and Astrophysics), 13

Association of Science and Technology Centers, 142

Asteroids, **3–7**, *4, 5, 6*
    classification of, 176–177
    close encounters with, 23–30, *24, 26, 29*
    from meteorites, 105–106
    Near Earth asteroids, 6
    near Earth space and, 185

Astrobiology, 22

Astrology, *vs.* astronomy, 8

Astrometry, 50

Astronomers, **7–9**, *8, 21*
    See also specific names

Astronomical units, 23, 176, 221

Astronomy
    careers in, **16–20**, *17*
    history of, **9–10**
    kinds of, **10–14**, *11, 12t*
    optical, 120

Astrophysics, 17

Atacama Large Millimeter Array (ALMA), 126

Atlantis, 128

Atmosphere
    of Earth, 10–11, 38
    of Jupiter, 77
    of Neptune, 115
    of Pluto, 149–150

of Saturn, 164
    of Triton, 117

Atmospheric probes, 154

Auger, Pierre, 31

Auroras, 180

B

Baade, Walter, 153

Bachelor's degrees, in astronomy, 19

Bacon, Roger, 190

Ballistic missiles, 108, 192

Basalt, in planetary crust, 38

Beidou Navigation Test Satellite-1, 61

Bell, Jocelyn, 151

Big Bang theory, 32–33
    cosmic microwave background radiation and, 131
    evidence for, 14
    Hubble, Edwin P. and, 1

Biomedicine, 21–22

BL Lac objects, 54

Black dwarf stars, 200

Black holes, *15*, **15–16**
    Chandra X-Ray Observatory and, 130
    collisions of, 16
    for galaxy energy, 54
    Hawking radiation and, 34
    Heavy stars and, 15–16
    Hubble Space Telescope on, 73
    study of, 200

Blazars, 129

Boeig Cyberdome, 141

Borrelly nucleus, 28

Brahe, Tycho, 81

Brazil, space program in, 62

A Brief History of Time (Hawking), 34

Brightness measurements, 148

Brilliant Pebbles (Defense system), 109

Brown dwarf stars, 13, 130
Brownian motion, 39
The Buck Moon (Hale), 191
Buhl Planetarium, 140
Buses (Spacecraft), **188–189**
Bush, George H.W., 109
Bush, George W., 109
Butler, Paul, 48

C

C-type asteroids, 4
Callisto, 76, 80
Canada, space program of, 61
Canadian Space Agency, 61
Carbonate minerals
    on Mars, 96
Careers
    in astronomy, **16–20**, *17*
    in space sciences, **20–22**, *21*
Carnegie Space Center, 140
Cassini, 46, 145, 158, 159, 168
Cassini, Giovanni Domenico, **23**
Cassini Division, 23
Cassini Program, 23, 46, 75
Cat's Eye Nebula, 72
Cepheid variables, 1, 2, 55, 56, 67,
    173
Ceres (Asteroid), 4
CGRO (Compton Gamma Ray Ob-
    servatory), 129–130
Chandra X-Ray Observatory, *11*, 13,
    129–130, *130*, 131
Chandrasekhar, Subrahmanyan, 129
Charon (Moon), 148–149
Chassignites, 96
Chicxulub Crater, 25, 29
China, space program in, *60*, 61, 62
CHON (Comet composition)
    29
Clarke, Arthur C., 137
    spaceguard surveys and, 26
Clarke orbits, 137, 182
Clementine (Project), 45
Close encounters, **23–26**, *24*, *26*,
    55
Clouds, *37*, 38
COBE (Cosmic Background Ex-
    plorer), 131
Cocconi, Giuseppe, 170
Colleges, space science employees
    in, 22
Collins, Michael, 193–194
Collisions, of black holes, 16
    See also Cosmic impacts
Columbus space laboratory, 59
Comet Wild-2, 46

Comets, **27–30**, *28*, *29*
    in Kuiper belt, **81–83**
    near Earth collisions of, 24
    in near Earth space, 185
    orbits of, 136
    Shoemaker-Levy 9, 174
Communications, 108
Composition, of comets, 28–29
Compton, Arthur Holly, 129
Compton Gamma Ray Observatory
    (CGRO), 129–130
Computer modeling, in astronomy,
    18–19
Congrieve, William, 190
Contact (Sagan), 163
Convection, 38, 107
Copernicus, Nicholas, 7, 9, **30**, 57
Coronal holes, 220
Coronal mass ejections, 178, 179,
    186, 220
Corrective Optics Space Telescope
    Axial Replacement (COSTAR),
    70
Cosmic Background Explorer
    (COBE), 131
Cosmic impacts, 23–27, *24*, *26*, *29*,
    36–37
Cosmic microwave background,
    33
Cosmic microwave background radi-
    ation, 131
Cosmic origins spectrograph, 71
Cosmic rays, **30–32**
Cosmological constant, 33–34
Cosmology, **32–35**, *33*
    evidence in, 14
    heliocentric system of universe, 9
    Hubble, Edwin P and, 14
    See also Big Bang theory
Cosmos (Sagan), 163
COSTAR (Corrective Optics Space
    Telescope Axial Replacement),
    70
Crab Nebula, *152*, 153
Crash-landers, 154
Craters
    impact, *24*, 95, 105–106, 118,
        216, 217
    lunar, 142–143
Crust, of Earth, 37–38
Curtis, Heber, 173

D

Dark matter
    Chandra X-Ray Observatory and,
        130
    of galaxies, 50

Data assistants, 19
Dating methods. *See* specific names
    of methods
De revolutionibus orbium coelesticum
    (Copernicus), 30
Debris, space, **182–183**, *183*
Deep Space I, 46
Defense, from Near Earth objects,
    186
Defense Meteorological Satellite
    Program, 108–109
Defense Space Communication Sys-
    tem, 108
Deimos (Moon), 92, 175–176
Desalguliers, Thomas, 190
*Dialogo sopra i due massini sistemi del
    mondo* (Galileo), 57
Dialogue on the two great world systems
    (Galileo), 57
Dinosaurs, cosmic impact and, 25
Discovery missions, 45, 128
DNA, extraterrestrial life and, 86
Doctoral degrees. *See* Ph.D.
Doppler technique, for extrasolar
    planets, 47–48
Dornberger, Walter, 192
Drake, Frank, *170*, 170–171
Drake equation, 172
Duncan, Martin, 82
Dwarf elipticals, 52

E

Earth, **35–39**, *36*, *37*
    astronomy careers and, 20–21
    electromagnetic spectrum in,
        10–11, 12*t*
    geocentric system and, 9
    impact craters and, *24*, 174
    magnetosphere in, 181
    meteorites on, 145
    near Earth space, 183–184
    orbits of, 182
    planetary defense of, 186
    See also Cosmic impacts; Low
        Earth Orbit
Earth Resources Observation Satel-
    lite, 62
Earthquakes, 37–38
Eclipses, solar, 202, *203*
Education, in astronomy, 17–18
Einstein, Albert, 1, 2, *39*, **39–40**,
    73
    on cosmological constant,
        33–34
    on special relativity, 31
    as theoretical physicist, 15
    See also General relativity

Electromagnetic spectrum
    Earth's atmosphere and, 10–11, 12*t*
    observatories and, 119, 127
    spacecraft and, 157
    telescopes for, 20
Electron volts, 177
Elements, formation of, 33
Elliptical galaxies, 51, 54
Elliptical orbits, 81
    of Mercury, 98
    near Earth encounters and, 23–24
    of Nereid, 118
    observatories and, 129
    of Saturn, 164
Enceladus (Moon), 66, 167
Energia, 58
Energy, solar, 91
Engineering, on space missions, 18
Eros (Asteroid), *4, 5*, 5–6
    discovery of, 24
    exploration of, 45
    NEAR Shoemaker on, 145
Erosion, on Earth, 36
ESA. *See* European Space Agency
Europa (Moon)
    exploration of, 76, 79
    extraterrestrial life on, 76
European Astronaut Centre, 59
European Space Agency (ESA)
    Comet Halley study by, 45
    observatories and, 131
    space program of, 59
European Space Research and Technology Centre, 59
Event horizons, of black holes, 15–16
Evolution
    Hubble constant and, 66
    impact cratering in, 25
    stellar, 13
Expansion rate, of universe. *See* Inflation, of universe
Exploration programs, **40–47,** *41, 43, 44*
Extrasolar planets, **47–50,** *49,* 74, 171
Extraterrestrial life
    on Europa, *76*
    search for, **84–90**
    SETI and, 87, 89–90, **170–173,** *171*
Extremophiles, 87

F

Faculty, astronomy, 18
Faint Object Camera, 70

Field equations, 34
Flares
    energy from, 220
    near Earth space and, 185
    solar particle radiation and, 177
    solar wind and, 180
Flocculent galaxies, 52
Fly-bys
    in planetary exploration, 144–145
    robotics and, 154
Freedom 7, 193
From Earth to Mars (Verne), 190
Fundamental physics, 14

G

Gagarin, Yuri, 193
Galilean satellites, 79–80
Galaxies, *33,* **50–56,** *51, 53, 54*
    black holes in, 16
    collisions of, 55
    distances to, 1
    elliptical, 51, 54
    flocculent, 52
    formation of, 33
    spiral, *51,* 51–52, 54
    study of, 13
Galileo Galilei, 7, 9, **56–57,** *57*
    on heliocentric system, 30
    Jupiter and, 76
Galileo (Satellite), 45, 59
    on Jupiter, 76–78, 145, 156–157
    robotics in, 112, 154, 155
Galle, Johann, 116
Gamma-ray bursts, 73
Gamma-ray Observatory (GLAST), 131
Gamma-rays
    Compton Gamma Ray Observatory and, 129–130
    observatories, 119, 121, 127, 129
    solar particle radiation and, 177
    spacecraft and, 158
Ganymede, 76, 80
Gemini, 156
General relativity, 15, 33–34, 39–40, 64–65
Generators, radioisotope thermoelectric, 157
Genesis, 181
Geocentric system of universe, 9, 30
Geology, of Earth, *36,* 36–37
Geomagnetic field, 220
Geomagnetic storms, 219*t*
Geostationary orbit, 137, 182

Geostationary Satellite Launch Vehicle, 62
German V-2 rockets, 108
Giotto, 45
GLAST (Gamma-ray Observatory), 131
Glenn, John, 107
Global change, space research in, 22
Global Positioning System (GPS), 108
Global Protection Against Limited Strike, 109
Global warming, 140
Globular clusters, 2–3, 55, 173
Glushko, Valentin P., 192
Goddard, Robert, 191
Gotoh, 139
Government jobs
    as astronomers, 18
    in space science, 22
Government space programs, **57–63,** *58, 60*
GPS (Global Positioning System), 108
Gravitational contraction, 116
Gravitational force, of black holes, 15
Gravity, **63–65,** *65,* **119,** *119*
Gravity assist techniques, 43
Great dark spot, 116–118
Great Observatories Program, 129–130
Great red spot, 77, *78*
Greenhouse effect, on Venus, 216
Gulf War, 108–109
Gyroscopes, 169

H

Habitable zone (HZ), 86
Halca spacecraft, 131
Hale, Edward Everette, 191
Hale-Bopp (Comet), 28–29, *29*
Hall, Asaph, 175
Halley (Comet), 28–29, 45
Halley, Edmund, 24, 28
Hawking, Stephen, 16, 34
Hawking radiation, 16, 34
HD209458, 49, *49*
Heavy stars, black holes and, 15–16
Heliocentric system of universe, 9, 30, 57
Heliosphere, 177
Henry Crown Space Center, *140*
Herschel, Caroline, 50, **66**
Herschel, John, **66**
Herschel, William, 50, **66,** 210
Hertzsprung, Protostar, 199

Hess, Viktor, 31

Hewish, Antony, 151

High Resolution Microwave Survey, 170

High-Resolution Spectrograph, 70

High school teaching, in physics and astronomy, 19

High-Speed Photometer, 70

Hot Jupiters, 48–49

HST. *See* Hubble Space Telescope

Hubble, Edwin P., 1, 9–10, **67–68**, *68*

   on galaxies, 50–52, 56

   on inflation, 34

Hubble constant, **66–67**

Hubble Deep Field, 71–72

Hubble Space Telescope (HST), 2–3, **68–75**, *69*, *72*

   for Cepheid variables, 67

   Charon from, 149–150

   Discovery launch of, 128

   Europe and, 59

   image clarity in, 121

   Jupiter from, 77

   Mars from, *93*

   Neptune from, 118

   operation of, 121, 126–129, 131, 133

   optics in, 150

   for protostellar disks, 13

   on quasars, 53

   repair of, 161, 170–171

   Uranus from, 212–213

Huygens, Christiaan, 23, **75**

Huygens probe, 46, 59

HZ (Habitable zone), 86

I

Impact, of asteroids, 6–7

Impact craters, *24*, *25*, 36–37, 95

   on Earth, 174

   on Mars, 95

   meteorite, 105–106

   on Triton, 118

   on Venus, 216, 217

India, space program in, 61–62

Inertia, law of, **119**

Inflation, of universe, 1–2, 10, 32–33, 34

   Hubble, Edwin P. on, 67–68

   Hubble Space Telescope and, 72–73

Infrared Astronomy Satellite, 128

Infrared detectors, 83

Infrared radiation

   near Earth space and, 184

observatories and, 119, 121, 128

   spacecraft and, 157

   Spacelab and, 197

Infrared Space Observatory, 131

Infrared telescopes, 13–14, 120, 122–126

Institute of Space and Astronautical Science (Japan), 59–60

Inter-Agency Space Debris Coordination Committee, 182

Interferometry, 18, 125–126

International Geophysical Year, 142

International Planetarium Society, 142

International Space Station (ISS)

   Brazil and, 62

   Canada and, 61

   Europe and, 59

   interior module, *65*

   Japan and, 60–61

   microgravity and, 107

   orbit of, 136

   robotics and, 162

   Russia and, 57–58

   space debris and, 182

International Ultraviolet Explorer, 128

Internships, in astronomy, 17–18

Interstellar medium, 13

Interstellar travel, 170

Ionosphere

   Spacelab and, 197

   Sun and, 184

Irregular galaxies, 52

Isotope ratios, 112

Israel, space program in, 62

Israel Aircraft Industries, 62

ISS. *See* International Space Station

J

Japan, space program in, 59–61

Jewitt, David, 83

Jupiter, **76–80**, *77*, *78*, *79*

   atmosphere and interior, 77–78

   asteroids and, 3

   Cassini, Giovanni Domenico, 23

   exploration of, 43–44, 45

   magnetic fields and rings, 78

   moons of, 79–80, *79*, 92

   study of, **76–80**, *78*, *79*

K

Kant, Immanuel, on Milky Way, 55

KAO (Kuiper Airborne Observatory), 13

Keck telescopes, 124–126

Kennedy Space Center (KSC), 195

Kepler, Johannes, 9, **81**

   on heliocentric system, 30

   on Mars orbit, 92

   See also Orbits

Kibo science laboratory, 60

Kirkwood gaps, 3

Korolev Rocket & Space Corporation Energia, 58

KSC (Kennedy Space Center), 195

Kuiper, Gerald Peter, 29, **83–84**, *84*

Kuiper Airborne Observatory (KAO), 13

Kuiper belt, 29, **81–83**, *82*, 176

   Oort cloud and, 132

   planetesimals in, 83, 147

   undiscovered planets and, 138

L

Lambert, Johann Heinrich, 55

Lapetus (Moon), 167

Large Magellanic Clouds, 54

Large Space Telescope. *See* Hubble Space Telescope

Lava-like meteorites, 4

Law of acceleration, 119

Law of action and reaction, 119

Law of inertia, 119

Law of universal gravitation, 63–64

Laws of motion, *119*, 119

LDEF (Long Duration Exposure Facility), **90–92**, *91*

Leavitt, Henrietta, 55

LEO. *See* Low Earth orbit

Leverrier, Urbain, 116

Liberty Bell, 193

Life in the Buck Moon (Hale), 191

Liquid propellant rockets, 191, 192

Long Duration Exposure Facility (LDEF), **90–92**, *91*

Long March rockets, 61

Low Earth Orbit (LEO)

   defined, 136, 157

   LDEF and, 91

   space debris and, 182

   trajectories and, *209*, **209–210**

Lowell, Percival, 93

Luna probes, 40, 143, 154, 155, 156

Lunar exploration. *See* Moon, exploration of

Lunar Orbiter, 40–41, *143*, 174

Lunar orbiters, 145

Lunar Prospector, 46, 145, 174

Luu, Jane, 83

M

M80 globular cluster, *2*

Magellan, 44–45

Magnetic fields

geomagnetic field, 220

of Jupiter, 78

of Mercury, 99–100

of Neptune, 117

of Saturn, 164

of Sun, 203

of Uranus, 211

Magnetosphere

of Earth, 181

of Jupiter, 78

of Saturn, 164

solar particle radiation and, 178

Main belt, asteroids in, 3

Map (Microwave Anisotrophy Probe), 131

Marcy, Geoffrey, 48

Mariner missions, 42, *43*, 143

on Mars, 94, 145

on Mercury, 98, 102

on Venus, 156–157, 214

Marius, Simon, 76

Mars, **92–98**, *93, 94, 96*

exploration of, 42, *44*, 45

extraterrestrial life on, 87–88

Mariner 9, 145

Mars Global Surveyor, 97

Mars Odyssey, 137

Pathfinder on, 145

Sojourner on, 145

Mars Express mission, 59

Mars Global Surveyor (MGS), 45–46, 97, 137

Mars Observer, 45

Mars Odyssey, 97, 137

Mars Pathfinder, 45, 145, 159, 162

Mars Sojourner, 45, 145, 159, 162

Marshall, George C., 195

Master's degrees, in astronomy, 19

The Mathematical Principles of Natural Philosophy (Newton), 119

Mathematics, in astronomy, 18–19

Mauna Kea Observatory, *120*, 121, 122, 124

Mayor, Michel, 48

Mentoring, in astronomy, 19–20

Mercury, **98–103**, *99, 100, 102*

atmosphere, 101

geology and composition, 101–102

interior and magnetic field, 99–101

motion and temperature, 98–99,

origins, 103

MESSENGER, 98, 102, 103

Messier, Charles, 50

Messier 31. *See* Andromeda

Messier 33, 54

Messier 100, 54

Meteorites, **103–106**

craters, 104–105

on Earth, 145

extraterrestrial life and, 88

on Mars, 96

study of, **103–106**, *104, 105, 106*

Meteoroids

LDEF and, 90–92

near Earth space and, 186

MGS (Mars Global Surveyor), 45–46, 97, 137

Microgravity, *107*, **107–109**

research and, 21

Spacelab and, 194

Microwave Anisotrophy Probe (MAP), 131

Military missiles, *108*, **108–109**

Milky Way, 54–55

Millisecond pulsars, 153

Milstar, 108

Mimas, 66

Minolta, 139

Minorities, in astronomy, 19–20

Mir space station, 58

Miranda (Moon), 84, 212

Mirrors, in telescopes, 123–125

Missiles

ballistic, 108

military exploration and, *108*

R-7, 192

Titan II, 193

V-2, 192

Molniya orbit, 137

Moon (Earth), **109–115**, *110,*

Apollo 11, 142, 160, 191–192, 193–194

crust, composition of, 114

exploration of, 40–42, *41, 62,* 111–112, *111,* 112*t, 113,* 142–143

isotope ratios in, 112

global and interior characteristics, 112

orbit and rotation, 110

origin of, 114

Lunar Orbiter, 174

lunar orbits, 110–111

Lunar Prospector, 174

mapping for, 174

radioisotope dating and, 112

Surveyor, 142

topography, 112–114

United States Ranger, 40, *43,* 142, 156, 158

Moons

of Jupiter, 79–80

of Mars, 92

of Neptune, 84, 117–118

of Pluto, 148–149

of Saturn, 84, 166–168

of Uranus, 84, 212–213

See also terms beginning with Lunar

Morrison, Philip, 170

Motion

Johannes Kepler and, *81*

Newton's laws of, 119

N

Nakhlites, on Mars, 96

NASDA (National Space Development Agency of Japan), 59

National Aeronautics and Space Administration (NASA)

astronomy careers at, 18

on extrasolar planets, 50

on extraterrestrial life, 88

space exploration by, 42–46

National observatories, 18

National Oceanic and Atmospheric Administration (NOAA)

solar particle radiation and, 178

weather space scales and, 200

National Optical Astronomy Observatories, 18

National Radio Astronomy Observatory, 18, 170

National Space Development Agency of Japan (NASDA), 59

National Virtual Observatory, 19

Near Earth Asteroid Rendezvous. *See* NEAR Shoemaker

Near Earth Asteroid Rendezvous (NEAR), 45

Near Earth asteroids (NEAs), 6, 24–26, *26*

Near Earth objects (NEOs), 24, 186

Near Earth space, 183–184

Near Infrared Camera and Multi-Object Spectrometer, 70

NEAR (Near Earth Asteroid Rendezvous), 45

NEAR Shoemaker, 5–6, *6,* 24, 145

NEAs. *See* Near Earth asteroids

Nebula

Cat's Eye, *72,*

Nebula (continued)
  crab, 153
  planetary, 200–201
  spiral, 1
NEOs (Near Earth objects), 24, 186
Neptune, **115–118**, *116*, *117*
  discovery of, 116–118
  exploration of, 44,
  moons of, 84, *117*, 117–118
  rings of, 118
  rotation and magnetic field, 117
Nereid (Moon), 84, 118
Neutrons, 152–153, 177, 200
New Horizons Pluto-Kuiper Belt
  Mission, 150
New Millennium missions, 46
Newton, Isaac, 9, 107, *119*, **119**
  law of universal gravitation,
    63–64
  Milky Way theories and, 55
Next Generation Space Telescope
  (NGST), 13, 131
NOAA. *See* National Oceanic and
  Atmospheric Administration
Non-scientific careers, in space sci-
  ences, 22
Nozomi probe, 60
Nuclear fuels, starshine from, 15–16
Nucleon, 177
Nucleus, of comets, 28

O

OAO (Orbiting Astronomical obser-
  vatories), 128
Oberon (Moon), 66, 212, 213
Oberth, Hermann, 191
Observational astronomers, 7–8
Observatories
  ground-based, **119–126**, *120*,
    121*t*, *122*, 124*t*
  national, 18
  space-based, **126–132**, *127*, *128*,
    130
  See also names of specific obser-
    vatories
Ofeq 3 satellite, 62
On the revolutions of the celestial orbs
  (Copernicus), 30
Oort, Jan Hendrik, 29, 133
Oort cloud, 29, **132–134**, *133*
  discovery of, 82, 133–134
  Kuiper belt, 132, 134
  planetesimals, 147
  undiscovered planets and, 138
Oppenheimer, Robert, 153
Optical astronomy, 120, 123–126
Optical interferometry, 125–126

Optical telescopes, 120, 123–126
Optics
  adaptive, 122
  in Hubble Space Telescope, 150
  mirror, 123–125
Orbital debris, 182
Orbiters, robotics and, 154
Orbiting Astronomical observatories
  (OAO), 128
Orbits, **134–137**, *135*, *136*
  of comets, 136
  of Earth, 182
  of Earth's Moon, 110–111
  geostationary, 137, 182
  Molniya, 137
  of Nereid, 118
  observatories and, 129
  of Pluto, 148
  of satellites, 137
  of space debris, 182
  study of, 81
  See also Elliptical orbits; Low
    Earth Orbit
Osiander, Andreas, 30
Overwhelmingly large telescope
  (OWL), 126

P

Palomar telescope, 123
Payload
  spacecraft buses for, 188
  Spacelab and, 197
Pendulum, 75
Perihelion, of comet Halley, 28
Ph.D., in astronomy, 17–19
Phobos (Moon), 92, 95, 175–176
Phoebe (Moon), 166–167
Photoelectric effect, 39
Photography
  of Mars, 42–43, 45–46
  of Venus, 42–43
Physical sciences, space research in,
  21
Physics, fundamental, 14
Pioneer, 43
  on Jupiter, 76
  on Saturn, 166
  on Venus, 214, 216
Planet-star distance, 47
Planet X, **137–138**
Planetariums, 19, **138–142**, *139*,
  140
  career options at, 141
Planetary defense, 186
Planetary exploration
  future of, **142–146**, *143*, *144*, *145*

  stages of, 144–145
  See also specific planets
Planetary geology, 173
Planetary motion, *81*
Planetary science, 17
Planetesimals
  Kuiper belt and, 83, 147
  study of, **146–147**, *147*
Planets
  extrasolar, **47–50**, *49*, 171
  geology of, 173
  orbits of, 136
  study of, 11, 13
  terrestrial, 101
  X, **137–138**
  See also names of specific planets;
    Small bodies
Plate tectonics. *See* Tectonism
Pluto, **147–151**
  atmospheric readings, 149–150
  discovery of, 207–208, *208*
  exploration of, 147–151, *149*,
    150
  orbit of, 148
  size and composition of, 148
  surface of, 149
Pluto-Kuiper Express spacecraft,
  150
Point masses, 63–64
Polar (Spacecraft), 62, 181
The Principia (Newton), 63
Progenitor star, 206
Prograde, 118
Progress spacecraft, 57
Project Ozma, 170
Project Phoenix, 171
Projectors, for planetariums,
  139–140
Prominences, 220
Protostar, 199
Protostellar disks, 13
Ptolemy, 9
Pulsars, **151–154**, *152*

Q

Quasars, 14, 53, 73, 129
Queloz, Didier, 48
Quinn, Thomas, 82

R

R-7 missiles, 192
Radarsat, 61
Radial velocity technique, 47–48
Radiation, Hawking, 34
  See also specific types of radiation

Radiation belts
  of Jupiter, 76
  study of, 21
Radio blackouts, 220*t*
Radio interferometry, 125–126
Radio observatories, 121, 122–123, 126
Radio telescopes
  for Earth's moon, 132
  operation of, 119, 122–123
  SETI use of, 171, 172
Radio waves, 119, 123, 184
Radioisotope dating
  of Earth's moon, 112
  of meteorites, 104
Radioisotope thermoelectric generators, 157
Radiometers, 169
Ranger missions. *See* United States Ranger
Reagan, Ronald, 109
Reflecting telescopes, 123–125
Refracting telescopes, 123
Relativistic time dilation, 31
Relativity. *See* General relativity; Special relativity
Religious controversy, on heliocentric view of universe, 30, 57
Remote Manipulator System (RMS), 161
Remote sensing satellites, 137
Rendezvous with Rama (Clarke), 26
Research
  in astronomy education, 17–18
  careers in, 20
Research assistants, 19
Retrograde motion, 30, 115
Rings (Planetary)
  of Jupiter, 78
  of Neptune, 118
  of Saturn, 166
  of Uranus, 213
RKK Energia, 58
Robotics
  space exploration and, **154–159,** *155, 157,* **160–163,** *161, 162*
  technology of, 160–163, **160–163,** *161, 162*
Rockets, 191, 192
  See also Missiles; names of specific rockets
Rocks, on Earth, 38
Rocky planets, 35–36
Römer, Ole, 23
Röntgensatellit (ROSAT), 13
Rosetta comet probe, 59
Rosviakosmos, 57

Rotation (Planetary)
  of Neptune, 117
  of Saturn, 165
RR Lyrae stars, 55
Russell, Henry Norris, 199
Russia, space program in, 57–59, *58*
  See also Soviet Union
Russian Aviation and Space Agency, 57

S

Sa galaxies, 52
Sagan, Carl, **163–164,** *164*
Sakigake, 45
San Andreas Fault, *36*
Sänger, Eugene, 192
Satellites, military, 108–109
Saturn, **164–168,** *165, 166, 166t, 167*
  Cassini on, 23, 145, 158, 159, 168
  exploration of, 44, 46, 164–168, *165, 166, 166t, 167*
  moons of, 84, 166–168
  physical and orbital properties, 164
  rings of, 166
Saturn (Rocket), 193
Sc galaxies, 52
Schiaparelli, Giovanni, 93
Schmidt, Maarten, 53
Science
  from space, 20–21
  in space, 21–22
  See also Space sciences; specific sciences
Science and Technology Agency (Japan), 60
Science centers, **138–142,** *139, 140*
  career options at, 141
SDI (Strategic Defense Initiative), 109
Sea of Tranquility, 42
Search for Extraterrestrial Intelligence (SETI), 89–90, *170,* **170–173,** *171*
Secchi, Angelo, 9–10
Selene project (Japan), 61
Sensors, **168–170,** *169*
SERENDIP, 171
Servicing and repair, of Hubble Space Telescope, 70–71
SETI. *See* Search for Extraterrestrial Intelligence
Seyfert, Carl, 54
Seyferts, 54

Shapley, Harlow, 55–56, *173,* **173**
  Visiting Lectureships Program, 173
Shavit (Rocket), 62
Shenzhou spacecraft, 61
Shepard, Alan B., 193
Sherogottites, on Mars, 96, 106
Shielding, solar particle radiation and, 178
Shoemaker, Eugene, 5, **173–174,** *174*
Shoemaker-Levy 9 comet, *21, 25,* 29, 174
Sidereus nuncius (Galileo), 57
SIM (Space Interferometry Mission), 50
Singularity, 15
SIRTF (Space InfraRed Telescope Facility), 130
Slipher, Vesto, 1, 207
Sloan Digital Sky Survey, 19
Small bodies, *175,* **175–177,** *176*
  difficulties classifying, 176
  discovery of, 175
  See also Planets
Small Magellanic Clouds, 54
SMART-1 spacecraft, 59
SNC meteorites, 96, 106
SOFIA (Stratospheric Observatory for Infrared Astronomy), 10
Soft-landers, 154
Solar and Heliospheric Observatory (SOHO) satellite, 13, 59, 181
Solar corona, 178, 203
Solar eclipse, 202, *203*
Solar energy, 91
Solar flares, impact of, 185
Solar particle radiation, **177–180,** *178,* 213
  impact of, 178–179
  solar particles, origins of, 177–178
Solar prominence, 178
Solar radiation
  near Earth space and, 184
  storms, 219*t*
Solar wind, **180–181,** *181*
  characteristics of, 180
  defined, 220
  Jupiter and, 78
  Mars and, 97
  Mercury and, 101
  near Earth space and, 185
  solar particle radiation and, 178
  undesirable consequences of, 180–181

Soviet Union
 lunar exploration by, 40–42
 missile development in, 108
 Venusian exploration by, 42–43
 See also Russia
Soyuz rockets, 57–58, *58*
Space
 environment, **183–188**, *184*, *185*, 187*t*
 interplanetary, **220–222**, 221*t*
 science from, 20–21
 science in, 21–22
Space debris, **182–183**, *183*
Space InfraRed Telescope Facility (SIRTF), 13, 130
Space Interferometry Mission (SIM), 50
Space probes, 168
Space sciences, careers in, **20–22**, *21*
Space shuttle, 161–162
 See also names of specific shuttles
Space shuttle STS-95, 107
Space surveillance, of space debris, 182
Space Telescope Imaging Spectrograph, 70
Space Telescope Science Institute (STSI), 18, 71
Space-time, black holes and, 15–16
Space Weather Scales, 220
Spacecraft
 buses, **188–189**, *189*
 gamma-rays and, 158
 infrared radiation and, 157
 orbit of, 136
 studies of asteroids, 5–6
 See also names of specific spacecraft and missions
Spacecraft buses, **188–189**, *189*
Spaceflight, history of, **190–194**, *191*
Spaceguard Survey, 26–27
Spacehab, 58
Spacelab, 59, **194–198**, *195*, *196*
Special relativity, 31
Spectrographic analysis
 for extraterrestrial life, 85
 of Kuiper belt, 84
 what is, 85
Spectrometers, 162, 169
Spiral galaxies, *51*, 51–52, *54*
Spiral nebula, 1
Spitz, 139
Spitzer, Lyman, 128
Standard candles, 66–67

Star formation, 13, 33, **198–201**, *199*, *200*
 cosmic rays and, 32
 galaxies and, 51–52
Starbursts, 52
Stardust (Mission), 46
The Starry Messenger (Galileo), 57
Stars, **198–201**, *199*, *200*
 age of, 2
 Hubble Space Telescope on, 74
 with multiple planets, 48
 study of, 13
 See also subjects beginning with the word Stellar; specific types of stars
Starshine, from nuclear fuels, 15–16
Stellar evolution, 13
Stellar-mass black holes, 15–16
Stellar systems, 13–14
Strategic Defense Initiative (SDI), 109
Stratigraphy, 174
Stratosphere, of Saturn, 164
Stratospheric Observatory for Infrared Astronomy (SOFIA), 10
STSI (Space Telescope Science Institute), 18, 71
Suisei, 45
Sun, 201–205, *202*, *203*, *204*
 eclipses, solar, 202, *203*
 evolution of, 204–205
 future of, 205
 heliocentric system and, 9, 30, 57
 study of, 13
Sun spots, 180, 203, 220
Sun-synchronous orbit, 137
Supergiants, red, 200
Supermassive black holes, 16
Supernova 1987A, 74
Supernovae, 74, 153, **205–207**, *206*
 black holes and, 15
 Chandra X-Ray Observatory and, 130
 defined, 201
 Hubble Space Telescope on, 73
Surveyor (Lunar lander), 41–42, 142

T

Teachers, astronomy, 18, 19
Technology Experiment Facility, 62
Tectonism, on Earth, 36, 37–38
Telescopes
 adaptive optics in, 122, 150
 electromagnetic spectrum and, 10–11, 12*t*, 20
 galaxies through, 50–51

Galileo and, 56–57
 ground-based, **119–126**
 Huygens, Christiaan and, 75
 infrared, 120, 122–126
 interferometry in, 125–126
 new technology in, 123–124
 Newtonian, 119
 optical, 120, 123–126
 OWL, 126
 radio, 119, 122–123
 reflecting, 123–125
 refracting, 123
 space-based, **126–132**
 See also specific types and names of telescopes
Temperature
 before inflation, 33
 of Pluto, 149–150
 of Triton, 149–150
Tenth planet, **137–138**
Terrestrial Planet Finder, 50
Terrestrial planets, 101
Theoretical astronomers, 8, 18–19
Theoretical physics, 15
Titan II Missiles, 193
Titan (Moon)
 discovery of, 75
 exploration of, 84, 166–168
 extraterrestrial life on, 76
Titania (Moon), 66, 212
Tito, Dennis, 59
Tombaugh, Clyde, 148, **207–208**, *208*
Training, in astronomy, 17–18
Trajectories
 defined, *209*, **209–210**
 orbits, 134
Tremaine, Scott, 82
Triton (Moon), 117–118
Tsiolkovsky, Konstantin Eduardovich, 190

U

Ultraviolet radiation
 near Earth space and, 184
 observatories and, 122, 127–128
 spacecraft and, 157
 Spacelab and, 197
Ulysses solar probe, 59, 181
Umbriel (Moon), 212
Uncompressed density, of Mercury, 100
United Kingdom Infrared Telescope, *17*

United States
  lunar exploration by, 40–42
  Martian exploration by, 42
United States Ranger, 40, *43*, 142, 156, 158
Universe
  age of, **1–3**, *2*, 72, 126, 131
  search for life in, **84–90**, *85*, *88*
  See also Cosmology
Universities, space science employees in, 22
Uranus, **210–213**, *211*,
  characteristics of, 211–212
  discovery of, 66, 210
  exploration of, 44
  moons of, 84, 212, *212*
  rings of, 213
U.S. Ranger. *See* United States Ranger

V

V-2 missiles, 192
Van Allen Radiation Belts, 129
Vanguard rocket, 108
Variable stars, 55
Vega (Spacecraft), 45
Venera missions, 42–43, 144, 216
Venus, **214–218**, *215*, 215, *217*
  atmospheric conditions, 214–216
  exploration of, 42, 44–45, 214–218, 215, *215*,

extraterrestrial life on, 87
greenhouse effect on, 216
lava flows on, *145*
Magellan on, 145
surface features, 216–218
Verne, Jules, 190
Vesta (Asteroid), 4
Viking lander, *44*, *85*
Viking missions, 42, 87–88, 95, 145, 154, 160
Virgo, 54
VLS-1 rocket, 62
Volcanism, on Earth, 36, 37
Volkoff, George, 153
Voyager, 44, 144, 175
  on Jupiter, 76–77
  mission of, 144, 156, 175
  on Neptune, 117–118
  on Nereid, 118
  on Saturn, 166, 167
  on Triton, 117
  on Uranus, 212–213

W

Water
  on Earth, 38
  on Mars, 95
Wave front sensor, *169*
Weather, space, **218–220**, *219*, **219–220**, 219t, *220*
Whirlpool galaxy, *53*

White dwarf stars
  age of universe and, 3
  defined, 200
Wide-Field Planetary Camera, 70
Wilson-Harrington (Comet), 28
Wind, 117
  See also Solar wind
Wind (Spacecraft), 181
Women astronomers, *8*, 19
Wright, Thomas, 55

X

X-ray pulsars, 153
X-ray telescopes, 13–14
X-rays
  black holes and, 16
  observatories and, 119, 121, 129, 131
  spacecraft and, 157, 158
  Spacelab and, 197
XMM-Newton, 131

Y

Yerkes telescope, 123
Yohkoh spacecraft, 131

Z

Zarya, 58
Zond, on Earth's Moon, 142
Zvezda Service Module, 58, *65*
Zwicky, Fritz, 153